Designer Oil Crops

Edited by
Denis J. Murphy

© VCH Verlagsgesellschaft mbH, D-69451 Weinheim (Federal Republic of Germany), 1994

Distribution:

VCH, P.O. Box 10 11 61, D-69451 Weinheim (Federal Republic of Germany)

Switzerland: VCH, P.O. Box, CH-4020 Basel (Switzerland)

United Kingdom and Ireland: VCH (UK) Ltd., 8 Wellington Court, Cambridge CB1 1HZ (England)

USA and Canada: VCH, 220 East 23rd Street, New York, NY 10010-4606 (USA)

Japan: VCH, Eikow Building, 10-9 Hongo 1-chome, Bunkyo-ku, Tokyo 113 (Japan)

ISBN 3-527-30040-6 (VCH, Weinheim)   ISBN 1-56081-827-1 (VCH, New York)

Designer Oil Crops

Breeding, Processing and
Biotechnology

Edited by Denis J. Murphy

Weinheim · New York · Basel · Cambridge · Tokyo

Denis J. Murphy
Department of Brassica & Oilseeds Research
John Innes Centre
Norwich
NR4 7UJ
United Kingdom

Tel: UK +60 35 25 71
Fax: UK +6 03 50 22 70
EMAIL: MURPHYD@UK.AC.AFRC.JII

Published jointly by
VCH Verlagsgesellschaft mbH, Weinheim (Federal Republic of Germany)
VCH Publishers Inc., New York, NY (USA)

Editorial Director: Dr. Hans-Joachim Kraus
Production Manager: Max Denk

The title illustration was reproduced with kind permission from the Bayer publication "research", edition 6, 1993.

Library of Congress Card No. applied for.

British Library Cataloguing-in-Publication Data: A catalogue record for this book is available from the British Library.

Deutsche Bibliothek Cataloguing-in-Publication Data:

Designer oil crops : breeding, processing and biotechnology /
ed. by Denis J. Murphy. – Weinheim ; New York ; Basel ;
Cambridge ; Tokyo : VCH, 1993
  ISBN 3-527-30040-6
NE: Murphy, Denis J. [Hrsg.]

Composition: Mitterweger Werksatz GmbH, D-68723 Plankstadt
Printing: betz-druck GmbH, D-64291 Darmstadt
Bookbinding: IVB, D-64646 Heppenheim
Printed in the Federal Republic of Germany.

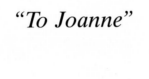

"To Joanne"

Preface

Recent spectacular advances in plant breeding and biotechnology have opened up new vistas for the production of novel varieties of oil crops. The potential gene pools available to plant breeders have been extended enormously following the development of wide crossing techniques, e.g. protoplast fusion and embryo rescue and the use of marker-based selection methods for the rapid identification of favourable traits in screening programs. At the molecular level, techniques are now available for the isolation of specific genes from donor organisms and their transfer into crop plants in order to facilitate the production of new and valuable products for human kind. These developments have lead to the formulation of the concept of designer oil crops. This concept involves the extension of classical breeding methods in order to allow the conscious design of a particular crop to fulfil a specific end use, whether this be for edible or for non-edible purposes.

In this monograph, the concept of designer oil crops will be examined further. A comprehensive survey of the major oil crops in Chapter 1 is followed by a detailed account of the actual and potential non-food uses of oils from a variety of plant species in Chapter 2. The role of breeding, both classical and modern, in enabling the improvement of oil crops is then examined in Chapter 3. In Chapter 4, there is a description of the complex biochemical processes that are responsible for the formation of storage oils, and in particular of their fatty acid components, which determine the quality and downstream uses of the oil. This is followed by a number of chapters on biotechnology-related issues, including methods of transforming oilseed crops with novel genes in Chapter 5, a general survey of oil crop biotechnology in Chapter 6, an account of the processing of some novel seed oils in Chapter 7 and a review of the release of transgenic oil crops into the environment in Chapter 8. Finally, in Chapter 9 there is a brief consideration of the future perspectives of designer oil crops.

These chapters have been designed to overlap with each other to some extent, with regard to their subject matter. This allows the various topics to be examined in depth by each of the specialist authors from their own unique perspectives. While each chapter tells its own story, together they form a comprehensive overview of the science and future potential of oil crops. It is hoped that this monograph will provide a useful introduction and/or update to students, scientists, educators, industrialists, farmers and members of the general public, who are interested in this exciting new area of agricultural endeavour.

July 1993,
Norwich, UK   Denis J. Murphy

Contents

Contributors

Dr. A. E. Arthur
Department of Brassica & Oilseeds
Research
John Innes Centre
Norwich NR4 7UJ
U.K.
Tel: +44 60 35 25 71
Fax: +4 46 03 25 98 82
EMAIL:
ARTHUR@UK.AC.AFRC.JII

Dr. F. P. Cuperus
Agrotechnological Research
Institute (ATO-DLO)
PO Box 17 NL-6700
AA Wageningen
The Netherlands
Tel: +3 18 37 07 50 97
Fax: +3 18 37 01 22 60

Dr. P. J. Dale
Department of Brassica & Oilseeds
Research
John Innes Centre
Norwich NR4 7UJ
U.K.
Tel: +44 60 35 25 71
Fax: +4 46 03 25 98 82
EMAIL:
DALE@UK.AC.AFRC.JII

Dr. J. T. P. Derksen
Agrotechnological Research
Institute (ATO-DLO)
PO Box 17 NL-6700
AA Wageningen
The Netherlands
Tel: +3 18 37 07 50 97
Fax: +3 18 37 01 22 60

Professor Dr. W. Friedt
Institut für Pflanzenbau
und Pflanzenzüchtung 1
Ludwigstr. 12
6300 Gießen, FRG
Tel: +64 17 02 59 83
Fax: +64 17 02 59 97

Professor J. L. Harwood
Department of Biochemistry
University of Wales College of
Cardiff
Cardiff CF1 1ST
U.K.
Tel: +44 22 28 74 1 08
Fax: +44 22 28 74 1 16

Dr. J. Irwin
Department of Brassica & Oilseeds
Research
John Innes Centre
Norwich NR4 7UJ
U.K.
Tel: +44 60 35 25 71
Fax: +44 60 32 59 88 2

Dr. W. Lührs
Institut für Pflanzenbau
und Pflanzenzüchtung 1
Ludwigstr. 12
6300 Gießen, FRG
Tel: +64 17 02 59 83
Fax: +64 17 02 59 97

Dr. D. J. Murphy
Head of Brassica & Oilseeds
Research Department
John Innes Centre
Norwich NR4 7UJ
U.K.
Tel: +4460352571
Fax: +44603259882
EMAIL:
MURPHY@UK.AC.AFRC.JII

Dr. B. G. Muuse
Agrotechnological Research
Institute (ATO-DLO)
PO Box 17 NL-6700
AA Wageningen
The Netherlands
Tel: +31837075097
Fax: +31837012260

Dr. R. A. Page
Department of Biochemistry
University of Wales College of
Cardiff
Cardiff CF1 1ST
Tel: +44222874108
Fax: +44222874116

Dr. J. Scheffler
Department of Brassica & Oilseeds
Research
John Innes Centre
Norwich NR4 7UJ
U.K.
Tel: +4460352571
Fax: +44603259882

Introduction

D. J. Murphy

Oil crops have been used as sources of both edible and non-edible products throughout recorded history. There is evidence from written records and archaeological excavations that oil crops such as linseed, olive, and sesame were in widespread use in classical civilizations at least six thousand years ago. Amongst the more important uses of such oil crops were a variety of non-edible applications, including as lubricants in simple machines, as sources of illumination in oil lamps, as cosmetics, soaps and as massage oils. One of the current major European and North American oil crops, rapeseed, was used extensively for non-edible purposes until late in the last century. For example, during the latter part of the Middle Ages, rapeseed oil was the most important lamp oil in Europe north of the Alps. Even earlier than this, wild Brassicas were collected (rather than being cultivated) for use in illumination and soap making. The demise of oil crops as major industrial feedstocks, began with the increasing large-scale exploitation of coal and fossil- oil reserves during the late 19th and early 20th century. This resulted in a cheap and abundant supply of fossil-derived mineral oil which, in addition to being used as a fuel, could also replace many of the other non-edible uses of vegetable oils. At the same time, the exponentially increasing human population made it necessary to utilise as many agricultural resources as possible for the cultivation of edible crops.

These two factors led to a drastic decrease in the non-edible uses of oil crops. Furthermore, the plentiful supply of readily available, cheap mineral oil stimulated the growth of entirely new petrochemical-based industries. These industries now supply us with products such as nylons and other textiles, paints, lubricants, plastics, detergents, cosmetics and pharmaceuticals. Refined fossil oils provide fuel for our homes, cars, planes, trucks and also power many electricity generating stations. Our civilisation is so firmly based on the ready availability of fossil oil, that a brief interruption in its supply in the early 1970s was enough to cause a severe disruption to the global economy. Unfortunately, however, the supply of fossil oil is not unlimited. This is a non-renewable resource and it is estimated that there is approximately 50-100 years supply left in readily accessible reserves. To some extent, it may be possible to use coal to substitute for mineral oil, although in many cases this would be prohibitively expensive. In addition, it has been estimated that our global coal reserves will only last another 200 years at present rates of consumption. Increasing use of coal as a substitute for fossil oil will of course reduce this timescale considerably. There are also many environmental drawbacks involved in the mining and refining of coal and coal-derived products. Added to this is the enormous contribution made by the combustion of coal and oil to increased atmospheric CO_2 concentrations and the attendant risks of unpredictable global climatic change.

These factors have focused attention on possible renewable sources of hydrocarbon products and in particular on oil crops. Most plant oils are already more complex than mineral oils when they are initially extracted. This eliminates the necessity for some of the costly and environmentally unfriendly processes required for the refining of mineral oils. With recent advances in oil crop breeding and biotechnology, even more sophisticated plant oils can be produced. Hence we can imagine the plants as non-polluting "bio- refiners" or "bio-reactors" replacing many of the visually and environmentally displeasing chemical refineries which litter the landscapes of developed and developing countries alike. Plant oils have the further advantage that they are CO_2-neutral, i.e. unlike fossil-oils, they contain recently-fixed CO_2 and the re-emission of this CO_2, if the plant oil or its products, are combusted, will not add to the net atmospheric CO_2 concentration.

The recognition that renewable sources of hydrocarbons for our petrochemical-based industries will eventually be required over the next 50-100 years has recently been combined with more short and medium term factors favouring the increasing use of oil crops for non-edible products. A major consideration here has been the proliferation of surpluses of edible agricultural products, particularly in the developed economies of Europe and North America. The problem of agricultural surpluses is caused largely by the vastly increased yields of the major food crops now possible in developed agricultural economies. These high crop yields are contingent upon the availability of elite relatively disease-free cultivars, a high degree of mechanisation and intensive input regimes, particularly of fertilizers and pesticides. Another important factor in stimulating agricultural yield has been the willingness of many countries or trading blocs to subsidise their agricultural sectors and hence to encourage intensive high-yielding farming practices which would otherwise be uneconomic. Much of the burden of such measures has fallen upon taxpayers in developed economies and it is now recognised that the system is ripe for reform. Such reforms are now in progress both within trading blocs, e.g. a restructuring of the Common Agricultural Policy (CAP) of the European Economic Community (EEC), and between the major trading blocs, e.g. reduction of farm subsidies under the General Agreement on Tariffs and Trade (GATT). Decreased agricultural production is being achieved both by lower subsidies and by the practice of "set-aside". "Set-aside" land is land that farmers are paid to keep fallow. It is estimated that the amount of "set-aside" land may increase to 30% of total arable land in some European countries by the end of this century. "Set-aside" is also a common practice in the United States.

Such large areas of set-aside land are, of course, undesirable for the agricultural sector as a whole. With less land, farmers will require less manpower, machinery, seed stocks, etc. This will result in a depression of the entire agricultural sector. A number of alternative uses of some set-aside land have been proposed. These include amenity developments, such as golf courses, theme parks or other recreational uses. Afforestation is also a possibility, particularly on relatively marginal land. Unfortunately, none of these proposals will maintain the agricultural sector at current levels and they run the risk of leading to the possible permanent loss of some of the set-aside land, which may be required for edible crop production in the future if circumstances were to change.

The combination of steadily depleting and non-renewable hydrocarbon reserves, large food surpluses and large areas of empty set-aside land provides a powerful incentive for the development of new types of oil crop. Such crops could serve as renewable sources of partially-refined hydrocarbon-based products for future generations. Their cultivation would help to maintain overall levels of agricultural production, without contributing further to increasing food surpluses. The recent perceived need for the expanded cultivation of industrial crops has coincided with a number of research breakthroughs in plant biotechnology and breeding. For example, it is now possible to transfer individual genes or even whole chromosomes from one plant or animal species to another totally unrelated species. This technique of "genetic engineering" has already been used to produce transgenic farm animals, such as sheep, which secrete valuable pharmaceuticals in their milk. The milk is collected in the usual way, and the pharmaceuticals purified by relatively straightforward methods. Such use of animals as pharmaceutical factories has caused some concern in the public at large. A more attractive alternative may be to use plants, such as oil crops, as the sources of such valuable pharmaceutical products. Recent research in California has shown that the genes encoding human antibodies can be successfully transferred to plants. These plants will synthesise the antibodies which can then be purified for research or medicinal uses. Novel molecular tagging methods have also been developed which can assist plant breeders to select desirable new cultivars for use in agriculture.

Such developments make it realistic to consider using genetic engineering and other advanced biotechnological techniques for the improvement of existing oil crops, or to enable the introduction of novel oil crops into temperate agricultural systems. The concept of "designer oil crops" is therefore rapidly becoming a reality. For example, during 1992, two new varieties of rapeseed were produced in the United States by genetic engineering. These two new rapeseed varieties are the beginning of what will eventually be an extensive portfolio of designer rapeseed cultivars producing different types of valuable edible or non-edible oils, each of which would be suited for a particular set of market applications. This technology is being pioneered in rapeseed, but in the future there is the possibility that it could be extended to other major annual oil crops such as soybean and sunflower. A more distant, but nevertheless realizable prospect is the extension of genetic engineering technology to the major high-yielding tropical perennial oil crops such as coconut and oil palm.

1 The Major Oil Crops

W. Lühs and W. Friedt

1.1 Introduction

Fats and oils derived from agricultural crops have for many centuries fulfilled the needs of mankind for both edible and industrial commodities. Productive oil crops are distributed all over world, and each of them has its own history of domestication. Consequently the oil crops differ in their cultural, economic and utilization characteristics. For example, rapeseed and linseed are adapted to relatively cool climates, whereas oil palm and coconut are predominantly suited for the rainfed growing conditions of the tropics. For oil palm, rapeseed, sunflower seed, and coconut the production of oil is the principal motivation for growing and marketing. However, the oil of cottonseed is a by-product of the more valuable fiber, and its availability is mainly influenced by the demand for cotton fiber. In the case of soybean, the protein-enriched meal usually is the more interesting commodity.

The storage oils of different agronomically important oil crops are basically characterized by their individual fatty acid patterns. This is due to the action of specific genes encoding enzymes involved in fatty acid biosynthesis. Consequently, fatty acid composition can be further modified by genetic manipulations, i.e., both by classical and modern breeding methods. Thus extensive breeding efforts have been made in the past in order to provide oil crops with an improved seedoil quality. In some instances, plant genotypes from long established crops have been bred to provide oils with drastically altered fatty acid compositions and, consequently, new physical and chemical properties. For example, for nutritional purposes, the removal of erucic acid in rapeseed oil was one of the most important breeding objectives almost 30 years ago. Furthermore, the reduction of linolenic acid in the seed oil of linseed, soybean and rapeseed, as well as the development of high oleic sunflower represent successful examples of conventional genetic approaches via mutagenesis and selection. An important requirement for successful selection in all these cases was the development of rapid screening techniques to enable a very large population of seeds to be analyzed for increase in oil content and desirable changes in fatty acid composition.

In this chapter, besides some general considerations on breeding and genetics of oil crops, an overview of the major oil crops is presented. More particularly, some of the more recent research activities, advances and opportunities of different genetic approaches in modifying fatty acid composition are discussed.

1.2 Production and consumption of vegetable oils and fats

World production of naturally occurring oils and fats was about 83.5 million metric tons (MT) in 1991/92 and is predicted to increase to more than 100 MT in the year 2000 [1,2]. Vegetable oils are one of the most valuable and essential agricultural commodities moving in international trade. The world production of vegetable oils presently amounts to about 64 MT, i.e., 75% of total world production. Some 20 MT are fats of animal origin such as tallow, lard, butter and marine oils. The primary demand for oils and fats (approximately 80%) is in the

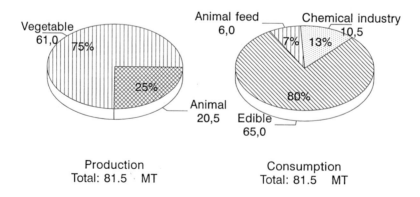

Production
Total: 81.5 MT

Consumption
Total: 81.5   MT

Figure 1.1. World production and consumption of oils and fats 1990–91 (adapted from [3,4]; MT, million metric tons).

food industry with an usage of about 65 MT. The remaining quantity of natural oils and fats sources serves - almost constantly over the past few years - as a raw material for animal feed and the chemical industry, 7% and 13% of world consumption, respectively (Fig. 1.1) [2–4].

There are only six major oil crops which are of international importance: i.e., soybean (*Glycine max*), oil palm (*Elaeis guineenis*), rapeseed (*Brassica* spp.), sunflower (*Helianthus annuus*), cottonseed (*Gossypium* spp.), and groundnut (*Arachis hypogaea*), together accounting for about 84% of world vegetable oil production. The relative share of these main vegetable oil sources has changed dramatically during the past 30 years. In particular, oil palm, rapeseed and soybean have contributed to the concomitantly rapid growth of vegetable oil supply during this period (Table 1.1). Production of coconut (*Cocus nucifera*), flax/linseed (*Linum usitatissimum*), castor (*Ricinus communis*) and palm kernel (*Elaeis guineenis*) accounts for a minor proportion of the total world oilseed production, although these oils are important feedstocks for industrial applications. Since oil crops belong to rather distantly related botanical families, such as *Fabaceae*, *Palmae*, *Brassicaceae*, *Compositae*, *Malvaceae*, *Linaceae* and *Euphorbiaceae*, deposition of oils and fats in different plant

Table 1.1 Contribution of individual crops to world vegetable oil production

Source	1961	1978/79	1991/92*	Relative rank 1961	1978/79	1991/92	% of 1991/92 total
		(in MT **)		1961	1978/79	1991/92	
Soybean	3.3	11.7	16.7	1	1	1	26.0
Palm	1.3	4.3	11.9	7	3	2	18.5
Rapeseed	1.2	3.7	9.4	8	4	3	14.6
Sunflower	2.0	4.7	8.0	4	2	4	12.4
Cottonseed	2.4	3.0	4.4	3	6	5	6.8
Groundnut	2.5	3.3	3.9	2	5	6	6.1
Coconut	2.0	2.8	2.8	5	7	7	4.4
Olive	1.4	1.6	2.3	6	8	8	3.6
Corn	na	0.4	1.6	na	11	9	2.5
Palm kernel	0.4	0.6	1.5	10	10	10	2.3
Linseed	0.9	0.7	0.7	9	9	11	1.1
Others	0.4	0.5	1.1	-	-	-	1.7
Total	17.8	37.3	64.3				

* October-September marketing year; **MT, million metric tons; na, not available; Sources: [1,5].

tissues is generally not limited to only a few taxonomic groups. With regard to the ecological distribution of commercial production, it has to be emphasized that each growing region of the world possesses at least one productive oil crop adapted to the range of climatic conditions existing in that area [6,7]. Another important aspect related to sources of vegetable oils is their relative efficiency with regard to the quantity of oil produced per hectare of land (Table 1.2).

Common vegetable oils and fats predominantly consist of triacylglycerols, i.e., glycerol esters of saturated or unsaturated fatty acids with a carbon number ranging from 12 to 22 (C_{12}-C_{22}). Soybean and sunflower oils are basically rich in linoleic acid (50–70%), whereas Canola ("double-low" rapeseed), peanut and olive are comparatively high in oleic acid (approximately 50–70%); palm and coconut oils are rich in saturated fatty acids (50 and 90%, respectively) (Table 1.3).

The bulk of vegetable oils, i.e., more than 90%, is directed to food uses [15], predominantly in the form of margarines, shortenings, salad oils and for frying

Table 1.2. Relative efficiency and ecological distribution of major oil crops [6,11]

Oil Crop	Ecological habitat	Oil (%)	Yield# (T/ha)	Oil Yield# (T/ha)
Oil palm (pulp)	tropical rain forest	20	30.0	6.0
Palm kernel	tropical rain forest	50	1.6	0.8
Soybean	humid temperate; humid subtropical	20	3.5	0.7
Rapeseed	cool/humid temperate; humid subtropical	45	4.0*	1.8
Sunflower	humid/dry temperate; humid/dry subtropical	45	3.5*	1.6
Cotton	humid/dry subtropical semi-arid tropical	20	2.5**	0.5
Groundnut	humid temperate; subtropical; semi-arid tropical	48	3.0**	1.4
Coconut (copra+)	tropical rain forest; tropical monsoon	64	3.5+	2.2
Linseed	cool/humid temperate; humid subtropical	40	2.5*	1.0

#under optimum agricultural management; T, metric tons; * Central Europe; ** seed; + meat (endosperm) after drying

Table 1.3. Typical fatty acid composition of major oil crops (adapted from [12–14]#)

Oil	I.V.##	Lauric C12:0	Myristic C14:0	Palmitic C16:0	Stearic C18:1	Oleic C18:1	Linoleic C18:2	Linolenic C18:3	Erucic C22:1	Others
						(% of total fatty acids)				
Soybean	125–138			11 (7–14)	4 (3–6)	23 (18–26)	54 (50–57)	8 (5–10)		
Palm	45–56		1 (0.5–2)	45 (41–47)	4 (3–6)	39 (36–44)	9 (6–12)			2
Palm kernel	14–24	50 (41–55)	16 (14–18)	8 (6–10)	2 (1–3)	14 (12–19)	2 (1–3)			8+ (5–11)
Rapeseed (high erucic)	97–110			3 (2–6)	1 (1–3)	24 (8–60)	15 (11–23)	8 (5–13)	35 (5–60)	14** (3–18)
Rapeseed (low erucic)	110–115			4 (3–6)	2 (1–3)	60 (50–66)	20 (18–28)	10 (6–14)	2 (0–5)	2** (0–4)
Sunflower	122–139			7 (3–10)	5 (1–10)	19 (14–35)	68 (55–75)			
Cottonseed	99–121		1 (0–2)	24 (17–29)	2 (1–4)	18 (13–44)	54 (33–48)	0.5 (0–2)		0.5 (0–2)
Groundnut	84–102			12 (6–16)	4 (1–7)	47 (35–72)	31 (13–45)			6* (2–11)
Coconut	7–13	49 (43–51)	17 (16–21)	9 (7–10)	2 (2–4)	6 (5–10)	2 (1–3)			15++ (9–18)
Linseed	169–196			6 (4–7)	4 (2–8)	18 (12–38)	14 (5–27)	58 (26–65)		

#As considerable variation occurs in commercial samples, ranges are given in parentheses; ## I.V. iodine value, classification into non-drying (I.V. <100), semi-drying (I.V. = 100–170) and drying oils (I.V. >170); *saturated fatty acids, 2 (1–3) % C20:0, 3 (1–5) % C22:0 and 1 (0.5–3) % C24:0; ** monounsaturated fatty acids, 13 (3–15) % C20:1 and 1 (0–3) % C24:1; + saturated fatty acids, 4 (2–6) % C8:0 and 4 (3–5) % C10:0; ++ saturated fatty acids, 8 (5–10) % C8:0 and 6 (4–8) % C10:0.

purposes. As nutrients, texturizers, and the most concentrated source of energy, fats and oils play an integral role in food. Major concerns with regard to nutritional quality include the impact of dietary oils and fats on blood cholesterol levels, inhibition of certain cancers, suppression of auto-immune disease, and obesity [16–18]. Nevertheless, suggestions for the increased use of vegetable oils as renewable sources for many industrial and non-food purposes have been increasing in recent years, in particular in view of the recognition that the resource of fossil-derived mineral oils is definitely limited (see Chapters 2 and 9). The needs of industry, however, are quite different to the major world oil production which generates an excessive supply of oils with a rather monotonic C_{16}/C_{18} fatty acid mixture. Apart from a few traditionally used oils and fats - like tallow, castor, linseed and tung oil - soybean and coconut/palm kernel oil currently are the most important feedstocks because they represent sources for detergent-range oleochemicals (C_{12}/C_{14}) and inexpensive C_{16}/C_{18} oleochemicals for a wide range of applications [2,10,19,20]. Both in nutrition and in industry, specific fatty acid profiles determining the suitability of vegetable oils for special applications are needed [21,22]. However, oil is not the only commodity of value available from oil-bearing plants. Stanton and Blumenfeld [23] compared the value of the various raw materials derived from various major oil crops. For instance during the decade 1981–90 the relative annual value of soybean oil ranged from 29–51%, with an average value of 36.4% as compared to 63.6% for the meal. Cotton is an even more extreme example; during the above period the cotton fiber determined 80.7% of the economic value, while the oil and meal component accounted for only 8.1% and 7.9%, respectively. Therefore, in these cases any attempt to change oil yield or fatty acid composition at the expense of protein or fiber content/quality should be considered carefully.

1.3 Breeding and genetics - general objectives and strategies

As discussed in more detail in Chapter 3, the choice of a breeding method to be used for an oil crop is based on several considerations: 1, mode of reproduction (self-pollinated vs. cross-pollinated, tolerance to inbreeding, seed multiplication method, etc.); 2, mode of gene action (monogenic vs. polygenic, dominant vs. recessive, heterosis, epistasis); 3, source of available diversity (cultivars, wild and related species); and 4, priority assigned in relation to other agronomic traits. With regard to the latter aspect the following objectives in oilcrop breeding are generally recognized in this order of importance: 1, yield improvement and adaptation to environmental stress; 2, increased oil content; 3, alteration of oil quality; and 4, quantitative and qualitative improvement of meal protein. Genetic progress has been made for all of these parameters and to a varying extent for each of the oilseed species. There has been a continuous increase in oil yield of the better cultivars, which has been accomplished through improved agronomy. For instance, in sunflower the increased oil content has been achieved by decreasing the proportion of the seed hull, and the rapid increase in seed yield was mainly due to the introduction of modern hybrid cultivars.

The value and utility of a vegetable oil primarily depends upon its fatty acid composition (see Chapter 4). Thus the aim of breeding programs for improved quality and versatility in oil crops is the acquisition and application of knowledge on the variation as well as the genetic and biochemical control of fatty acid composition. In nearly all important annual oil crops conventional approaches have resulted in new genetic strains with marked alterations of fatty acid composition (for an overview see Fig. 1.2). These strains or lines were developed as a result of selection of available germplasm of the species or as a result of an artificial induction of genetic variability followed by selection, e.g., chemically induced mutants of soybean, rapeseed, sunflower and linseed (Table 1.4). Frequently, these genetic changes are inherited by one or a limited number of genes according to Mendelian expectations. The availability of non-destructive determination of oil content by nuclear magnetic resonance (NMR) spectrometry or other methods, such as gas liquid chromatography, as well as rapid screening methods like the half-seed technique [24,25], has provided plant breeders with the means of tailoring seed oils for edible and industrial requirements (see Fig. 1.2). Consequently, a variety of plant species and cultivars with distinct and widely different fatty acid compositions are now available, e.g., low erucic rapeseed (Canola), high oleic safflower and sunflower as well as low linolenic linseed [21,26].

Further goals with regard to oil quality are, for example: 1, higher levels of stearic acid to reduce the need for hydrogenation in the manufacture of margarine and frying oil; 2, reduced level of palmitic acid, regarded as the important cardiovascular risk factor amongst the saturated fatty acids; 3, developing a commercial source of high oleic canola with more than 80% oleic acid in triacylglycerols to improve stability of frying oils as well as suitability for industrial applications, and 4, decreased linolenic acid in soybean and rapeseed in order to improve oxidative stability [21,26,27]. Some of these goals are achievable by traditional breeding methods of selection and mutagenesis.

More rapid breeding progress can be achieved by an application of biotechnology, i.e., cell and tissue culture techniques and molecular methods, as a promising additional tool [21,26,28–30]. Considerable steps towards

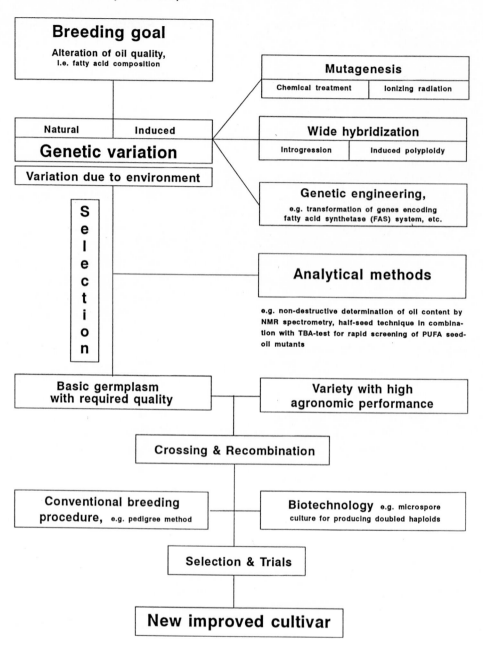

Figure 1.2. Scheme illustrating the procedure of plant breeding, based on the definition of a breeding goal (oil quality in this case), and exploiting natural or induced variation via conventional or modern breeding methods and techniques, finally leading to the new, improved crop cultivar.

Table 1.4. Mutants of different oil crops and the proposed models for genetic control of fatty acid composition, respectively.

Crop	Genotype	Palmitic	Stearic	Oleic	Linoleic	Linolenic
				% total fatty acids		
Soybean						
Standard		11	4	24	53	7
"Century"[a]	*Fan Fan*	11	3	19	58	8
"C1640"[b]	*fan fan*	12	3	22	60	3
"FA8077"[a]	*Fas Fas*	9	4	44	38	5
"A6"[b]	*fas fas*	8	30	21	35	6
Rapeseed (Canola)						
"Westar"		4	2	59	19	9
"Stellar"		4	2	60	26	2
"Cascade"[a]		5	1	65	18	9
"X-82"[b]	M_2 (half-seed)			84	4	3
"X-82"[b]	M_3 (half-seed)			79	5	5
"X-82-B"[b]	M_4 (line)	4	3	80	4	5
Sunflower						
Standard		5–6	3–5	20–40	50–69	
High oleic	$OL_1\ OL_2\ OL_3$*	3	2	88	7	
High oleic	$OL_1\ OL_2\ ol_3$*			84–90	1–8	
Intermediate	$OL_1\ ol_2\ OL_3$*			43–52	37–47	
Low oleic	$ol_1\ OL_2\ OL_3$*			27–46	45–64	
---	$ol_1\ OL_2\ OL_3$*			---	---	
---	$ol_1\ ol_2\ ol_3$*			---	---	
Linseed						
Standard		5	4	21	13	56
"Glenelg"[a]	*Ln1Ln1Ln2Ln2*	7	4	35	14	40
"M1589"[b]	*ln1ln1Ln2Ln2*	7	5	44	25	19
"M1722"[b]	*Ln1Ln1ln2ln2*	8	5	36	28	23
"Zero"[b]	*ln1ln1ln2ln2*	9	6	34	51	2

*alleles in homozygous condition; --- ibid.; [a] = original material; [b] = mutant; for references see text.

practical application of such novel techniques have been reported for different oil crops - in particular rapeseed and linseed. In these species it is now possible to obtain haploid plants reproducibly through microspore or anther culture which results in a time gain of several years for the breeder (see Fig. 1.2).

Interspecific hybridization is another interesting supplementary technique in plant breeding; it can help to create new genetic variation. Wide crosses have been successfully used, for example in the *Cruciferae* family and in the genus *Helianthus*, by application of the embryo rescue technique [31–33]. On this basis, related species can be used as gene sources to improve various agronomically important characters. For species recalcitrant to sexual hybridization, the protoplast fusion technique is an elegant method to achieve hybrids, provided that the regeneration of intact plants is feasible [30].

The most radical changes in proportion of special fatty acids are those desired for vegetable oils and fats destined to serve as industrial feedstocks. For instance, it is a subject of investigation in several laboratories to create a temperate-zone, annual crop like rapeseed as a source of high contents of lauric acid or trierucoylglycerol [27, 34–39; Chapter 6]. Evidently such goals will not be realized satisfactorally by classical means of plant breeding - not even assisted by biotechnology. In these cases sophisticated molecular methods will be required for transferring specific foreign genes between distant species (see Chapter 6). The necessary prerequisites for an application of gene technology in oilseed breeding are now becoming available, i.e., efficient vectors and transformation systems have been elaborated and a number of target genes have been isolated. Therefore, the transfer of numerous relevant genes, especially for seed oil quality, to cultivated species can be anticipated for the near future - provided that entire, reproductive plants can be regenerated from the manipulated cell(s) or tissue. Based on a better understanding of biochemistry and physiology of seedoil synthesis and accumulation as storage oil bodies, genetic engineering is recognized as a novel and efficient pathway for creating novel genetic variation for the plant breeder [see ref 21 and Fig. 1.2].

1.4 Major vegetable oils and fats

In the following description the major oil crops are considered in order of their importance as world agricultural commodities and sources of oil for edible and/or industrial use. More details on cultivation, breeding and utilization of the whole group of oil-bearing crops have been compiled by Röbbelen *et al* [9] and Salunkhe *et al* [10].

1.5 Soybean

1.5.1 Importance, distribution and utilization

The soybean (*Glycine max*) is the world's most important single source of oil and high protein meal. The plant is a grain legume originating in north-eastern China which was domesticated probably during the Chou dynasty (ca. 3100–3700 B.P.) or even earlier [40]. From that area of China, the soybean spread to southern China, Korea, and other countries in south-eastern Asia. Soybeans were introduced to the North American continent by the end of the 18th century, but until the 1920s they remained a very minor crop grown mostly for forage. During the past 50 years, however, it has become a major grain crop in the USA [41,42]. World production of soybean currently comprises about 106 MT. The USA accounts for about 50% of the world production, while the other major soybean-producing countries, viz., Brazil, Argentina and China, account for another 35–40% of the total world production. The remaining production is scattered among various countries in Asia and South America, with almost negligible quantities produced in Africa and Europe [1,11]. The EEC soybean production, predominantly grown in Italy, amounts to about 1.5 MT at a high average yield of approximately 3.0 metric tons per hectare [11,43]. The United States is the world's largest soybean exporter, while the EEC always has imported the largest quantities for supplying protein for human and animal feeding purposes [44].

Soybean is mainly grown for its seed, which is used commercially as human food and livestock feed as well for the extraction of oil. It is the world's most important grain legume both in terms of total production and international trade [7,45]. The highest domestic consumption exists in Asia, where it has been used as a basic food since ancient times [40,46]. However, in the western world the value and utility of soybean is derived mainly from its major seed components, viz., protein and oil. The protein is of greater importance because the monetary value of the protein fraction (average 60–65%) exceeds that of the oil fraction [23,44]. Protein and oil make up about 60% of the commercially produced soybean's gross composition. The protein and oil content average 38–42% and 18–22% (on dry weight basis), respectively. Since the oil-free meal contains about 45–50% protein, soybean meal represents a major protein component in livestock feed. Additionally, in recent years soybean meal has been used in a variety of forms, such as "textured vegetable protein" in human foodstuffs [10,47]. Soybean oil, which accounts for about 35–40% of the value of the seed, dominates the global supply of traded vegetable oils. Over the past 20 years, the demand for edible vegetable oil has continued to rise steadily.

Even though the production and use of other oils, especially rapeseed and palm oil, have increased dramatically in the same time, production of soybean oil (currently some 16.5 MT) has maintained its first rank in world vegetable oil production [1,2]. Most of the oil is used in the preparation of food products, such as salad and cooking oils, margarine, shortenings and frying fats. During degumming, an important by-product of soybean oil processing, the phospholipid lecithin, is obtained. This is used in many products as an emulsifying agent [10,16,47–50]. Because of its relatively low cost and dependable supply, considerable quantities of soybean oil are used also in the manufacture of non-food products (see Chapter 2).

1.5.2 Botany - taxonomy and general description

The soybean is a member of the genus *Glycine* which belongs to the legumes (*Fabaceae*). According to Hymowitz [45] the legume family, comprising about 650 genera and 18,000 species, is by far the most versatile plant family. The genus *Glycine* is divided into two subgenera, *Glycine* and *Soja*. Taxonomically, at present fifteen wild perennial species, distributed to the Australasia region, are grouped in the subgenus *Glycine*. The most common chromosome number is $2n=2x=40$. The species *G. tabacina* is known to contain both tetraploid ($2n=4x=80$) and diploid individuals; *G. tomentella* is diploid, tetraploid, and aneuploid ($2n=38,40,78,80$) [41,45,51–53]. The subgenus *Soja* includes the cultivated soybean (*G. max*) and its wild ancestor, *G. soja*, both of which are annual and diploid [51,54,55].

Soybeans grown for seed production are normally bushy, erect, usually not more than 75cm in height, much branched, with well developed roots. Modern cultivars are less branched and classified into both determinate and indeterminate growth types adapted to different growing regions [56,57]. Roots carry nodules containing species-specific *Bradyrhizobium* bacteria, which, when present in the soil and well developed, render the plant generally independent of nitrogen fertilization. There is evidence that *Bradyrhizobium* strains and soybean genotypes vary in nitrogen fixation and utilization, respectively [56,58,59].

The soybean plant produces numerous short, hairy pods containing about three round, usually yellow seeds. Flowers are born on short racemes originating in leaf axils, each inflorescence bearing up to twenty very small, purple or white flowers, which are typically leguminous in shape. Many flowers are shed without forming pods. Self-pollination is the rule, and normal outcrossing is estimated at less than 1%. Pods normally shatter when ripe, releasing the seed, thus the rate and degree of shattering is an important varietal characteristic [7,56,57].

Soybeans are highly photoperiodic, i.e., cultivars differ in their day- length requirements for flowering and seed setting. Temperature is probably the next most important factor [7,56,57]. Although basically a sub-tropical short-day plant, soybean has been adapted to a wide range of conditions, growing also in tropical or cool-temperate regions with day length varying from 12 to 16 hours. Particularly in the northern, cooler regions, growth and development of soybean is limited by photoperiod and/or temperature. Daylength sensitivity has long been considered to be a limiting factor of soybean production in Europe and Canada, since delayed flowering often resulted in the failure of the crop to reach maturity before the first frost. Identification of insensitive genotypes and development of day-neutral "Fiskeby" type material by Holmberg [60] in Sweden has resulted in remarkable progress in eliminating the short-day requirements for flowering in soybean. However, the local adapted variety itself often shows a relatively narrow range of latitude, i.e., sensitivity to day length becomes important when genotypes are grown outside of their area of adaption. Although plants appear to be generally unaffected by a fairly wide variation in day temperatures before flowering, the optimum range for most cultivars is usually close to 30 °C. Thus a daytime temperature of 25 °C or less delays flowering. However, short-season varieties with substantial cold toler- ance have been bred for cultivation in Russia and Canada (e.g., "Maple Arrow"), requiring only 22 °C from emergence to flowering, and 15 °C from flowering to maturation [7,56,57].

1.5.3 Breeding

1.5.3.1 Sources of genetic variation

The cultivated soybean, as it evolved since its domestication, constitutes an extremely diverse species mostly due to photoperiod response. Several thousand local varieties or landraces were introduced from northeast China and Korea into the United States between 1910 and 1931. This introduced germplasm provided the basic breeding material for the improved cultivars currently cultivated in the USA [42].

G. max can be crossed readily with its wild ancestor, *G. soja*, to obtain fertile progeny. Although *G. soja* has many agronomically unacceptable characteris- tics, it had been used as a source of genes for small seed [57]. In a research program with regard to diversity in oil quality [61,62], wild *Glycine* species showed an extended array of fatty acid compositions. However, with a range of 11–27% linolenic acid none of the related *Glycine* species contained lower levels of linolenic acid than *G. max*. Nevertheless, the perennial *Glycine* species represent a potentially valuable source of germplasm for soybean breeding as they possess many interesting characteristics such as drought, heat and cold tolerance, apparent daylength insensitivity as well as disease and pest resistances. However, the incorporation of these useful alien traits from wild species of the subgenus *Glycine* to cultivated soybean has not been feasible so far because of extremely low crossability and early pod abortion. Interspecific hybrids which had been obtained, remained highly sterile even after chromosome doubling, and attempts to develop backcross progenies have not been very successful either [41,51,53,63–65].

1.5.3.2 Breeding procedures

Soybean breeding methods are dominated by the self-pollinating property of the plant. Current cultivars are generally pure lines developed by crossing of two or more lines followed by self-fertilization to the F_4 or a later generation (see Chapter 3). Homozygous lines are isolated and tested to determine those with desirable performance and cultivar potential [56,57]. Since success in improving quantitatively inherited characteristics such as yield, oil or protein content depends on having a diversity of genes for the character to be improved in the parent lines, recurrent selection has also been used for the gradual enhancement of these traits in soybean - though this method is commonly used in breeding of cross-pollinated species. A limitation of recurrent selection for soybean improvement has been the number of cross-pollinations that can be achieved among selected individuals for creating genetic variation due to recombination which is the indispensable basis for subsequent selection. Emasculation and pollination of soybean by hand is tedious and time-consuming. However, the discovery of genetic male sterility has been successfully applied to recurrent selection schemes to facilitate natural crossing [56,57,66–68].

1.5.3.3 Breeding objectives

Productivity

The establishment of soybean as a major agricultural commodity has been accomplished predominantly through genetic improvement of cultivars. During the development of breeding, particularly in North America, it became a general practice to group soybeans according to their photoperiodic behaviour and area of local adaptation. At present, thirteen maturity groups are distinguished. They are designated starting with 000, 00 and 0 for the earliest ripening groups adapted to the longer days and short summer seasons of southern Canada and northern United States. These are followed by Roman numerals from I to X where the latter stands for the latest maturity group, which is considered for cultivation in the short-day tropical regions on either side of the equator [56,57]. Thus, soybean breeding has been traditionally focused on improving crop productivity, i.e., the increase of seed yield. The rapid growth in production and importance of soybean would not have been accomplished without the progress that was made in the breeding of highly productive, disease- and pest-resistant cultivars, adapted for specific ecological areas and for the mechanization of cultivation practices and harvest [56,57].

Improvement of seed components

Basic breeding efforts devoted to the genetic improvement of soybean quality have intensified over the past 20 years [69–72], although quality traits are not adequately emphasized in cultivar development. This is due to the fact that until now the price paid for commercial soybean is not based on seed composition. However, a number of quality factors are under investigation that may become selection criteria in the near future. Over the past few years, progress has been made in the development and release of improved germplasm with modified seed composition. This development must be seen with regard to special quality traits fitting niche markets due to consumer and dietary demands [57,71–73]. Respective breeding objectives receiving particular attention are enhancing protein content and quality by increased methionine and decreased antinutritional factors (e.g., trypsin inhibitors), as well as improving oil quality by reducing linolenic acid.

Since protein and oil are the seed components which give the soybean crop its value, there have been attempts by breeders to increase their concentration and improve their quality. Besides considerable modifications by environment there is much variation for protein (30–50%) and oil content (14–24%) in the soybean germplasm collection. However, most of the primitive germplasm is not agronomically acceptable immediately [10,71,74,75]. In addition, these traits are usually polygenic which makes it difficult to incorporate them into existing high yielding or locally adapted cultivars without loss in seed yield. Furthermore, attempts to increase oil yield will always result in less protein and *vice versa*. Sucrose metabolism by the developing soybean seed plays a major role in the synthesis of protein and oil. Thus, the biochemical regulation of the partioning of carbon from sucrose into these seed components is probably the basis for the generally negative correlation between oil and protein content [66,76].

Protein quality: Because of the higher value of the protein component of soybeans, breeding for enhanced protein content and quality should receive more consideration than in the past. However, it has been shown that selection for increased protein content tends to reduce overall seed yield as well as oil content. Inheritance of protein content is complex, and genetic potential may be altered through recurrent selection [66,68,77,78].

Soybean meal is a major protein component in livestock feed, since it contains lower levels of crude fiber and almost the entire spectrum of amino acids including those considered essential. Although the sulfur-containing amino acids, methionine and cystine, are present in limiting amounts, soybean meal is a valuable source of lysine, the amino acid of greatest concern in the cereals and most other seeds. Soybean seed, or the meal obtained from it, must be treated with heat to inactivate the trypsin inhibitor, an anti-nutritional factor that prevents proper digestion of the protein. However, besides having several forms of trypsin inhibitors, soybean and wild *Glycine* species also display a large variation in their total content [10,47,79–81]. This may allow for the selection of low-trypsin-inhibitor cultivars in the future.

Oil quality: Due to its abundant supply and favorable fatty acid composition, soybean oil is probably the world's most widely used vegetable oil. While improving crop productivity is still the main objective of soybean breeding programs, increased emphasis is directed to improving the seed oil characteristics. Soybean oil contains a relatively high level of linoleic acid (approximately 50%) which makes the oil suitable for both edible and industrial applications. But it also has an undesirably high content of linolenic acid ($C_{18:3}$, 7–10%) causing inadequate oxidative stability and flavor quality. Industrial processes have been developed to partially hydrogenate and deodorize soybean oil, but these processes are expensive and result in formation of undesirable *trans* geometrical isomers of its fatty acids. Therefore, it is becoming more important to improve oil stability through genetic changes in triacylglycerol composition and structure [82–85]. Recently, the potential of breeding for improving soybean oil quality has been comprehensively reviewed [71,72]. Besides the modification of polyunsaturated fatty acids (PUFAs), other breeding objectives, such as those concerning the alteration of saturated fatty acids are also recognized. In particular, an important goal is the reduction of palmitate in soybean oil because of nutritional concerns about excessive proportions of saturated fats in the human diet, along with an increase in stearic acid in order to maintain a suitable margerine-type oil. Of course, it will be necessary that this oil will have the processing qualities needed by the food industry to substitute palmitate-rich tropical fats.

Discussions on soybean oil quality are predominantly focused on PUFA content. In the past, a number of different means of increasing genetic variability in fatty acid composition were considered, e.g., the use of mutagenic agents or interspecific hybridizations. The latter approach was dismissed quite quickly, since the results of Hymowitz *et al* [61] and Chaven *et al* [62] did not favor wild *Glycine* species as sources for developing a low linolenic acid soybean. However, several sources of low linolenic acid material with approximately 3–4% linolenate, i.e., with a 50% reduction in linolenate content compared to traditional soybean cultivars have been identified in available germplasm or have been developed by mutagenesis [71,86]. With regard to the inheritance of low linolenic content in such lines, e.g., the chemically induced mutants "C1640" and "A5", considerable differences have been established. In "C1640" the low linolenate level is controlled by a recessive allele at the *Fan* locus [87]; the decrease in linolenic acid is accompanied by a corresponding increase in linoleate ($C_{18:2}$). Contrary to the *fan* mutant, the "A5" mutant has decreased levels of both linoleate and linolenate, suggesting a defect in oleate desaturation. The genetic basis for the low linolenic, high oleic phenotype of "A5" could not be attributed to a single nuclear mutation. Rather, it behaves like a quantitative character with partial maternal effects [71,88]. In experiments with extreme temperature regimes, low linolenic lines with the *fan* allele were relatively stable, while the other low linolenic lines were quite sensitive to environmental variation [71,86]. Recently, Fehr *et al* [89] described a second gene locus *Fan2*(A23) involved in the inheritance of reduced linolenic acid content. Via combination with the "A5" mutant line two recombinant lines, "A16" and "A17", with about 2% linolenate in the seed oil were derived. Several tests suggest that soybean oil from such low linolenic lines is stable and has a longer shelf life than soybean oil produced commercially so far [82,84].

With regard to the breeding objective to reduce palmitic acid ($C_{16:0}$) in soybean oil, several lines with lower levels have been reported [71,72]. By recombining two different mutagen-derived lines, Fehr *et al* [90] recently produced a new genotype, which had approximately 4% palmitate, as compared to 10–12% in commonly grown cultivars. The segregation data revealed that the low level of palmitic acid is controlled by different alleles (*fap1* and *fapx*) at two independent loci. Furthermore, the alleles at each locus act in an additive way. The oil of chemically induced high-stearic mutants range from approximately 15% to 30% stearate as compared to 4% for usual cultivars. Genetic studies indicate that the high stearic trait is determined by different recessive alleles at the *Fas* locus [91].

1.6 Oil palm

1.6.1 Importance, distribution and utilization

The oil palm (*Elaeis guineensis*) was domesticated in West Africa, its original area of distribution with humid-tropical climate. Palm oil has a long history of food use, playing an important part in the African village economy dating back for over 5,000 years [10,92]. The oil palm was introduced to Malaysia in 1878, while the development into a plantation crop only happened early in the 20th century. This was followed by plantations in the Belgian Congo (now Zaire) and then in other parts of West Africa during the colonial era [92,93].

The fruit-coat fat of the oil palm has become a major vegetable oil in the past few decades. Production has increased in a rapid manner, more than quadrupling from 1970 to 1990. In 1991/92 world production of palm oil was reported at 11.9 MT [1] taking second place after soya oil. The bulk of palm oil (about 80%) is produced in South-East Asia, notably in Malaysia and Indonesia [1,11]. Concomitant with the production of palm oil, the by-product palm kernel oil also became important, because the similarity of its oil composition to that of coconut oil [94,95]. The oil palm considerably out-yields other oil crops in oil per hectare. In Malaysia, the world's leading producer with more than 50% market share (6.2 MT in 1991/92) [1,11], yields of 5 to 7 metric tons (T) of oil per hectare are achievable under good agricultural management. Usually it can be assumed that for each 100 T of refined palm oil that are produced, there will be 13–14 T of palm kernel oil [10,96,97].

Ong [93] and De Vries [98] described the outstanding development of the Malaysian palm industry, which nowadays is producing and exporting to a large extent processed, i.e., neutralized, bleached, deodorized and/or fractionated palm oil. In order to open new markets for palm oil, substantial resources have been invested in the processing industry. The palm oil industry in Malaysia is based on two main technologies, physical refining (distillation) and fractiona-tion, which have enabled the industry to offer a wide range of partially and fully processed products of crude palm oil to the world market. Refined palm oil is commercially fractionated into the solid palm-stearin fraction and the valuable liquid palm-olein fraction [10,93,94,98].

Since the oil palm produces fruits throughout the year, the fresh fruit branches are harvested and worked up regularly. The crude palm oil contains relatively high amounts of undesirable free fatty acids dependent on the maturity of fruits and the method of oil extraction. Steam sterilization and immediate processing of the palm fruits in oil mills is indispensable because of rapid enzymatic reactions which lead to an increase in free fatty acids, off-flavors and discoloration contributing to deterioration of the oil. Good plantation oil consistently runs lower than 5% in free fatty acids; whereas, native-produced oil was not uncommonly as high as 15, 25 or even 50% in free fatty acid content [10,99–101]. The palm fruit produces two distict oils: crude palm oil from the pulp (about 50% oil on fresh weight basis) and palm kernel oil from the kernel. The oil content of the dried kernels ranges from 44–53% with an average of 50%. The harvested palm fruit bunch provides 20–24% refined palm oil and 2–4% palm kernel oil. Both oils differ appreciably in their composition, physical properties and usability (Table 1.2). Palm kernel oil containing a considerable proportion of lauric acid ($C_{12:0}$, approximately 50%) is classified as a lauric oil - a typical natural source of medium-chain saturated fatty acids important for the oleochemical industry. Palm oil contains nearly equal amounts of saturated C_{16}/C_{18} and unsaturated C_{18} fatty acids [10,95]. Palm oil is very versatile and suitable for food and technical uses. Because of its tendency to cloud at low temperature, edible uses tend to be restricted to hardened fats such as margarines, baker shortenings or cooking fats, frying, confectionery and creams. Only about 10% of palm oil is used for technical applications [10,93,102; see also Chapter 2).

1.6.2 Botany - taxonomy and general description

The genus *Elaeis* belongs to the palm family (*Palmae*), an important member of the monocotyledonous group of angiosperms. It is included in the tribe *Cocoideae* - together with the genus *Cocos*. Within the genus *Elaeis* three species are distinguished: the economically important oil palm (*E. guineensis*) originally native to Africa; a South American relative, *E. oleifera* (syn. *E. melanococca*); and a less well-known species of American origin, *E. odora*. *E. guineensis* and *E. oleifera* hybridize readily, producing fertile offspring and thus suggesting a close relationship in spite of their different areas of origin [96].

E. guineensis is one of the larger palm species; it has a stem that can reach a height of 25–35m topped by 35 to 60 pinnate leaves. The amount of growth is connected with the number of leaves produced each year. The young tree aged 3–5 years bears the first fruit bunches and the adult stage begins at 6 to 9 years. 12 to 15 year-old oil palms are most productive. The lifespan of the oil palm is extraordinary, since the first palms brought to Indonesia have an age of 140 years and are still bearing fruits. However, an oil palm has an economically productive life of 20–30 years, and replanting is usually carried out after about 24 years [92,96].

The palm plant is monoecious, which makes cross-pollination necessary because it produces separate male and female inflorescences from axillary buds at different times, i.e., the pure male and pure female cycles rarely overlap in one plant. The inflorescence is initiated at the same time as the leaf, about 33 months prior to flowering. Almost 6 months after pollination, a fruit bunch weighing 15 to 20kg and consisting of approximately 1,000 to 1,500 fruits is produced. Both the vegetative and the generative characters, e.g., the successive cycles of sexual differentiation, the number and size of the fruit bunches, are under genetic control but also strongly influenced by environmental factors [96]. The mature fruit is a deep orange-red drupe containing the pulp (mesocarp), the shell (endocarp) and the kernel. The fibrous mesocarp is rich in oil and yellowish-orange colored due to its high content of carotene. The amount of oil in mesocarp averages 40–60% of fresh weight, but seems also to be greatly affected by environment. Two alleles determine the constitution of the shell defining the three naturally occurring fruit types varying in shell content and thickness: *dura* with thick shell (2–8mm, 20–40% of the fruit); *pisifera* is shell-less; and *tenera*, the thin-shelled hybrid (0.5–4mm, 5–20%). The endocarp encloses the kernel consisting of the embryo and the fatty endosperm [10,92,97].

1.6.3 Breeding

1.6.3.1 Sources of genetic variation

Inter-population crosses between *dura* and *pisifera* types have been effectively applied in improving mesocarp oil production. The *pisifera* fruit type is rare, and since it is usually female-sterile, it is of no interest for cultivation. The most common fruit in natural palm groves is the *dura* type with thick shells. However, the hybrid type *tenera* is preferable to *dura* as planting material due to the higher proportion of oil-bearing mesocarp, 60–90% per fruit weight as compared to 20–65%, respectively [92,97].

Since the chromosome number for both species, *E. guineensis* and *E. oleifera*, is $2n=32$, interspecific hybrids are feasible accompanied by almost normal chromosome pairing. The South American species *E. oleifera* is very attractive for oil-palm breeding because of its low vertical growth rate, its excellent oil quality, and as a source of resistance to several palm diseases. However, the oil yield of *E. oleifera* is low, primarily because of the low oil content of the mesocarp [92,96,103].

1.6.3.2 Breeding procedures

The progress in breeding work on oil palm has been comprehensively reviewed by Hardon *et al* [92] and by Gascon *et al* [96]. *E. guineensis* is an outbreeder normally propagated by seed. Natural oil-palm populations exhibit a wide variation and are mainly composed of heterozygous individuals. In general, breeding of tropical perennial oil crops is very consuming in time, labor and space. For instance, in breeding programs applying reciprocal recurrent selection [56] the duration of one improvement cycle is in the order of 12–14 years due to the monoecious character and the slow development of the palm plant. The plants are planted usually at a density of about 140 trees per hectare (ha), so that the land area needed to evaluate a single progeny is roughly 0.5ha [92,96].

With the shift from *dura* to *tenera* planting material, yield was increased by 25–30%. Thus, all the seed now produced for planting material is of *tenera* type obtained by artificial fertilization of *dura* mother-trees with pollen from *pisifera* trees. Some 80 to 100 million seeds are produced annually, enabling the planting or replanting of 300,000ha of palm groves each year. For the production of improved *teneras* breeding programs must provide *duras* and *pisiferas* with high combining ability [92,96,97].

Fruit characteristics, such as pulp proportion of the fruit, the quantity of kernel and the average weight of the fruit, possess relatively high heritability (i.e., degree of genetic determination of a trait), thus improvement should be feasible - even through simple mass selection. Therefore, it is not surprising that application of mass selection programs in the early history of palm breeding led to ecotypes with better quality fruit bunches; e.g., the improved *dura* ecotype "Deli" with few, large bunches, which is considered to have the best fruit composition so far. However, most of the economically interesting traits, such as total yield of fruit bunches as well as oil content of the pulp, are quantitatively inherited and modified by environmental factors. Consequently, they display continuous variation. Selections carried out within ecotypes gave little progress towards yield enhancement because the locally adapted material frequently had a narrow genetic basis and showed a certain degree of inbreeding after selection. However, ecotypes of different origin displayed a larger variation in many important characteristics. While crosses between ecotypes of different origin complementing each other in terms of yield components, such as number, size and average weight of bunches, gave good results, reciprocal recurrent selection (RRS) has been adopted as a breeding procedure to improve the oil palm's yield [96]. The suitability of RRS is not generally accepted, since the observed superiority in inter-population ("inter-origin") crosses may also partly be explained by the avoidance of further inbreeding. After Hardon *et al* "it is obviously impossible ever to realize the maximum potential for specific combining ability between two populations because only few of the total number of palms can be tested adequately on their cross-performance" [92]. This is considered as a limitation of the RRS breeding scheme, since only the parents of the progenies with the best performance should be used in the next cycle for further crossing. Malaysian breeding programs are aiming at widening genetic variation in basic populations by inter-crossing material of various origins, but selection is based on family and individual palm performance emphasizing general combining ability. The percentage of fruit in the bunch depends particularly on an effective pollination, which is greatly affected by the environment. Thus, in South-East Asia, investigations are under way to enhance the fruit-to-bunch ratio via increased pollination by a more efficient weevil pollinator (*Elaeidobius kamerunicus*) imported from West Africa [92,98].

In the past few years, biotechnology has become a valuable supplementary tool to oil-palm breeding [104,105; see also Chapters 3 and 6]. Because of the long generation time and the observed variation within hybrid seed progenies it is apparent that there is a considerable scope for improvement if the best of the hybrid palms could be propagated vegetatively *in vitro*. Such multiplication techniques possess many advantages [92,104,105]. Nevertheless, the selected clones and their plantlets must be subjected to field trials since abnormalities have been observed in some clones. Active research is being carried out to understand their causes in order to prevent such abnormalities in tissue-culture-derived plant material [93,96].

1.6.3.3 Breeding objectives

Productivity

Over the past 40 years, the average oil yield of Malaysian oil palm plantations has been increased from 1.3 to over 5T/ha/year; both breeding and improved cultivation practices have contributed equally to the four-fold yield increase [97]. Although the oil palm has the highest oil yield compared to other oil crops, one of the most important objectives of breeding is still reducing the cost of production via increasing the productivity of oil palms. It is noteworthy that tissue culture has become a valuable supplementary tool to achieve this goal. Since the palm is strongly heterozygous, individual plants display a wide range of variation, particularly in yield. The establishment of homogeneous plantations using multiplicated plantlets (clones) from the best trees will probably improve oil yield. However, the effect will obviously be limited by genotypically determined tissue-culture ability [93,104,105].

The vertical growth of the trees determines the operation and harvest costs of a palm plantation. Since harvesting trees higher than about 14m is almost impossible, and with regard to an assumed cultivation time of 20 years, the goal is to develop trees with a mean annual upward growth of 0.65 to 0.7m. Resistance to diseases and tolerance to drought are important goals in breeding programs aimed at a better adaptation to each respective environment. In particular, drought tolerance is an important objective since many palm oil producing countries have a marked dry season of 2–3 months [92].

Improvement of fruit components

Crossing the *dura* and *pisifera* to give the thin-shelled *tenera* fruit type has resulted in only small differences in the yield of fruits, fruits per bunch or mesocarp oil, but large differences exist in fruit composition. Improved partitioning of dry matter within the fruit has resulted in a 30% increase of oil yield due to the higher proportion of mesocarp at the expense of shell [97].

Oil quality: There is a considerable difference in fatty acid composition between *E. guineensis* and *E. oleifera*, since the latter contains about 75–80% unsaturated fatty acids, i.e., 55–70% oleic and 10–25% linoleic acid [103,106,107]. One trend in producing countries to allow for both edible and non-food usability of the oil, is directed towards increasing the proportion of unsaturated fatty acids, particularly oleic acid [93,96,108]. In programs intended to improve palm oil as a source of oleic acid (approximately 40%), selection on *E. guineensis* germplasm as well as breeding and exploiting interspecific hybrids between *E. guineensis* and *E. oleifera* has shown the possibility of increasing the oleate content [106,108–110].

1.7 *Brassica* species

1.7.1 Importance, distribution and utilization

Rapeseed is one of the major global sources of vegetable oil, which comes from several *Brassica* species belonging to the family *Cruciferae* (syn. *Brassicaceae*). These oilcrop species include *B. napus* (rape), *B. rapa* (syn. *campestris*) with turnip rape and the Indian sarson ecotypes, as well as *B. juncea* (Indian or Brown mustard), or other related species occasionally included with rapeseed

[111–113]. *Brassica* vegetables and oilseeds were undoubtedly among the crops domesticated by early man. Domestication must have occurred whenever and wherever the usability of locally adapted cruciferous weeds was recognized. There is evidence that some vegetable types were in wide-scale use in the neolithic age, and references to its use or that of close relatives appear in the earliest writings of Asian and European civilizations. Early records in ancient Sanskrit literature suggest that rapeseed, notably the Yellow sarson type ("*Siddhartha*"), was cultivated in India as early as 4000 B.P.. The introduction to China and Japan must have happened around some 2000 years ago. Other Brassicas, such as turnips (*B. rapa* ssp. *rapifera*), cabbages or kales (*B. oleracea*), represent an important ancient group of vegetables. For example, the Greeks and Romans were interested in developing an array of cultivated forms, e.g., stemkales and headed cabbages. *B. nigra* (Black mustard) is mentioned in Greek literature for its medicinal value. The cultivation of rapeseed in Europe, north of the Alps, is considered to have begun as early as the 13th century. In the latter part of the Middle Ages it became the most important source of lamp oil, until it was replaced by petroleum towards the end of the 19th century. The oil was also widely used in soap making [111,114].

Brassica oil crops are of particular importance because they are well adapted to both the temperate and relatively cool climates, especially where moisture is adequate. There are both winter and spring (summer) forms of *B. napus* and *B. rapa*, while *B. juncea* only occurs as summer form. In Canada, as well as North and Central Europe, rape and turnip rape are grown commercially as rapeseed. In the semitropics of Asia, notably India, *B. rapa* ecotypes (Yellow sarson, Brown sarson and toria) and *B. juncea* are major vegetable oil sources. In China, all three species are in cultivation, but the winter form of *B. napus* dominates. *B. napus* winter cultivars are generally the highest-yielding ones under good growing conditions. Under a summer growing regime, *B. juncea* is more resistant to shattering, has better resistance to blackleg disease (*Leptosphaeria maculans*), and tends to have higher yields than either of the other two *Brassica* oilseed species, particularly where moisture is limiting. *B. rapa* is favored where early maturity is desired because of short growing seasons [112,113,115].

In *Brassica* oilseeds, the occurrence of two components traditionally distinguished them from other major oilseeds. Both components, erucic acid and glucosinolates, were considered anti-nutritional for humans and for animals. Firstly, the seed oil formerly contained approx. 25–50% erucic acid (cis-13-docosenoic acid, $C_{22:1}$), referred to as HEAR (high-erucic acid rapeseed). It was claimed that erucic acid had an adverse effect on experimental animals when fed as a very large proportion of the diet. Secondly, the meal contained considerable amounts of glucosinolates, a group of chemically related thioglucosides, some of which had goitrogenic and other anti-nutrional properties [10,111,116–118]. However, in Asian countries, notably India and China, the oil of domestic *Brassica* crops has been used commonly for edible purposes. In contrast, in the western world the use of rapeseed was almost exclusively for the production of oil for non-edible applications, e.g., initially for domestic usage and as a lubricant for steam engines, and later on as a feedstock for the chemical industry. The situation changed after plant breeders successfully altered the chemical composition of rapeseed [119–121]. Firstly, the introduction of oils virtually free of erucic acid enhanced its attractiveness as an edible oil. Secondly, low-erucic acid rapeseed (LEAR) cultivars with very low glucosinolate content in the meal additionally increased its potential as protein supplement in animal feeds. Currently, in the major rapeseed producing areas (Canada, Europe) production has been shifted almost completely to *B. napus* and *B. rapa* cultivars with minor quantities of both erucic acid and glucosinolates in the oil and the meal, respectively [122]. In Canada, this double-low seed type is called "Canola" to facilitate identification of the seed and the products (oil and meal) made from it. Canola is currently defined as having less than 2% erucic acid in the oil and less than 30 micromoles of the aliphatic glucosinolates per gram of oil-free meal [121]. Recently, *B. juncea* has also been developed to Canola quality, which could have a significant impact on the Indian subcontinent where *B. juncea* is a major oil crop [121,123,124].

Largely due to the major improvements in seed quality described above, *Brassica* oilseed crops have become the third most important world source of vegetable oil (9.4 MT in 1991/92) after soybean and palm. Furthermore, rapeseed also ranks third in the production of oilseed meal after soybean and cotton. The major rapeseed producing countries are China, India, Canada, Germany and France [1,11,43]. Generally, rapeseed containing about 40–45% oil and 20–25% protein, is grown and processed for its oil [10,126,127]. It is important to be aware that the *Brassica* oilseed crops yield three types of seed quality, viz., HEAR, LEAR, and Canola, which differ substantially in relation to human and animal nutrition as well as industrial usage [118]. LEAR/Canola oil is utilized primarily for food purposes as margarine, shortening, cooking and salad oil. In Canada, Canola oil accounts for about 60% of the domestic production of refined oil [121,128]. Although the by-product, Canola meal, has about 40% protein (on a dry matter, oil free basis) and a relatively well-balanced amino acid composition, it has not received the same acceptance as soybean meal in the animal feed industry [10,122]. *Brassica* oil of the HEAR type with proportions of erucic acid substantially higher than 55% is still being sought by breeders, biotechnologists and chemists for use in industrial processes. High oleic acid rapeseed (HOAR) oils (over 75% oleate) are suitable for both edible and industrial applications. Thus, besides the steady growing Canola market, the opportunity exists to supply speciality *Brassica* oils for several niche markets [122].

1.7.2 Botany - taxonomy and general description

Within the *Cruciferae* (*Brassicaceae*) the genus of the greatest economic importance is *Brassica*. The genomic chromosome number in *Brassica* ranges from $x=7$ to $x=19$ [129,130]. The botanical relationship among the most common *Brassica* species is presented by the well-known triangle of U [131]. The three amphidiploid species, *B. napus* (genome AACC, $2n=4x=38$), *B. juncea* (AABB, $2n=4x=36$), and *B. carinata* (BBCC, $2n=4x=34$), have most likely originated in nature from diploid ancestors through unidirectional hybridization followed by spontaneous chromosome doubling. Additionally, it is recognized that *B. nigra* (BB, $2n=16$) is more distantly related to both *B. rapa* (AA, $2n=20$) and *B. oleracea* (CC, $2n=18$) [114,130,132,133]. Nevertheless, there is cytological evidence that the three basic diploids are themselves secondary polyploids, which probably originate from a common archetype with a lower basic chromosome number ($x=6$). Thus, interchromosomal similarities (homoeology) exist within as well as between the diploid genomes [114,130,134].

The botany and taxonomy of *Brassica* species is comprehensively described elsewhere [111,112,117,129,135]. During domestication, *Brassica* species have been utilized and modified for almost every part of the plant, including roots, stems, leaves, terminal and axillary buds, and seeds. *B. oleracea*, *B. rapa*, and *B. juncea* are highly polymorphic. *B. oleracea* includes important vegetables, forage and fodder crops, such as cabbage, cauliflower, broccoli, marrow-stem kale, kohlrabi, brussels sprouts. *B. rapa* has three well-defined groups: oleiferous forms which are widely cultivated in Canada and northern Europe (turnip rape) as well as the Indian subcontinent (Yellow sarson, Brown sarson, toria); rapiferous forms (turnip) which have a worldwide distribution; and the leafy forms (Chinese cabbage, pak-choi) which are favored Chinese/East Asian vegetables. *B. juncea*, oleiferous forms are cultivated in India, Middle East, southern China, eastern Europe and in other moisture-deficient areas. Leafy forms of *B. juncea* are important vegetable and forage crops in China. *B. napus* is important as an oilseed and forage crop predominantly in Europe, but to a lesser extent also in China, Japan and North America. *B. napus* also has a rapiferous form (rutabaga or swede) which has minor importance as fodder or vegetable. *B. nigra* (Black mustard) is used mostly as condiment in India and some European countries. *B. carinata* (Ethiopian mustard) has local importance in the Ethiopian region as oilseed and forage crop [111,135].

With regard to floral characteristics *Brassica* crops show considerable differences. *B. napus*, *B. juncea*, and *B. carinata* are predominantly self-pollinated crops, although under field conditions, an average of about one-third outcrossing is observed, particularly in *B. napus*. Most *B. rapa*, *B.*

oleracea and *B. nigra* strains, on the other hand, are forced to natural cross-fertilization due to a sporophytic self-incompatibility (SI) system. However, self-pollinating lines occur in *B. rapa*, such as the Indian Yellow sarson types, as well as some cauliflowers and broccolis [115,136–138].

1.7.3 Breeding

1.7.3.1 Sources of genetic variation

The biological and chemical diversity available in *Brassica* crops has been comprehensively described in excellent monographs [116,117]. In comparison to *B. napus* and *B. carinata*, which are not known to occur wild in nature, and are considered to be of rather recent origin, the polyphyletic origin of *B. juncea* has given rise to large morphotypic variation and goes back to ancient times - at least in the Middle East and India, where the oldest forms are found [114,123,135]. Additionally, the development of oleiferous *B. juncea* types was obviously due to the existence of oily *B. rapa* forms participating in the interspecific cross [114]. It is still not clear which parental combinations of *B. rapa* and *B. oleracea* forms gave rise to the amphidiploid *B. napus*, and whether natural hybridization occurred more than once resulting in different *B. napus* forms.

However, it is most likely that *B. napus* originated in the Mediterranean region, the area of overlap between the more widely distributed *B. rapa* and *B. oleracea* [130,133,139–142]. Due to its origin, *B. napus* contains the genetic variability of only those sub-species or varieties of *B. oleracea* and *B. rapa* which were involved in the original cross(es). Thus, genetic variation seems to be rather limited in *B. napus*. In particular, for winter rape only three main types are described as local land races. They have evolved in Europe under different climates by natural selection displaying variation in vegetative growth and winter hardiness. The first released cultivar "Lembkes", selected in Germany from a Mecklenburg land race in the early 20th century, was extensively exploited in French, Swedish, German and Polish breeding programs. The genetic base is nowadays even narrower because the introduced double-low quality again originates from single sources, the spring cultivars "Liho" and "Bronowski". Consequently, there is still a need to introduce new genetic variation, since most of the European winter cultivars share common parentage [115,136].

An impressive strategy to broaden the genetic basis is the production of resynthesized rapeseed by crossing the original ancestors, *B. oleracea* and *B.*

rapa. For such hybridizations a variety of biotechnological tools, e.g., embryo rescue techniques as well as protoplast fusion, have been used to circumvent existing incompatibility barriers [30–32,143,144]. Some resynthesized rape forms have already been used in rapeseed breeding [136,145–147; see also Chapter 6]. Diederichsen and Sacristan [148] reported the successful transfer of clubroot resistance from *B. oleracea* to rapeseed by crossing *B. oleracea* and *B. rapa*; the resynthesized rapeseed progeny were included in a practical breeding program. Mithen and Magrath [149] derived synthetic lines of *B. napus* resistant to blackleg disease (*Leptosphaeria maculans*) via embryo culture. Recently, crosses between *B. rapa* ssp. *trilocularis* (Yellow sarson) and several selected cauliflowers (*B. oleracea* convar. *botrytis* var. *botrytis*) have been carried out in order to create new oilseed rape germplasm with a high erucic acid content. The offspring display desirable variation in the content of major fatty acids, so that it is possible to produce breeding lines with an erucic acid content of 60% or even more [30,150].

1.7.3.2 Breeding procedures

The majority of *B. napus* and *B. juncea* cultivars are pure lines derived from breeding schemes designed for self-fertilizing crops, i.e., pedigree selection or modifications thereof. In the self-incompatible *B. rapa* species, recurrent selection as a standard procedure for improving populations has been employed most frequently. Backcrossing has been successfully used to transfer simple inherited traits, such as low erucic acid and glucosinolate content into adapted breeding materials of both self-fertilized and cross-fertilized *Brassica* species [115,120,136]. Nevertheless, significant levels of heterosis for seed yield, that can be more than 30%, have been reported for both *B. napus* and *B. rapa* [120,151–154]. In particular, the strongest heterotic effects occurred in experimental crosses between material of the most distant geographical and genetical origin [155,156]. Thus, extensive efforts are underway to develop cytoplasmic genetic male-sterile and restorer lines as the most promising system for the production of hybrid cultivars [120,121].

Rapeseed (*B. napus*) is currently among the crop species most amenable to improvement through biotechnology [30,157; see also Chapter 6]. For instance, it is possible to obtain haploid and subsequently doubled haploid (DH) plants reproducibly through anther and/or microspore culture. The principal advantage of the haploidy technique is the rapid fixation of segregating genotypes, occurring in lower frequency, in which recessive genes coding for specific traits are combined in the homozygous condition. Thus, utilization of microspore culture can allow a substantial acceleration of the breeding cycle [158]. Due to the generally high response of *B. napus* genotypes, many oilseed rape breeders are now testing DH-lines in their breeding programs and a large number of lines are already in advanced stages of field testing, thus some of them may soon be released as licensed cultivars.

In addition to the applied aspects, rapeseed is one of the most interesting species to be genetically transformed using recombinant DNA technology, especially with regard to oil quality [21,38,39,159,160; see also Chapter 6].

1.7.3.3 Breeding objectives

Productivity

Unlike soybeans, peanut, and most other oilseeds, progress in breeding *Brassica* oil crops was always connected with drastic improvements of seed quality, which was followed by a relatively quick acceptance by growers and the processing industry. However, the research towards seed quality does not ignore the importance of high yields. According to the morphological and agronomical characteristics, the yield of rapeseed consists of the number of siliques per unit area, the number of seeds per silique and the 1,000-seed weight. The improvement of productivity also includes several agronomic parameters, such as early maturing, resistance to lodging and shattering, as well as resistance to weeds, insects and particularly to the major diseases (*L. maculans*, *Sclerotinia sclerotiorum*, *Verticillium dahliae*, etc.). In the future, hybrid cultivars have to show their potential to raise the yield of both, *B. napus* and *B. rapa* [120,121].

Improvement of seed components

The gross composition of *Brassica* oil crop species varies widely depending on both genetic (i.e., species, variety, cultivar) and environmental factors (e.g., temperature). The oil content ranges from 36 to 50% (on a dry matter basis), while the oil-free meal contains 33–48% protein [10,111,127,161,162]. The improvement of oil content is an important goal due to the primary economic value of the oil component and its relatively high heritability [163], plus the ease and speed with which oil content can be measured by non-destructive NMR techniques. Although oil and protein content are negatively correlated, improvements can be achieved through selecting for the sum of the two seed components [119,127,163]. Low glucosinolate rape meal still bears several problems. Besides the presence of undesirable compounds like sinapic acid esters, phytic acid and phytates, phenolic acids and tannins, the comparatively high crude fiber content (approximately 15% of dry oil-free meal) is disadvantageous [10,164,165]. Due to the small size of the seeds, the hull, which accounts for about 10–20% of the seed weight, imparts most of the fiber content to the meal [111,126]. Currently, the most promising route to reducing fiber and hull content genetically, is to breed cultivars with a yellow seed coat, like pure yellow-seeded cultivars occurring in the sarson subspecies of *B. rapa* or in *B. juncea*. Since yellow seed coats are significantly thinner than brown or black ones, the development of pure yellow-seeded *B. napus* cultivars with agronomically acceptable performance still remains an important goal in quality breeding aiming in an increased oil and protein content [112,120,122, see also Chapter 3].

Oil quality: The value and suitability of rapeseed oil for nutritional or industrial purposes is again determined by its fatty acid composition. The identification of naturally occurring zero-erucic mutants in both *B. napus* and *B. rapa*, was indeed the first discovery heralding the era of mutant-derived quality improvement in oil crops [166–168]. Canola quality rapeseed almost lacking nutritionally undesirable long chain fatty acids meets all the requirements of a prime edible oil [122,128]. However, the quality of the oil can still be improved further by decreasing the linolenate content from on average 10% to less than 3%, to give an enhanced shelf life [122,169–172]. In the last two decades, improvements in the C_{18} fatty acid composition in rapeseed (*B. napus*) were achieved by selecting altered linoleate/linolenate genotypes after chemical mutagenesis. The fatty acid profiles of these lines indicated that nearly all of the linolenic acid was being directed to linoleic acid and that the level of oleate increased only insignificantly [168,169,173–175]. In 1988, the spring rapeseed cultivar "Stellar", which produces an oil containing less than 3% linolenate, was released for commercial production in Canada, although its agronomic performance is not satisfactory [176]. Recently, Kräling and Röbbelen [172] reported the transfer of desirable alleles controlling the low linolenic trait into winter rapeseed by different breeding strategies.

Due to strong environmental and marked maternal influences, only low correlations have been found between the contents of polyenoic fatty acids determined from half-seeds and their progeny. The most important factor influencing the biogenesis of the unsaturated fatty acids is the temperature prevailing during seed development [170,171]. Only recently was research carried out or reported on mutants with reduced levels of PUFAs by blocking oleic acid desaturation. The development of Canola cultivars with reduced levels of PUFA accompanied by higher oleate content would produce a dietary oil with additional markets [17]. For industrial applications, a very high content of oleic acid (80–90%) is preferred because this is most suitable for certain consecutive chemical reactions [20,177, Chapter 2].

Following gamma radiation treatment of "Regent" spring rape, Wong *et al* [178] identified a single M_4 family, FA677, with substantially increased oleate content, viz., 79%. By combining the high oleic with a low linolenic trait through crossing and subsequent progeny selection, recombinant genotypes with very high oleate (>85%) and low linolenate levels (<3%) were found [179]. Auld *et al* [180] used chemical mutagenesis via EMS-treatment for the same objective. Starting materials in this case were the spring HEAR cultivar "R-500" (*B. rapa*,Yellow sarson) and "Cascade" winter rapeseed (*B. napus*). In the latter, the mutant "X-82" was identified as the most promising line, displaying 84.4% oleate and 6.6% PUFA in the M_2 generation, vs. 26.7% PUFA in "Cascade". The best M_3 selections of "X-82" had oils with >86% oleate , <3% linoleate and <3% linolenate. Furthermore, it was possible to select four M_4 lines with approximately 80% oleic acid and satisfactory seed yield.

Traditional varieties of *Brassica* oilcrop species typically contain appreciable levels of long-chain fatty acids, viz., eicosenoic ($C_{20:1}$), erucic acid ($C_{22:1}$) and nervonic acid ($C_{24:1}$). Studies on the inheritance of erucic acid in *Brassica* oil crops revealed that erucic acid is under the control of the embryonic genotype. In monogenomic species like turnip rape (*B. rapa*) the erucic acid synthesis is governed by one gene locus. Consequently, in the amphidiploid species (*B. napus*, *B. juncea*) two major gene loci are participating. Multiple alleles occur at each locus acting in largely additive manner. Homozygous genotypes with various alleles produce levels of erucic acid ranging from less than 0.1% to about 60%. In the winter forms of *B. napus*, alleles are present which in single dose give about 15–18% erucate [181]. The accumulation of fatty acids of preceding chain elongation steps, viz., oleate and eicosenoate, is probably influenced by the same major gene loci [182,183]. With regard to oleic acid operating as substrate for desaturation at the same time, it is assumed that in HEAR (*B. napus*) at least two minor gene loci are involved additionally, i.e., controlling the desaturation of oleic acid to form linoleate and linolenate, respectively [181,183].

In resynthesized rapeseed (*B. napus*), produced with the aim of genetic modification of oil quality [30,150,184], the effect of alleles responsible for a specific erucic acid content is probably also influenced by the level of ploidy and genetic background, i.e., the parental *B. oleracea* (CC) and *B. rapa* (AA) genomes. Due to the procedure of resynthesis commonly used, i.e., hybridization of the monogenomic parents followed by artificial diploidization, synthetic *B. napus* forms are homozygous lines, so that a change in their fatty acid composition, as compared to that of their parents, could also be explained by interactions between the non-homologous genes of the A and C genomes. These interactions have to be attributed to epistasis (non-allelic gene actions) rather than dominance [113,184]. Lühs and Friedt [150] have suggested that it may be possible to achieve recombinants with new allele combinations via introgression of resynthesized germplasm into conventional high erucic acid breeding material. However, the possibility to increase the erucic acid synthesis by accumulation and combination of desirable alleles is obviously limited due to the restriction of the subsequent triacylglycerol synthesis [30]. Cruciferous seed oils seem to contain erucic acid almost exclusively in the *sn*-1 and *sn*-3 positions of the glycerol backbone [185–187], so that a maximum of approximately 60–63% erucic acid is currently achievable [150,188,189]. Mutants in rapeseed with a desirable erucic content beyond the theoretical limit (>66%) may be unlikely to occur. With regard to well-known industrial and non-food uses (see Chapters 2 and 6) trierucoylglycerol (trierucin) would make the processing of comparatively pure erucic acid much easier and more attractive commercially. However, in the near future it might be possible to introduce the property for producing trierucin-enriched seed oil into *B. napus* with sophisticated techniques of genetic engineering [35,37,190,191].

1.8 Sunflower

1.8.1 Importance, distribution and utilization

The cultivated sunflower (*Helianthus annuus*) is one of the most important oil crops of the world. This New-World crop probably originated in the Southwest of North America and was domesticated in the central-western area of the present USA before 5000 B.P.. After the discovery of America, sunflower was introduced to Europe in the 16th century. Initially, the plant was grown as an ornamental and later for food and medicinal purposes. Sunflower subsequently reached Russia where it became an oil-bearing crop plant during the 19th century [192–195]. Sunflower performs well in most temperate climatic regions, with significant production in each of the six crop-growing continents. Europe and the former USSR (now CIS) are the largest producers of sunflower seed, accounting for about 55% of the world's total production of about 22.5 MT. Among the individual countries the former USSR and Argentina are the leading producers followed by France, USA, China and Spain [1,11]. The oil, with about 8 MT, ranks fourth in world production of vegetable oils [1,2]. Sunflower seed derives almost 80% of its economic value from the oil, while defatted meal is the main by-product after oil extraction. The meal, containing about 40–45% highly digestible protein, is suitable for human food or livestock feed [10,23,196]. The oil is generally considered a premium edible oil because of its high linoleic acid content followed by oleic acid. Since over 90% of the fatty acid composition is made up by oleic and linoleic, the reciprocal proportions of both have been an important concern of breeders and processing industry. Although sunflower is primarily grown for extraction of its oil, there is limited production of sunflower of non-oilseed type used for human confectionery or as bird feed [10,197].

1.8.2 Botany - taxonomy and general description

The genus *Helianthus*, belonging to the *Compositae* family, comprises probably over 100 species; about 50 species found in North America have been described in more detail. The basic chromosome number is $x=17$; diploid, tetraploid, and hexaploid species with annual or perennial growth habit have been identified [192,198,199]. Besides several sunflowers being utilized as ornamentals, the genus *Helianthus* comprises two crop plants: *H. annuus* ($2n=2x=34$), the common sunflower, cultivated predominantly for its oily seeds; and *H. tuberosus* ($2n=6x=102$), Jerusalem artichoke (topinambour), a perennial species which is grown occasionally for its tubers. The cross *H. annuus* with *H. tuberosus* has been widely exploited in Russia for the improvement of the disease resistance of cultivated sunflower [7,192]. *H. petiolaris* ($2n=2x=34$), an annual wild species, has been crossed with *H. annuus*, and out of backcross progeny a cytoplasmic male sterility (CMS) system was developed [200].

With regard to morphology, *H. annuus* itself is a highly variable species. There are weedy and wild sunflower forms, which show profuse branching, have smaller heads, and are usually taller than the cultivated sunflower. Modern sunflower cultivars, reaching a plant height between 1.5 and 2.5m at flowering, have a stem with 20 to 30 leaves and only one apical inflorescence or head. The leaves represent the largest part of the total assimilating area, and the heliotropic movement of the foliage leaves increases the photosynthetic efficiency. The inflorescence is also heliotropic until the beginning of flowering, after which it remains fixed facing eastwards. The flowering head has normally a maximum diameter of 15 to 30cm, which consists of the mostly yellow and sterile ligulate or ray flowers, and the fertile disc- or tube flowers. Sunflower is generally cross-pollinated, and wild populations or open-pollinated cultivars of sunflower are highly self-incompatible. In contrast, cultivated hybrid cultivars often possess high levels of self-fertility.

The 1–2 rows of ray flowers form the visual apparatus, which attracts bees, bumble-bees and other pollinating insects. Flowering of the complete disc florets proceeds from out- to inside and takes 8–12 days under favorable conditions. The number of disk flowers is variable; usually there are 600–1,200 flowers, rarely 3,000 to 4,000. The number of fruits (achenes) varies between 250 to 1,500/head, arrayed in a spiral pattern in the large circular structure initially surrounded by the ray flowers [7,193]. The average size (length 7–25mm, width 4–13mm) of the achenes decreases from the outer positions towards the centre of the head. Physiological maturity, defined as maximum dry matter accumulation, occurs between the 30th and 40th day after start of flowering. At this stage, the achenes, with about 35% moisture content, have attained maximum dry weight and oil accumulation [201–203]. There is considerable variation in oil content and fatty acid composition with regard to position of the seeds within the head as well as the size of individual heads [7,204,205]. At maturity, the heaviest achenes with the highest oil content are found in the outer circular positions. Smaller heads, with a diameter less than 20cm, have been shown to carry seed with up to 40% more oil than seed from heads of 30cm size. There are usually unfilled achenes at the centre of the head. The optimum temperature range for seed production is 20–25 °C. Water stress (soil moisture deficiency) and temperature, which remains above 30 °C at flowering and seed filling, are believed to reduce seed yield. Additionally, the oil content is reduced under these conditions, whereas the protein content is increased due to the distinct accumulation pattern and the generally inverse relationship of the two seed storage components [7,202,206,207]. The achene consists of the shell (pericarp) and the seed (kernel). The 1,000-seed weight varies from 40–60g (oilseed type) to more than 100g (non-oilseed type). The level of oil content is considerably influenced by the proportion of the shell. The oil content of the fruit averages 40–50%, and of the kernel 50–60%, respectively. The protein content of the fruit and kernel varies from 15 to 22% and from 20 to 30%, respectively [10,208].

Although originally a plant of subtropical and temperate zones, sunflower is highly adaptable to a wide range of environments due to breeding and selection. Further, it is a spring-annual crop, which can tolerate drought better than many annual crops because it is more efficient in extracting soil moisture. With regard to the response to environmental conditions, first of all, sunflower is relatively sensitive to light deficiency due to shading. Apart from genotypic variation, the plant can be considered more or less day-neutral, i.e., the effect of photoperiod is negligible compared to the influence of other environmental factors, particularly temperature. For instance, high temperature is the most likely factor which reduces the time to maturity. Sunflower grows well within a temperature range of 20–25 °C, however, maximum relative growth rates have been observed at temperatures around 28 °C. Low temperatures (ca. 15 °C) extend the duration of single developmental phases and reduce net photosynthesis, whereas frost damages the plant to some degree at all stages of development [7,207,209–211].

The components of oil yield per area are: number of heads (i.e., plants) per area, number of seeds (i.e., achenes) per head, 1,000-seed weight, and oil content of achenes. Optimum plant density is 4 to 8/m^2 influenced by the rainfall or irrigation available (300 or more than 500mm/year, respectively). The sunflower is able to compensate for plant losses to a certain extent by increasing the single head diameter and, consequently, the number of achenes per plant. Under long season conditions, vegetation period is about 130–150 days. Total dry matter can reach 12–20T/ha under experimental plot conditions, indicating the potential of further improvement. But on average, only 8–11T/ha total dry matter and 2–3T/ha of achenes are obtained and the average yield of oil amounts to 1.2–1.4T/ha. Thus, because of its outstanding carbon fixation capacity sunflower requires a much shorter vegetation period than winter rapeseed (*B. napus*) for the production of a comparatively high quantity of biomass, seed or oil [211].

1.8.3 Breeding

1.8.3.1 Sources of genetic variation

In sunflower breeding programs, a wide range of germplasm resources, such as plant introductions, open-pollinated varieties, populations, and breeding lines, are utilized. However, wild species represent the most diverse source of genetic variation for agronomic type, disease resistance, and seed characteristics [193,212]. Genetic diversity is one of the reasons making sunflower a widely adapted crop plant with a high potential. Utilization of wild species for improving cultivated sunflower towards agronomically and morphologically desirable traits as well as disease and pest resistance is reviewed by Seiler [199]. The wild species also have the potential to contribute additional cytoplasms for hybrid seed production [199,213,214]. Currently, commercially grown sunflower hybrid cultivars utilize a comparatively narrow genetic basis since it relies on one source of CMS only, the well-known *H. petiolaris* (PET) cytoplasm [200]. The wild species continue to serve as a source of considerable variation in oil content and quality [199]. Interspecific hybridization is considered as a suitable route to incorporate wild germplasm into the cultivated sunflower, although difficulties may arise due to differences in chromosome number (2*x*, 4*x*, 6*x*) and cross incompatibility [199,215–217]. In general, interspecific crossability of cultivated *H. annuus* and most annual species is fairly high. However, earlier extensive hybridization programs revealed that almost complete incompatibility exists between cultivated sunflower and diploid perennial species, e.g., *H.*

angustifolius, *H. divaricatus*, *H. giganteus*, *H. nuttallii* and *H. mollis* [216]. Partial compatibility with *H. annuus* was found for some tetraploids, such as *H. decapetalus*, *H. hirsutus*, and hexaploid species, e.g., *H. resinosus*, *H. tuberosus* [216]. In these cases, postzygotic incompatibility - i.e., between the embryo and the endosperm - was believed to prevent efficient and direct recovery of interspecific hybrids. In the past few years biotechnology has become an important supplementary tool with many applications in sunflower breeding, especially providing opportunities to develop interspecific hybrids via embryo rescue, more rapidly and with more precision than conventional techniques [30,33,217–219]. Evaluation of wild species and of the progenies of their hybrids with cultivated sunflower are underway and it may be anticipated that valuable lines will be identified and extracted from these progenies [199,217,220].

1.8.3.2 Breeding procedures

At present, the major breeding method in sunflower is the development of hybrids utilizing the CMS systems. This tendency had been stimulated by the higher yields of hybrids as opposed to populations, and the successful use of a complete, practicable CMS system (see Fig. 1.4). Hybrid cultivars were introduced during the early 1970s. They showed about 25% higher yields than open-pollinated varieties, greater uniformity in plant height and flowering, and a greater degree of self-compatibility. The inbred parental breeding lines are commonly developed by progeny selection and tests of combining ability. An application of the "haploidy technique", which produces haploids and spontaneously doubled haploids, would allow the breeder to develop inbred breeding lines more rapidly. In addition, the haploidy technique facilitates the selection for characters controlled by recessive genes, e.g., genes incorporated from alien species. The production of haploid sunflower plants derived from embryogenic microspores is generally possible although in sunflower the application of anther and microspore culture is rather difficult, and the response so far is low in comparison to rape or flax [217,221]; for an overview of biotechnology in sunflower breeding see Figs. 1.3 & 1.4. Some countries are not able to produce or distribute hybrid seeds economically due to the relatively high costs. In addition, populations (open-pollinated varieties) may be more adaptable to extreme environmental conditions. Therefore, mass and recurrent selection methods are still effectively practiced breeding methods in many parts of the world [193,212].

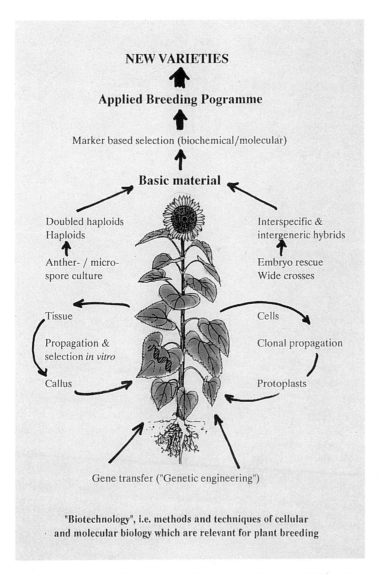

Figure 1.3. Scheme illustrating principle applications of biotechnology, i.e. cellular and molecular methods, in breeding of sunflower and other oilseed species.

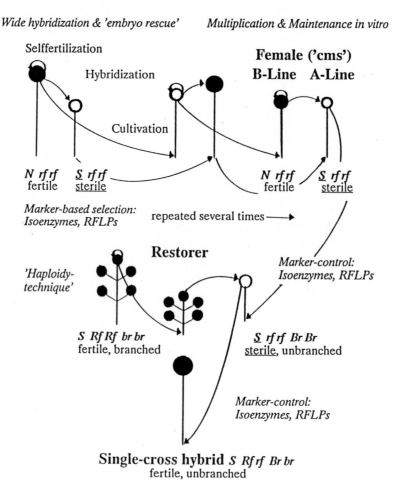

Wide hybridization & 'embryo rescue'   *Multiplication & Maintenance in vitro*

Possible applications of biotechnology in sunflower breeding - Explanations: 'cms' = cytoplasmic male sterility, A-Line = male sterile female, B-Line = fertile analogon (maintainer); Gene symbols: *N* = "normal" cytoplasm, *S* = sterility-inducing cytoplasm, *Rf/rf* = Restorer/Nonrestorer Gene, *Br/br* = unbranched/branched growth habit; RFLPs = restriction fragment length polymorphisms

Figure 1.4.   Breeding scheme for the development of parental inbred lines and commercial hybrids of sunflower and possible applications of biotechniques in the breeding procedure.

1.8.3.3 Breeding objectives

Productivity

Sunflower combines high yield potential with considerable adaptability to a wide range of growing regions, e.g., the desert conditions of the Mid-East, the African tropics, and the short-season as well as the relatively cool climates of Russia, Canada and Central Europe. Thus, the breeding objectives also may vary greatly with production area, length of season, relative prevalence of certain diseases or environmental stress conditions, economical considerations due to supply and demand, and last but not least the grower's preference. However, the basic objective of sunflower breeding is to increase yield [7,193,212]. With regard to this goal the breeder's efforts need to focus on different components of an integrated system as follows: 1, physiology of yield and genetics of yield components [207,211]; 2, plant ideotype and enhanced harvest index [222,223]; and 3, resistance to diseases, pests and environmental stress conditions [224,225]. With regard to the well-known narrow genetic base of sunflower hybrid cultivars, further expansion of sunflower production may be particularly limited by epidemic diseases, since sunflower is very sensitive to various fungal pathogens. Consequently, hybrids resistant to the most dangerous pathogens in Central and Northern Europe, e.g., *Sclerotinia sclerotiorum* and *Botrytis cinerea* have to be developed for economic cultivation under these temperate climatic conditions [217].

Improvement of seed components

Russian breeders, in particular the academic Pustovoit, are usually credited with the development of the oilseed type of sunflower generally grown today. The oil content of sunflower seeds (achenes) depends on both the proportion of hull and the percentage of oil in the kernel. Initially, oil content was increased mainly by reducing hull percentage. Over a longer period of selection the oil content of sunflower seeds has been raised from 20–30% to 40–50% by reduction of hull content from 40–50% to 20–25%. Consequently, a wide genetic variation exists nowadays, including high oil cultivars or hybrids with low hull (20–25%) content and 57–67% oil in the kernels [10,193,208,226].

Oil quality: In sunflower oil a strong inverse relationship between oleic and linoleic acid is observed, which is caused by genotype and temperature during seed formation. For the most part sunflower seed oil is an important source of edible oil rich in linoleic acid (50–70%), whereas oleic acid represents between 15 and 45% of the total fatty acids, depending on environment [125,203,227]. However, large genotypic variations in proportions of oleate/linoleate exist between *Helianthus* species and genotypes [199,208,226], as well as between seeds within heads. For instance, the latter can be exploited in the selection of inbred lines with genetically stable higher than normal levels of linoleate, even under high temperature during seed development [205,228]. Since linoleate is normally inversely proportional to temperature during seed maturation, under hot summer conditions the content of this PUFA may drop to <50%. This is undesirable for manufacture of polyunsaturated products such as margarines, where a linoleic acid content of at least 62% is required [228]. Modern high-oleic sunflower hybrids nowadays also provide an oleic-rich oil (about 85%), which has an improved utility as frying oil with increased oxidative stability, and is also a valuable feedstock for the oleochemical industry [20,28,177,229; see also Chapter 2]

This development traces back to a high oleic mutant, which was derived from dimethyl sulphate-induced mutation [230]. After further selection for high oleic acid content in progeny of the original mutant, the open pollinated variety "Pervenets" with an average oleic acid content of 75% was released. Individual half-seed analyses in this germplasm has shown inter- and intra-plant variation for oleic acid content of individual seeds ranging from 19 to 94%. Nevertheless, it was finally possible to select true-breeding progenies from "Pervenets" with a quite stable, almost temperature-insensitive oleate content higher than 83% [227]. Although the high oleic material used in different studies was derived from the same germplasm, some differences have emerged from examinations of genetic control. Urie [227] has suggested, that the high oleic trait is inherited by a single dominant gene (*Ol*) with embryo genotype control. Miller *et al* [231] have defined a second gene (*ml*) modifying the oleic acid content determined by the major gene *Ol*. The high oleic genotype is represented by the recessive *ml* gene in the homozygous condition in combination with the gene *Ol*. Dominant alleles of the modifier gene (*Ml*) appear to be present in common sunflower cultivars more or less frequently, correspondingly the recessive allele *ml* seems to be rare [231]. It should not be neglected, however, that Miller and coworkers also observed the presence of considerable maternal effects on oleate content. More recently, Fernández-Martínez *et al* [232] have confirmed the previous results obtained with similar material [227] indicating an almost complete absence of maternal influence and a high degree of dominance of the factors conditioning high oleic acid content. Genetic analysis of segregating progeny support the hypothesis that the high oleic trait is determined by three dominant complementary genes (Ol_1, Ol_2 and Ol_3). The dominant gene Ol_1 probably is the result of mutation in the Russian "Pervenets" material. The different alleles of Ol_2 and Ol_3 are present in the genetic background of the respective sunflower lines used as cross parents [232,233]. With regard to biochemical mechanisms, it is interesting to note that the high oleic character is expressed exclusively in the tissue of the developing seed. The block in oleate synthesis is confined to the

ER (endoplasmic reticulum) and, due to reduced oleoyl-PC (phosphatidyl-choline) desaturation, both membrane lipids and triacylglycerol composition of the seed are affected [234].

1.9 Cotton

1.9.1 Importance, distribution and utilization

Cotton (*Gossypium* spp.) is an important world crop, which has been cultivated in the semi-arid tropics and subtropics since prehistoric times. Nowadays it is grown extensively (more than 50% of total area) in temperate regions [235,236]. The oldest record of cotton textiles dates back to 4700 B.P. and was found in India, where the technology for spinning and weaving of flax (*Linum usitatissimum*) already existed. Recent reviews of cotton-growing history were provided by Lee [235], and Percival and Kohel [237]. Cotton is cultivated and processed mostly for its seed fibers, representing 80% of total value, while the seeds are also an important source of oil [23]. Among the oilseeds, cotton is considered to be an important commodity because of its use in food and animal feed throughout the world. The seed oil is used for edible purposes, and the meal is a protein-rich cattle feed [10,16,235]. The world production of cotton oil in 1991/92 was 4.4MT ranking fifth in world's vegetable oil production [1]. The world production of cotton seed was some 38MT. The major cotton producing countries are China, United States, the former USSR, Pakistan and India [1,11].

1.9.2 Botany - taxonomy and general description

The genus *Gossypium*, represented by more than 40 diverse species, is a member of the family *Malvaceae*. Four species with spinnable seed fibers on the seed coats, called "lint", have been domesticated and represent the cultivated cottons [237,238]. They include two diploid species ($2n=2x=26$) having their center of origin in Africa-Asia, *G. arboreum* and *G. herbaceum*, and two amphidiploid (allotetraploid) species ($2n=4x=52$) being native to the New World, *G. hirsutum* and *G. barbadense*. *G. arboreum* (tree cotton) remains an important crop in India, whereas *G. herbaceum*, probably the earliest cotton cultivated, has today only local significance in the drier areas of Africa and Asia. *G. barbadense*, commonly known as extra-long-staple, Egyptian long-staple, or Pima cotton, contributes about 8% of the world production of fiber. *G. barbadense* cultivars have extra-long, strong, and fine fibers, supplying high quality lint. *G. hirsutum*, known widely as Upland cotton, accounts for about 90% of the world cotton production. Upland cotton cultivars have shorter and weaker fibers, but are usually higher yielding than those of *G. barbadense* [56,239].

All species of *Gossypium* are exclusively propagated via seeds. The original growth habit of most species is that of a woody perennial shrub or tree with short-day photoperiodic behaviour. Nowadays the cultivated forms are grown as annuals, still indeterminate in growth habit, displaying a predominantly day-neutral flowering response [235,239]. *Gossypium* is originally adapted to arid-tropical climates. The crop is susceptible to cold temperature and needs more than 160 growing days above 15 °C. Cotton requires at least 500mm of water to grow a crop of minimum acceptable yield [236].

The cotton plant has a completely self-compatible flower, thus self pollination is readily accomplished. The cotton bolls are fruit capsules with 3–5 chambers (locules), each with 7 to 10 ovules. The fibers originate as cells growing outward from the epidermis of the fertilized ovule. A boll normally contains 20–30 mature seeds, which have an ovoid shape and an average size of about 10mm [239,240]. The seed hairs are of two types: spinnable long hairs ("lint") used in the manufacture of textile products, and short hairs ("linters") remaining as fuzz on the seed after ginning. The length of the lint hairs is the main criterion for fiber quality, which varies up to 5cm according to the *Gossypium* species and cultivar. The commercial cottonseed normally consists of about 10–15% linters, 35–40% hull and 50–55% kernel. Linters, the short fibers removed from seeds before crushing, are utilized as source of industrial cellulose.

The cottonseed contains approximately 15–22% oil and 17–21% protein varying due to variety and growing region [10]. However, Cherry and Leffler [241] reported the following values: 23–26% oil and 26–28% protein on a moisture, lint-free seed basis. Palmitic (22–26%), oleic (15–20%) and linoleic (49–58%) are the predominant fatty acids, while linolenic content is negligible. Noteworthy is the content of approximately 1% cyclopropene fatty acids, viz., sterculic and malvalic acid, which are characteristic for seed oils of *Malvaceae* [241]. The *Gossypium* species are characterized by darkly pigmented glands present on the stem, leaves, flowers and bolls of cotton plants which produce glanded seeds. The pigments deposited in these glands are predominantly the polyphenol (terpenoid compound) gossypol, well-known to be responsible for the undesirable dark color of cottonseed oil and the limited meal quality; gossypol is toxic to non-ruminant animals [10,241].

1.9.3 Breeding

1.9.3.1 Sources of genetic variation

At present, 37 diploid (2n=26) and 6 tetraploid (2n=52) species of *Gossypium* are distinguished [237,238]. However, none of the wild diploid species bears usable fiber, and thus these species are of no primary interest. The natural allotetraploid species combine the 26 larger chromosomes (A-genome, Old World species) with the 26 smaller chromosomes (D-genome, New World species). For more detail, particularly regarding time and place of the origin of the allotetraploids, the reader is referred to Percival and Kohel [237], Endrizzi *et al* [242] and Endrizzi [243]. In attempting to transfer useful properties from diploids into the cultivated tetraploids, artificial allo- or autopolyploids can be created from the diploids or from crosses between cultivated tetraploids and certain of the diploids [238,244]. Evaluated interspecific hybrids involving the tetraploids *G. hirsutum* and *G. barbadense* displayed heterosis towards highly desirable traits, e.g., yield or several fiber properties. However, the hybrid vigor was sometimes accompanied by undesirable excessive vegative growth or a lack of fiber length uniformity and maturity [245,246]. Although germplasm and primitive stocks of *G. hirsutum* and *G. barbadense* might be used directly in breeding programs, most of the primitive types of tropical cottons, especially of *G. hirsutum*, display a photoperiodical flowering response. Consequently, they fail to flower under the long-day regime of the temperate-zone growing seasons, and thus have to be converted to day-neutral behaviour before being used in a breeding program [56,238,239].

1.9.3.2 Breeding procedures

Cotton is commonly considered to be a self-pollinating crop, although natural crossing may range from 0–60%. Both wild and domesticated *Gossypium* species are able to self-pollinate without any loss of fertility and vigor. The methods used predominantly are derived from those used for self-pollinators, i.e., crossing within and among genotypes with subsequent selection. Additionally, modifications in breeding practice are made attempting both to maintain an adequate level of heterozygosity and to exploit the heterosis due to partial cross-pollination - usually 3–30%. So the evidence at hand suggests that cotton cultivars are usually not strictly pure lines [56,238,239].

1.9.3.3 Breeding objectives

Productivity

Lint yield and fiber quality have always been the primary objectives of cotton breeding [56,238,239]. The productivity of the cotton plant is influenced by important agronomic characters like climatic adaptation, earliness of maturity and resistance to stress environments (drought and heat tolerance, salt tolerance) as well as disease resistance (*Fusarium* and *Verticillium* wilt, bacterial blight, boll rots) and pest resistance (root knot nematode, boll weevil, cotton bollworm) [56]. Due to the comparison of high-quality Pima cotton and high-yielding Upland cotton, desirable fiber properties may be unfortunately associated with low fiber yield. However, some of these negative correlations have been broken in cotton breeding assisted by modern methods of fiber technology [56,235,247].

Improvement of seed components

The lack of interest in cottonseed quality is due to the natural predominance of cotton breeding conducted with the economically more important lint yield and quality [241]. One exception has been the effort devoted to breeding glandless cultivars with a low gossypol content in the seed. The presence of the glands and their gossypol contents are genetically controlled. Glandless genotypes like the primitive cultivated *G. hirsutum* stock "Hopi" are used for crosses with agronomically adapted cultivars. Glandlessness does not generally affect lint yield or fiber quality [245,248], but infesting insects seem to have a preference for glandless cotton [56]. Glandless cottonseeds are highly desirable for their nutritional value. Due to the elimination of gossypol, the color of the oil and the utility of the meal is distinctly superior to that from common cottonseeds. Additionally, the meal is no longer restricted as feed to ruminant livestock [240,241,249]. Almost nothing has been done genetically to improve other chemical seed properties including proportion and quality of oil and protein. Although considerable variation of the chemical composition of *Gossypium* seeds has been described [10,241,248], there have been no attempts to use such variability in breeding programs systematically [239,245].

1.10 Groundnut

1.10.1 Importance, distribution and utilization

The groundnut (peanut, earthnut), *Arachis hypogaea*, is an annual legume native to South America. Archaelogical evidence has shown that groundnut was used in Peru as early as 4500 B.P., although the domestication must have occurred before that time. The Portuguese probably brought groundnut from Brazil to West Africa and to India. Africa is regarded as a substantial secondary centre of diversity. It is believed that peanut seed came to North America with the transportation of slaves from Africa [7,250]. Nowadays the oil crop is grown throughout the tropical and warm temperate regions of the world between latitudes 40°N and 40°S. Although more than 100 countries in the world cultivate groundnut on a significant scale, about 55% of the 23.4MT world production (groundnuts in shell) is concentrated in India and China. Yields obtained in industrial countries, e.g., the USA, and in developing nations differ substantially. India has about 40% of the world production area contributing about 30% of the world production, whereas the USA, with a three-fold higher yield, has only about 4% of the world production area contributing 9% of world total production [1,11,251].

Edible oil is the major product of groundnuts in the world. With 3.9MT groundnut oil ranks sixth in world vegetable oil production [1]. About two-thirds of the world's peanut production is crushed to extract cooking oil. In the USA, on the other hand, about 70% of the peanuts harvested are used for peanut butter, snack food, confectionery, and various other edible products, while the oil for edible purposes is usually obtained from low-grade peanuts unsuitable for consumption as whole seed. The oil content of the kernels varies between 42 and 52%; most of the commercially grown peanut cultivars contain an average of about 50% oil [10,48,252]. Since refined peanut oil includes about 80% unsaturated fatty acids (oleate, linoleate) accompanied by relatively good oxidative stability and shelf life, it is widely used to prepare oleomargarine, mayonnaise and salad dressing. Furthermore it is regarded as a good cooking and frying oil, since foods fried in peanut oil have a popular flavor and good keeping quality [10]. Besides being rich in edible oil, groundnut meal is also an important source of animal feed and dietary proteins, particularly in developing countries of Asia and Africa. The protein content of the seed usually amounts to 22–30% (average 26%), while the peanut meal contains about 50% protein [10,46,252,253]. However, in western countries the meal accounts only for about 10–20% of the market value of groundnuts [23].

1.10.2 Botany - taxonomy and general description

The genus *Arachis* is a member of the family *Fabaceae*, which differs from other legumes in its geocarpic behaviour, i.e., after fertilization a long carpophore, a fruitstalk commonly known as peg, turns downwards and thrusts the immature pod (ovary) into the ground for further ripening [7,252]. The genus is subdivided into six or seven sections containing about 70 species and is far from complete description. In South America, the primary center of variation, perennial species are generally found in high rainfall areas, while annuals are found in semiarid regions. The section *Arachis* consists of at least 15 annual and perennial diploids ($2n=2x=20$), and two annual tetraploids ($2n=4x=40$): *Arachis hypogaea*, only known under cultivation, and its nearest wild relative, *A. monticola* [252,254–258]. The cultivated peanut (*A. hypogaea*) is considered to be an amphidiploid, which has originated from a hybrid (genome AABB) between *A. batizocoi* (BB, $2n=20$) and another member of the section as donor of the A genome ($2n=20$), possibly *A. cardenasii* [255–259].

The groundnut plant is a herbaceous legume, which normally has a rapidly growing, well-developed tap-root with many lateral roots and nodules in the root axes. The plant shows indeterminate growth and flowering habit, i.e., the prolongation of the floral axis is not arrested at the beginning of flowering. According to Nigam *et al* [251] and Coffelt [252] flowering behaviour shows sensitivity towards light (quantity and quality), photoperiod, temperature, and moisture (realtive humidity). The yellowish flowers arise from axillary buds produced near the base of the stem and the branches. Groundnuts are self-pollinated, with natural cross-pollination varying from 0–10% [254]. The rates of outcrossing as well as the number of flowers and the length of the flowering period depend on genotype and environmental conditions. Groundnut plants usually display an overproduction of flowers, since about 40% of the flowers fail to initiate peg development and another 40% abort before pod development [7,252,254]. The pods, attached to the branches of the plant by the peg, develop and mature in the ground usually 2–8cm below the soil surface. The mature pod is an indehiscent legume, which consists of the shell (pericarp) containing one to five seeds, each enclosed in a papery seed coat or testa (endocarp). Of the pod weight, 20–26% is contributed by the shell and 74–80% by the kernels. The latter weigh 0.2–2.0g [10,252].

The groundnut is a major crop in tropical and subtropical areas. A temperature range of 25–30 °C appears to be optimum. Compared to other oilseeds, groundnut plants are relatively drought-resistant making them especially important in semi-arid regions. However, most of the world's groundnut cultivation is located to rainfed areas under low input conditions. A precipitation of 500–1,000mm will allow commercial groundnut production, but the crop can grow on as little as 300–400mm. Rainfall is undesirable once the pods are mature, because some groundnut varieties have only a very weak fresh seed dormancy and germinate quickly under suitable conditions [7,251].

There is a wide variation in the types and strains of peanut cultivated in particular localities, but from a practical point of view two main commercial growing types are distinguished [7,10]: the bushy upright with an erect central stem and vertical branches (bunch type), and the recumbent, trailing with numerous creeping laterals (spreading or runner type). The first type is more suitable for mechanized production, while the second is more commonly grown under peasant farming systems.

A. hypogeae is classified into 2 subspecies, ssp. *hypogaea* and ssp. *fastigiata*, and each of them further divided into 2 botanical varieties. The subspecies *hypogaea* has no infloresences on the main stem, a spreading or erect growth habit, a longer maturation period, and pronounced fresh seed dormancy. In contrast, the subspecies *fastigiata* is always erect, with inflorescences on the main axis, earlier maturity, and only weak fresh seed dormancy, germinating immediately as they become mature [7,252]. Besides the differences in growth habit, the botanical varieties form a rough basis for classification of peanuts in market types primarily due to pod and seed size characteristics determining the utility for distinct consumer products - e.g., small-seeded types for oil and peanut butter, or large-seeded types for direct use as snack food. However, only three of the botanical varieties are grown commercially: var. *hypogaea* (the "Virginia Bunch" and "Virginia Runner" type) in the former subspecies and vars. *fastigiata* (the "Valencia Bunch" type) and *vulgaris* (the "Spanish Bunch" type) in the latter [10,251,254].

1.10.3 Breeding

1.10.3.1 Sources of genetic variation

According to Nigam *et al* [251] the Genetic Resources Unit of the ICRISAT (International Crop Research Institute for Semi-Arid Tropics) maintains a world collection of over 12,000 accessions of cultivated groundnut and wild *Arachis* species. Besides the cultivated groundnut, there are about 70 wild *Arachis* species which are all native to South America, providing additional sources of variation for many traits, such as resistance to many diseases and insect pests [255,257,260]. Those species belonging to the section *Arachis* are cross-compatible with *A. hypogaea*. Although most of these are only diploid

($2n=20$), several techniques have been developed to facilitate their use in breeding programs. For instance, successful introgression could be accomplished by previous doubling of the diploid donor genome followed by hybridization and subsequent backcrossing to *A. hypogaea* [244,251,255,257]. Although interspecific crosses within the section *Arachis* have been made, it is considered doubtful if crosses outside the section *Arachis* are feasible [255,257,261].

1.10.3.2 Breeding procedures

The breeding methods applied to groundnut are the same as those used in improving other self-pollinated crop species. The pedigree and bulk system are used more commonly by groundnut breeders than the single-seed decent (SSD) method. Nevertheless, modifications of pure-line breeding systems are often practised. A disadvantage of traditional pedigree method is the severe restriction on the amount of recombination among linked genes during selecting among homozygous lines generated from biparental crosses. Hence, the use of recurrent selection for self-pollinators has been suggested to give rise to population improvement, and to overcome this limitation of progress. The application of backcrossing procedures has its importance due to the increased emphasis on breeding for disease resistance [251,252,254,262].

1.10.3.3 Breeding objectives

Productivity

Despite the more than doubling of yields obtained in the USA over the last 30 years, increased productivity, i.e., yield potential and yield maintenance, is still one of the principal goals of groundnut breeding. The dramatic yield increases have resulted from a combination of factors including improved cultivars, better agronomic practices and conditions, e.g. new pesticides and herbicides, as well as mechanized cultivation [251,254,262]. In particular, selection for yield has resulted in cultivars with enhanced partitioning of the total assimilates to the fruits during the fruit filling period. This suggests the opportunity for further yield improvement due to structural alterations in the peanut plant, i.e., an increase of the number of fruiting sites available [263]. In contrast, the average yield in developing countries remains around 1.0T/ha (groundnuts in shell), i.e., yields are only about one third of the yields in developed countries. Besides the lack of improved agronomic practices, production technology and infrastructure as well as generally low input, the main reasons for low productivity in Asia and Africa are biotic and abiotic stresses, such as diseases and insect pests, unpredictable and unreliable rainfall as well as drought [251].

Improvement of seed components

Various physical, sensory and biochemical parameters comprise the quality of groundnut seed. Physical and sensory factors determine market-grade or confectionary traits including pod size and several seed characteristics: intact testa and resistance of seed slitting, size and shape of the seed, as well as seed color, texture and flavor [251,254]. However, an important goal in most groundnut breeding programs is the improvement of seed components, viz., oil quality and/or protein quality. Since peanut is major part of the diet in developing countries, the enhancement of these characteristics is eminently interesting [46,252,253]. According to Nigam *et al* [251] over 8,000 germplasm lines have been screened at ICRISAT for oil and protein content ranging from 31 to 55% and from 16 to 34%, respectively. These results indicate the opportunity for further improvement, although it must be considered that the negative correlation between oil and protein content will hinder efforts to increase both characters simultaneously [10].

Oil quality: The oil quality trait comprises oil content and fatty acid composition, e.g., iodine value and ratio of oleic to linoleic (o:l ratio), as criteria for oil stability and flavor. Several studies have shown that in a diverse collection of germplasm, oil content ranges from 46 to 63% in wild species and 43 to 56% in cultivars. The variability in oil yield, especially in the wild species, indicates the opportunity for increased oil content in peanut [262,264]. The data from breeding programs dealing with the genetic manipulation of oil quality clearly indicate a wide genetic variability for various fatty acids in general, and for oleic (41–80%) and linoleic acid (4–42%) in particular [107,254,265]. The oil of genotypes with as high as 80% oleate may exhibit superior keeping and frying quality, since peanut oils having high levels of linoleate are more susceptible to oxidative rancidity [10,266]. "Virginia" botanical-type peanuts (ssp. *hypogeae* var. *hypogeae*) generally produce seed oils with higher oleic acid content than seed oils from "Spanish" or "Valencia" types. Additionally, production location has been shown to have a significant effect on the oleate:linoleate (o:l) ratio of peanut oil [48,107,267,268]. Crosses among pure breeding cultivars of these types, differing in oleic and linoleic levels, have shown that the genotype of the developing embryo determines the o:l ratio and that only a few additive genes are involved [252,269]. Moore and Knauft [270] showed that the high-oleic character in peanut line F435 is determined by two recessive genes (ol_1, ol_2).

Protein quality: Similar quality enhancement could be made by increasing protein content and improving the amino acid balance, although neither have received much attention due to the predominant use of the oil. The protein content of the peanut kernel may range from 16 to 36%. Typically for proteins of grain legumes, the major quality deficit is the low content of the sulfur-containing amino acids, especially methionine [10,253,262].

Unfortunately, much of the world's groundnut meal is contaminated with aflatoxins produced during growth of the mold *Aspergillus flavus*. Because of these highly toxic compounds, a considerable amount of groundnut protein is not suitable for feed or food use. However, besides prevention of mold contamination several sources of resistance referring to distinct mechanisms have been recognized [10,251,252].

1.11 Linseed/flax

1.11.1 Importance, distribution and utilization

Flax (syn. linseed, *Linum usitatissimum*), widely adapted to warm and cool temperate climates, is cultivated both for its stem fiber (flax) and for its seedoil (linseed). In addition, linseed is traditionally used in therapeutics because of the laxative properties of mucilaginous carbohydrates (pentosans) associated with the seed coat [10,271–275]. The cultivation of flax goes back to the dawn of civilisation, as shown by the remains of prehistoric Swiss lake dwellers. There are strong indications that at least the fiber type of *Linum* originated in an area east of the Mediterranean region, notably nearby India, and that it spread northwards and westwards. At this time (8000 B.P.) cultivation of domestic flax was widespread in the Near East region. The ancient Egyptians had a high regard for linen, and they used it not only for clothing but also for the wrapping of mummies. They also embalmed bodies with linseed oil. After flax production spread to Europe, it was grown extensively for linen manufacture until the beginning of the 20th century, but since then another fiber crop, namely cotton, has taken its place. Since flax fibers are stronger, more durable, and more resistant to moisture than cotton or wool, linen fabrics still have the image of a superior textile. As the demand for flax fiber declined, more and more linseed was used as a source of drying oil [274,276]. Some dual-purpose cultivars have been bred [273], but both the seed and fiber obtained were of lesser quality. However, the shorter flax fibers extracted from linseed cultivars by mechanical means are used for several technical applications [274]. *L. usitatissimum* is now grown in many parts of the world, either for its stem fiber or for its seed oil, depending on the cultivar used as well as the cultural and climatic conditions. The taller, only slightly branched flax type is usually cultivated in cool, temperate regions of the former USSR, the largest producer of flax fiber. To a minor extent it is also grown in France, Belgium, the Netherlands and other northern and eastern European countries. The shorter, branched and more quickly maturing linseed type prefers the warmer climates of Canada, Argentina, India, China and the former USSR, contributing together about 75% of the total 2.65MT linseed produced in 1991 - reflecting a strongly decreased area (some 4.0 million ha) planted with linseed. In contrast to Canada and Argentina, the EEC has increased its linseed production, i.e., about 200–300,000ha were planted with linseed in 1992. Major growing countries were UK (140,000ha) and Germany (90,000ha). The world production of linseed oil has a potential of about 0.7MT [1,11].

Linseed has about 35–45% oil, which is unique among the major vegetable oils in that it contains high levels (approximately 40–65%) of linolenic acid [10,273,277]. The high susceptibility to autoxidation gives linseed oil a rapid drying property, which explains the traditional usage in the paint and varnish industry. Raw and cold-processed linseed oil may be used for nutrition, too. In India, about 35–40% of the oil is consumed as cooking oil [10]. However, the high linolenic acid content causes rancidity and a short shelf life, and since the commercial production is low compared to other oil crops, linseed oil is rarely used for edible purposes. Attempts are being made to achieve a drastic reduction of the linolenic acid content, which would provide an edible-quality linseed oil. Recently, the cultivars with so-called "Linola" quality bred by Green *et al* [278,279] have been patented. However, the impact on the edible oil market is doubtful, because the oil must compete with sunflower and safflower oil which have a very similar fatty acid composition to Linola.

The protein content in linseed ranges from 20 to 24%. After oil extraction a meal is obtained, which is considered a valuable protein concentrate for livestock. The meal needs to be processed to remove the mucilage, and inactivate toxic constituents such as the enzyme linamarase, which hydrolyzes the cyanogenic glucoside linamarin forming poisonous hydrocyanic acid (prussic acid). However, the protein quality is inferior to other vegetable protein sources due to a low lysine level [10].

1.11.2 Botany - taxonomy and general description

The genus *Linum* belongs to the family *Linaceae*, and is further divided into five taxonomic sections, viz., *Linum, Dasylinum, Cathartolinum, Linastrum,* and *Syllinum*. The latter section is morphologically quite distinct from the others, being far more broad-leaved and prostrate in growing habit, and having considerable portions of ricinoleic acid in the seed oils [280]. According to Durrant [276] the genus *Linum* comprises of nearly 200 species, which are spread over the temperate and warm temperate region of the northern hemisphere, mostly in Europe and Asia, but with about 50 species in America. The haploid chromosome number in the genus *Linum* displays a wide range (n= 8, 9, 10, 12, 14, 15, and 16), but $n=9$ and $n=15$ are the most common chromosome numbers [274,276,281–283]. According to Pirson [284] there is cytological evidence that $n=15$ species are secondary balanced polyploids, probably originating from a wild ancestor with a basic chromosome number of $x=8$. *L. usitatissimum* is the only species of agricultural importance, although the closely related *L. angustifolium* (both $2n=30$) has been formerly cultivated in some areas. *L. usitatissimum* has non- or semidehiscent capsules and annual growth habit which makes it suitable for cultivation [276,281].

The linseed or flax plant is an annual herb with a thin, erect and wiry stem, about 60 to 120cm high. The fiber-type cultivars are generally slender, tall growing, non-tillering, and slightly branched. In contrast, the cultivars grown for seed purposes are usually smaller, much branched and profusely tillering. The small flowers with bright blue but sometimes pink or white petals grow on terminal panicles. Despite the presence of nectaries, the flowers are usually self-pollinated, although cross-fertilisation is possible and may occur at a rate of <10% [284a]. The fruit is a round capsule, divided into five chambers, with each containing up to 2 yellowish or brown seeds. The thousand-seed weight, ranging between 3 to 16g, is negatively correlated with the number of seeds per capsule [285,286].

1.11.3 Breeding

1.11.3.1 Sources of genetic variation

With regard to oil quality, natural variability of *L. usitatissimum* is comparatively limited as demonstrated by several investigations [277,287–289]. In contrast, wild *Linum* species show a wide range of variation towards linolenic acid content, with several wild species having very low levels (3%) of linolenic acid [280,290]. Besides oil content, varying from 23 to 46%, an extensive variation amongst wild *Linum* species, ranging from 4 to 68% linolenic and correspondingly 9 to 83% linoleic acid was found [291,292]. Furthermore, several of the wild species possess many agronomically desirable features such as resistance to diseases and drought, winter hardiness, etc. [293]. Interspecific crosses within the $n=15$ or $n=9$ groups have been successful, but crosses between $n=9$ species and *L. usitatissimum* or any other $n=15$ species have more or less failed. Unfortunately, characteristics of interest that are not found in *L. usitatissimum* are often present only in species with chromosome numbers different from $n=15$ [281,282,293]. The exploitation of these species via interspecific hybridization is hampered due to strong reproductive barriers [294]. Although the embryo rescue method has generally proved its usefulness [295], linseed and *Linum* species are much more recalcitrant to an application of this *in vitro* culture technique. Nevertheless, hybridizations of *L. usitatissimum* to other high-linolenic *Linum* species were successful and appear to open new opportunities for increasing the linolenic acid content of cultivated linseed [29,283,291,292].

1.11.3.2 Breeding procedures

Linseed (syn. flax, *L. usitatissimum*), regarded as a self-pollinating crop, has a considerable heritage in terms of classical breeding techniques. The pedigree method has been used most widely in developing improved linseed cultivars, although other methods such as single-seed descent and bulk breeding method may be used, too. In every case, cultivars of linseed represent pure, i.e., true-breeding lines [274,296]. Further cultivar improvement by breeding is feasible, particularly by application of biotechnology, i.e., tissue and cell culture techniques. Linseed can now be considered as one of the crop species most amenable to improvement through these techniques [283,297]. For instance, it is possible to obtain haploid and doubled haploid plants reproducibly through anther- or microspore-culture, which allows the rapid fixation of rarely segregating genotypes and a substantial abbreviation of the breeding cycle [29,30]. Furthermore, flax is one of the first species, which has been genetically transformed using recombinant DNA technology and of which transgenic lines have been already released entering official registration trials [275,298–300].

1.11.3.3 Breeding objectives

Productivity

The primary aims of linseed breeding are the improvement of seed yield and oil content as well as protection from yield losses due to lodging and diseases, e.g., wilt (*Fusarium oxysporum* f. *lini*) and rust (*Melampsora lini*). Linseed yields vary considerably, with national averages ranging from 0.3T/ha in India to about 1.3T/ha in Canada [1,11]. However, in cool-temperate regions up to 2.0T/ha are attainable within a duration of 140–160 days. With regard to yield and its components (plant density, 1,000-seed weight, seeds per capsule, number of capsules per plant) linseed/flax production systems have to be considered as complex, since they do not only require the provision of new, improved cultivars by breeding. Successful plant systems also need sophisticated agronomical techniques, e.g., tillering and lodging are highly influenced by planting density (for linseed: 400–800 plants/m^2) and nitrogen supply. The yield of modern linseed cultivars can be estimated at about 3T/ha under optimum growing conditions. However, it has to be kept in mind that the realization of this potential is usually limited by economic and ecological restrictions [29,285,286,301].

Improvement of seed components

Although the protein, the accompanying substances (mucilage) and the fibers (only of technical quality) are associated products that can also be used, it is the amount and quality of extracted oil that determines the value of linseed. The seed has an oil content of about 35–45%, which is influenced by variety, seed size, climate and maturity. Drought and higher temperatures during the sensitive seed-filling period accelerate maturity, and consequently seed size and oil content are reduced. Linolenic acid content is influenced by temperature, as the proportion of this PUFA is enhanced by cooler temperatures during ripening due to the stimulation of the desaturation of the seed oils. Despite this strong environmental modification, however, linseed oil can be further optimized by breeding [161,271,302–304].

Oil quality: With regard to genetic selection for linseed lipid composition, in principle, the choice may be conducted in the direction of industrial oil production, as well as in the direction of obtaining oil suitable for food. Thus, flax breeders expended some effort in identifying low-linolenic genotypes which might enable linseed oil to get a place in the expanding edible oil market [272,288,289,305]. Linseed normally produces a seed oil rich in linolenic acid, which drastically reduces the shelf life of the oil and makes it virtually unsuitable for edible use. Since natural variation of fatty acid composition of cultivated flax is insufficient, mutation breeding seems to be the most promising approach for reducing the linolenate content in linseed oil, so far [289]. An exceptionally successful example in this respect was the mutagenesis program of Green and Marshall [288]. Following ethyl-methanesulfonate (EMS) treatment of seeds of the high linolenic linseed cultivar "Glenelg", the two mutants M1589 and M1722 with about 50% of wild-type linolenate level were isolated. By recombining the two mutants, the line "Zero" with less than 2% linolenic acid in the seed oil was obtained [278]. Biochemically, the mutation events display a genetic block in the conversion of linoleate to linolenate in the developing seed, due to a reduced rate of linoleoyl-PC desaturation [306,307]. As a consequence, the oleic acid content is nearly the same as in cv. "Glenelg", but the level of the nutritionally desirable linoleic acid in the low linonenic genotype "Zero" is greatly increased. Recently, a Canadian breeding program with a similar intention identified an EMS-induced low-linolenic mutant in the cultivar "McGregor" [305,309]. Genetic analyses demonstrated that the low linolenic trait is determined by two genes (*Ln1* and *Ln2*) located on separate chromosomes and acting additively [308,309]. Although it is indicated that there is no dominance and the fatty acid composition of the F_1 generation is largely determined by the developing embryo, the contribution of maternal effects in reciprocal crosses between lines which differ in oil quality should not be neglected. This is due to the fact that the oil accumulated in the endosperm and seed coat accounts for about 20% of the mature seed oil content, and that the fatty acid composition of these tissues can differ significantly from that of the embryo [310].

Acknowledgements: We wish to thank Mrs Ingeborg Scholz, Dr. F. Ordon and Mr A. Thierfelder for help in preparing the manuscript. Own scientific work reported herein was supported in part by research grants from *Bundesministerium für Forschung und Technologie*, Bonn, *Bundesministerium für Ernährung, Landwirtschaft und Forsten*, Bonn, *Gemeinschaft zur Förderung der privaten deutschen Pflanzenzüchtung*, Bonn, and *Deutsche Forschungsgemeinschaft*, Bonn.

References

1 ISTA, *Oil World Statistics Update*, ISTA/Mielke GmbH, Hamburg, Germany, **1992**.
2 Eierdanz, H., in: *Proceedings of the World Conference and Exhibition on Oilseed Technology & Utilization*, Applewhite, T. (ed.), AOCS, Champaign, Illinois, **1993** (in press).
3 Baumann, H., Bühler, M., Fochem, H., Hirsinger, F., Zoebelein, H., Falbe, J., *Angew. Chem. Int. Ed. Engl.*, **1988**, *27*, 41–62.
4 Staal, L., *Industrial Crops and Products*, **1992**, (in press).
5 Haumann, B.F., *J. Am. Oil Chem. Soc.*, **1988**, *65*, 702–717.
6 Godin, V.J., Spensley, P.C., *Oils and Oilseeds. No. 1. Crop and Product Digests*, **1971**, The Tropical Products Institute, London.
7 Weiss, E.A., *Oilseed crops*, **1983**, Longman, London.
8 Chan, E., *Oléagineux*, **1983**, *38*, 371–377.
9 Röbbelen, G., Downey, R., Ashri, A. (eds.), *Oil Crops of the World*, McGraw-Hill Publ. Comp., New York, **1989**.
10 Salunkhe, D.K., Chavan, J.K., Adsule, R.N., Kadam, S.S., *World Oilseeds: Chemistry, Technology, and Utilization*, **1992**, Van Nostrand Reinhold, New York.
11 FAO, *Production Yearbook 1991*, *45*, Food and Agriculture Organization, Rome, **1992**.
12 Pardun, H., *Analyse der Nahrungsfette*, Verlag Paul Parey, Berlin, **1976**.
13 Podmore, J., in: *Recent Advances in Chemistry and Technology of Fats and Oils*, Hamilton, R.J., Bhati, A. (eds.), Elsevier Applied Science, New York, **1987**, 167–181.
14 Rossell, J.B., *Fat Sci. Technol.*, **1991**, *93*, 526–531.
15 Röbbelen, G., in: *Proceedings of the World Conference on Biotechnology for the Fats and Oils Industry*, Applewhite, T.H. (ed.), AOCS, Champaign, Illinois, **1988**, pp. 78–86.
16 Dziezak, J.D., *Food Technol.*, **1989**, *43*, 66–74.
17 Marsic, V., Yodice, R., Orthoefer, F., *INFORM*, **1992**, *3*, 681–686.
18 Gurr, M.I., *Prog. Lipid Res.*, **1992**, *31*, 195–243.
19 Pryde, E.H., Rothfus, J.A., in: *Oil Crops of the World*, Röbbelen, G., Downey, R., Ashri, A. (eds.), McGraw-Hill Publ. Comp., New York, **1989**, pp. 87–117.
20 Zoebelein, H., *INFORM*, **1992**, *3*, 721–725.
21 Friedt, W., Lühs, W., in: *Proceedings of the World Conference and Exhibition on Oilseed Technology & Utilization*, Applewhite, T. (ed.), AOCS, Champaign, Illinois, **1993**, (in press).
22 Stein, W., in: *Improvement of Oil-Seeds and Industrial Crops by Induced Mutations*, Proceed. Advisory Group Meeting, Intern. Atomic Energy Agency (IAEA), Vienna, **1982**, pp. 233–242.
23 Stanton, J.M., Blumenfeld, J.K., *INFORM*, **1992**, *3*, 1019–1022.

24 Thies,W., McGregor, D.I., in: *Oil Crops of the World*, Röbbelen, G., Downey, R., Ashri, A. (eds.), McGraw-Hill Publ. Comp., New York, **1989**, pp. 132–164.

25 Marquard, R., *Fat Sci. Technol.*, **1987**, *89*, 95–99.

26 Scowcroft,W.R., *INFORM*, **1990**, *1*, 945–951.

27 Knauf,V.C., *Trends in Biotechnol.*, **1987**, *5*, 40–47.

28 Friedt,W., *Fat Sci. Technol.*, **1988**, *90*, 51–55.

29 Friedt,W., in: *New Crops for Temperate Regions*, Anthony, K.R.M., Meadley, J., Röbbelen, G. (eds.), Chapman & Hall, London, **1993**, pp. 222–234.

30 Thierfelder, A., Lühs, W., Friedt, W., *Industrial Crops and Products*, **1993**, *1*, 261–271.

31 Nishiyama, I., Sarashima, M., Matsuzawa, Y., *Plant Breeding*, **1991**, *107*, 288–302.

32 Plümper, B., *Proceed. 8th Intern. Rapeseed Congress*, GCIRC, Saskatoon, Canada, **1991**, Vol. 4, pp. 1034–1039.

33 Kräuter, R., Steinmetz, A., Friedt, W., *Theor. Appl. Genet.*, **1991**, *82*, 521–524.

34 Battey, J.F., Schmid, K.M., Ohlrogge, J.B., *Trends in Biotechnol.*, **1989**, *7*, 122–126.

35 Taylor, D.C., Thomson, L.W., MacKenzie, S.L., Pomeroy, M.K., Weselake, R.J., in: *Proceed. 6th Crucifer Genetics Workshop*, McFerson, J. R., Kresovich, S., Dwyer, S.G. (eds.), USDA-ARS Plant Genetic Resources Unit, Cornell University, Geneva, NY, **1990**, pp. 38–39.

36 Hills, M.J., Murphy, D.J., *Biotechnol. Genet. Engineer. Rev.*, **1991**, *9*, 1–46.

37 Wolter, F.P., Bernerth, R., Löhden, I., Schmidt,V., Peterek, G., Frentzen, M., *Fat Sci. Technol.*, **1991**, *93*, 288–290.

38 Wolter, F.P., *INFORM*, **1993**, *4*, 93–98.

39 Voelker,T.A.,Worrell, A.C., Anderson, L., Bleibaum, J., Fan, C., Hawkins, D.J., Radke, S.E., Davies, H.M., *Science*, **1992**, *257*, 72–74.

40 Hymowitz, T., Newell, C.A., *Econ. Bot.*, **1981**, *35*, 272–288.

41 Hymowitz, T., Singh, R.J., in: *Soybeans: Improvement, Production, and Uses*, Wilcox, J.R. (ed.), Agronomy Monograph 16, Am. Soc. Agron., Madison, Wisconsin, USA, **1987**, 2nd edition, pp. 23–48.

42 Hymowitz, T., in: *Advances in New Crops - Proceedings of the First National Symposium New Crops: Research, Development, Economics*, Janick, J., Simon, J.E. (eds.), Timber Press, Portland, Oregon, USA, **1990b**, pp. 159–163.

43 Askew, M.F., *INFORM*, **1992**, *3*, 935–938.

44 Smith, K.J., Huyser, W., in: *Soybeans: Improvement, Production, and Uses*, Wilcox, J.R. (ed.), Agronomy Monograph 16, Am. Soc. Agron., Madison, Wisconsin, USA, **1987**, 2nd edition, pp. 1–22.

45 Hymowitz, T., in: *Advances in New Crops - Proceedings of the First National Symposium New Crops: Research, Development, Economics*, Janick, J., Simon, J.E. (eds.), Timber Press, Portland, Oregon, USA, **1990a**, pp. 154–158.

46 Singh, U., Singh, B., *Econ. Bot.*, **1992**, *46*, 310–321.

47 Mounts, T.L, Wolf, W.J., Martinez, W.H., in: *Soybeans: Improvement, Production, and Uses*, Wilcox, J.R. (ed.), Agronomy Monograph 16, Am. Soc. Agron., Madison, Wisconsin, USA, **1987**, 2nd edition pp. 819–866.

48 Salunkhe, D.K., Sathe, S.K., Reddy, N.R., in: *Chemistry and Biochemistry of Legumes*, Arora, S.K. (ed.),   Edward Arnold Publ., London, **1983**, pp. 51–109.

49 Szuhaj, B.F., *J. Am. Oil Chem. Soc.*, **1983**, *60*, 306–309.

50 Schneider, M., *Fat Sci. Technol.*, **1992**, *94*, 524–533.

51 Hymowitz, T., Palmer, R.G., Singh, R.J., in: *Chromosome Engineering in Plants: Genetics, Breeding, Evolution*, Tsuchiya, T., Gupta, P.K. (eds.), Part B, Elsevier, Amsterdam, **1991**, pp. 53–63.

52 Singh, R.J., Kollipara, K.P., Ahmad, F., Hymowitz, T., *Genome*, **1992a**, *35*, 140–146.

53 Singh, R.J., Kollipara, K.P., Hymowitz, T., *Theor. Appl. Genet.*, **1992b**, *85*, 276–282.

54 Palmer, R.G., Newhouse, K.E., Graybosch, R.A., Delannay, X., *J. Heredity*, **1987**, *78*, 243–247.

55 Singh, R.J., Hymowitz, T., *Theor. Appl. Genet.*, **1988**, *76*, 705–711.

56 Poehlman, J.M., *Breeding Field Crops*, Van Nostrand Reinhold, New York, **1987**.

57 Fehr, W.R., in: *Principles of Cultivar Development - Crop Species*, Fehr, W.R. (ed.), Macmillan Publ. Comp., New York, **1987**, Vol. 2, pp. 533–576.

58 Gresshoff, P.M., in: *Advances in New Crops - Proceedings of the First National Symposium New Crops: Research, Development, Economics*, Janick, J., Simon, J.E. (eds.), Timber Press, Portland, Oregon, USA, **1990**, pp. 113–119.

59 Sinclair, T.R., Soffes, A.R., Hinson, K., Albrecht, S.L., Pfahler, P.L., *Crop Sci.*, **1991**, *31*, 301–304.

60 Holmberg, S.A., *Agri Hortique Genetica*, **1973**, *31*, 1–20.

61 Hymowitz, T., Palmer, R.G., Hadley, H.H., *Trop. Agric.*, **1972**, *49*, 245–250.

62 Chaven, C., Hymowitz, T., Newell, C.A., *J. Am. Oil Chem. Soc.*, **1982**, *59*, 23–25.

63 Broué, P., Douglass, J., Grace, J.P., Marshall, D.R., *Euphytica*, **1982**, *31*, 715–724.

64 Newell, C.A., Delannay, X., Edge, M.E., *J. Heredity*, **1987**, *78*, 301–306.

65 Singh, R.J., Kollipara, K.P., Hymowitz, T., *Crop Sci.*, **1990**, *30*, 871–874.

66 Burton, J.W., in: *Soybeans: Improvement, Production, and Uses*, Wilcox, J.R. (ed.), Agronomy Monograph 16, Am. Soc. Agron., Madison, Wisconsin, USA, **1987**, pp. 211–247.

67 Burton, J.W., Koinange, E.M.K., Brim, C.A., *Crop Sci.*, **1990**, *30*, 1222–1226.

68 Xu, H.J., Wilcox, J.R., *Euphytica*, **1992**, *62*, 51–57.

69 Howell, R.W., Brim, C.A., Rinne, R.W., *J. Am. Oil Chem. Soc.*, **1972**, *49*, 30–32.

70 Hammond, L.G., Fehr, W.R., Snyder, H.E., *J. Am. Oil Chem. Soc.*, **1972**, *49*, 33–35.

71 Burton, J.W., *Fat Sci. Technol.*, **1991**, *93*, 121–128.

72 Wilson, R.F., *INFORM*, **1993**, *4*, 193–200.

73 Anon., *INFORM*, **1992**, *3*, 1112–1114.

74 Marquard, R., Schuster, W., *Fette, Seifen, Anstrichm.*, **1980**, *82*, 137–142.

75 Mohamed, A.I., Rangappa, M., *Plant Foods Human Nutrit.*, **1992**, *42*, 87–96.

76 Smith, A.J., Rinne, R.W., Seif, R.D., *Crop Sci.*, **1989**, *29*, 349–353.

77 Hanson, W.D., *Crop Sci.*, **1991**, *31*, 1600–1604.

78 Imsande, J., *Agron. J.*, **1992**, *84*, 409–414.

79 Nielsen, N.C., in: *Advances in New Crops - Proceedings of the First National Symposium New Crops: Research, Development, Economics*, Janick, J., Simon, J.E. (eds.), Timber Press, Portland, Oregon, USA, **1990**, pp. 106–113.

80 Mohamed, A.I., Mebrahtu, T., Rangappa, M., *Plant Foods Human Nutrit.*, **1991**, *41*, 89–100.

81 Kollipara, K.P., Hymowitz, T., *J. Agric. Food Chem.*, **1992**, *40*, 2356–2363.

82 Mounts, T.L., Warner, K., List, G.R., Kleiman, R., Fehr, W.R., Hammond, E.G., Wilcox, J.R., *J. Am. Oil Chem. Soc.*, **1988**, *65*, 624–628.

83 Neff, W.E., Selke, E., Mounts, T.L., Rinsch, W., Frankel, E.N., Zeitoun, M.A.M., *J. Am. Oil Chem. Soc.*, **1992**, *69*, 111–118.

84 Liu, H.R., White, P.J., *J. Am. Oil Chem. Soc.*, **1992**, *69*, 528–537.

85 Wolff, R.L., *J. Am. Oil Chem. Soc.*, **1992**, *69*, 106–110.

86 Rennie, B.D., Tanner, J.W., *J. Am. Oil Chem. Soc.*, **1989**, *66*, 1622–1624.

87 Wilcox, J.R., Cavins, J.F., *Theor. Appl. Genet.*, **1985**, *71*, 74–78.

88 Graef, G.L., Fehr, W.R., Miller, L.A., Hammond, E.G., Cianzio, S.R., *Crop Sci.*, **1988**, *28*, 55–58.

89 Fehr, W.R., Welke, G.A., Hammond, E.G., Duvick, D.N., Cianzio, S.R., *Crop Sci.*, **1992**, *32*, 903–906.

90 Fehr, W.R., Welke, G.A., Hammond, E.G., Duvick, D.N. and Cianzio, R., *Crop Sci.*, **1991** *31*, 88–89.

91 Graef, G.L., Fehr, W.R., Hammond, E.G., *Crop Sci.*, **1985**, *25*, 1076–1079.

92 Hardon, J.J., Rao, V., Rajanaidu, N., in: *Progress in Plant Breeding -1*, Russell, G.E. (ed.), Butterworths, London, **1985**, pp. 139–163.

93 Ong, A.S.H., in: *Fats for the Future*, Cambie, R.C. (ed.), Ellis Horwood Publ., Chichester, UK, **1989**, pp. 285–300.

94 Sen Gupta, A.K., *Fat Sci. Technol.*, **1991**, *93*, 548–554.

95 Haumann, B.F., *INFORM*, **1992**, *3*, 1080–1093.

96 Gascon, J.P., Noiret, J.M., Meunier, J., in: *Oil Crops of the World*, Röbbelen, G., Downey, R., Ashri, A. (eds.), McGraw-Hill Publ. Comp., New York, **1989**, pp. 475–493.

97 Corley, R.H.V., Lee, C.H., *Euphytica*, **1992**, *60*, 179–184.

98 De Vries, R.J., *J. Am. Oil Chem. Soc.*, **1984**, *61*, 404–407.

99 Berger, K.G., *J. Am. Oil Chem. Soc.*, **1983**, *60*, 206–210.

100 Swoboda, P.A.T., *J. Am. Oil Chem. Soc.*, **1985**, *62*, 287–292.

101 Siew, W.L., Mohamad, N., *J. Am. Oil Chem. Soc.*, **1992**, *69*, 1266–1268.

102 Berger, K.G., Ong, S.H., *Oléagineux*, **1985**, *40*, 613–621.

103 Meunier, J., *Oléagineux*, **1975**, *30*, 51–61.

104 Jones, L.H., *J. Am. Oil Chem. Soc.*, **1984**, *61*, 1717–1719.

105 Jones, L.H., in: *Agricultural Biotechnology: Opportunities for International Development*, Persley, G.J. (ed.), CAB International, Wallingford, UK, **1990**, pp. 213–224.

106 Ong, S.H., Chuah, C.C., Sow, H.P., *J. Am. Oil Chem. Soc.*, **1981**, 1032–1038.

107 Hamilton, R.J., in: *Recent Advances in Chemistry and Technology of Fats and Oils*, Hamilton, R.J., Bhati, A. (eds.), Elsevier Applied Science, New York, **1987**, 109–166.

108 Tan, B.K., Ong, S.H., Rajanaidu, N., Rao, V., *J. Am. Oil Chem. Soc.*, **1985**, *62*, 230–236.

109 Wuidart, W., Gascon, J.P., *Oléagineux*, **1975**, *30*, 406–408.

110 Noiret, J.M., Wuidart, W., *Oléagineux*, **1976**, *31*, 465–474.

111 Appelqvist, L.-Å., Ohlson, R. (eds.), *Rapeseed - Cultivation, Composition, Processing and Utilization*, Elsevier Publ. Comp., Amsterdam, **1972**.

112 Downey, R.K., in: *High and Low Erucic Acid Rapeseed Oils*, Kramer, J.K.G., Sauer, F.D., Pigden, W.J. (eds.), Academic Press, New York, **1983**, pp. 1–20.

113 Chopra, V.L., Prakash, S., in: *Oilseed Brassicas in Indian Agriculture*, Chopra, V.L., Prakash, S. (eds.), Vikas Publishing House, New Delhi, **1991**, pp. 29–59.

114 Prakash, S., Chopra, V.L., in: *Oilseed Brassicas in Indian Agriculture*, Chopra, V.L., Prakash, S. (eds.), Vikas Publishing House, New Delhi, **1991**, pp. 60–85.

115 Downey, R.K., Rakow, G.F.W., in: *Principles of Cultivar Development - Crop Species*, Fehr, W.R. (ed.), Macmillan Publ. Comp., New York, **1987**, Vol. 2, pp. 437–486.

116 Vaughan, J.G., Macleod, A.J., Jones, B.M.G, *The Biology and Chemistry of the CRUCIFERAE*, Academic Press, New York, **1976**.

117 Tsunoda, S., Hinata, K., Gómez-Campo, C. (eds.), *Brassica Crops and Wild Allies*, Japan Scientific Soc. Press, Tokyo, **1980**.

118 Kramer, J.K.G., Sauer, F.D., Pigden, W.J., *High and Low Erucic Acid Rapeseed Oils*, Academic Press, New York, **1983**.

119 Stefansson, B.R., in: *High and Low Erucic Acid Rapeseed Oils*, Kramer, J.K.G., Sauer, F.D., Pigden, W.J. (eds.), Academic Press, New York, **1983**, pp. 143–160.

120 Downey, R.K., Röbbelen, G., in: *Oil Crops of the World*, Röbbelen, G., Downey, R., Ashri, A. (eds.), McGraw-Hill Publ. Comp., New York, **1989**, pp. 339–362.

121 Downey, R.K., *Plant Breed. Abstracts*, **1990**, *60*, 1165–1170.

122 Downey, R.K. and Bell, J.M., in: in: *Canola and Rapeseed - Production, Chemistry, Nutrition and Processing Technology*, Shahidi, F. (ed.), Van Nostrand Reinhold, New York, **1990**, pp. 37–46.

123 Kirk, J.T.O., Hurlstone, C.J., *Z. Pflanzenzüchtg.*, **1983**, *90*, 331–338.

124 Love, H.K., Rakow, G., Raney, J.P., Downey, R.K., *Can. J. Plant Sci.*, **1990**, *70*, 419–424.

125 Schmidt, L., Marquard, R., Friedt, W., *Fat Sci. Technol.*, **1989**, *91*, 346–349.

126 Anjou, K., Lönnerdal, B., Uppström, B., Åman, P., *Swedish J. agric. Res.*, **1977**, 7, 169–178.

127 Arnholdt, B., Schuster, W., *Fette, Seifen, Anstrichm.*, **1981**, 83, 49–54.

128 Ackman, R.G., in: *Canola and Rapeseed - Production, Chemistry, Nutrition and Processing Technology*, Shahidi, F. (ed.), Van Nostrand Reinhold, New York, **1990**, pp. 81–98.

129 Harberd, D.J., *Bot. J. Linn. Soc.*, **1972**, 65, 1–23.

130 Prakash, S., Hinata, K., *Opera Bot.*, **1980**, 55, 3–57.

131 U, N., *Japan. J. Bot.*, **1935**, 7, 389–453.

132 Attia, T., Röbbelen, G., *Can. J. Genet. Cytol.*, **1986**, 28, 323–329.

133 Song, K.M., Osborn, T.C., Williams, P.H., *Theor. Appl. Genet.*, **1988**, 75, 784–794.

134 Röbbelen, G., *Chromosoma* (Berlin), **1960**, 11, 205–228.

135 Hanelt, P., in: *Rudolf Mansfelds Verzeichnis landwirtschaftlicher und gärtnerischer Kulturpflanzen (ohne Zierpflanzen)*, Schultze-Motel, J. (ed.), Springer, Berlin, New York, **1986**, Vol. 1, pp. 272–332.

136 Thompson, K.F., *Adv. Applied Biol.*, **1983**, 7, 1–104.

137 Shivanna, K.R., in: *Oilseed Brassicas in Indian Agriculture*, Chopra, V.L., Prakash, S. (eds.), Vikas Publishing House, New Delhi, **1991**, pp. 117–137.

138 Becker, H.C., Damgaard, C., Karlsson, B., *Theor. Appl. Genet.*, **1992**, 84, 303–306.

139 McNaughton, I.H., in: *Evolution of Crop Plants*, Simmonds, N.W. (ed.), Longman, London, **1976**, pp. 53–56.

140 Song, K., Osborn, T.C., Williams, P.H., *Theor. Appl. Genet.*, **1990**, 79, 497–506.

141 Hosaka, K., Kianian, S.F., McGrath, J.M., Quiros, C.F., *Genome*, **1990**, 33, 131–142.

142 Song, K., Osborn, T.C., *Genome*, **1992**, 35, 992–1001.

143 Glimelius, K., Fahlesson, J., Landgren, M., Sjödin, C., Sundberg, E., *Sveriges Utsädesförenings Tidskrift*, **1989**, 99, 103–108.

144 Chopra, V.L., Narasimhulu, S.B., in: *Oilseed Brassicas in Indian Agriculture*, Chopra, V.L., Prakash, S. (eds.), Vikas Publishing House, New Delhi, **1991**, pp. 257–301.

145 Olsson, G., Ellerström, S., in: *Brassica Crops and Wild Allies*, Tsunoda, S., Hinata, K., Gómez-Campo, C. (eds.), Japan Scientific Soc. Press, Tokyo, **1980**, pp. 167–190.

146 Kräling, K., *Plant Breeding*, **1987**, 99, 209–217.

147 Chen, B.Y., Heneen, W.K., *Hereditas*, **1989a**, 111, 255–263.

148 Diederichsen, E., Sacristan, M.D., in: *Proceed. 8th Intern. Rapeseed Congress*, GCIRC, Saskatoon, Canada, **1991**, Vol. 1, pp. 274–279.

149 Mithen, R.F., Magrath, R., *Plant Breeding*, **1992**, 108, 60–68.

150 Lühs, W., Friedt, W., *G.C.I.R.C. Bulletin*, **1993**, (in press).

151 Schuster, W., *Z. Pflanzenzüchtg.*, **1969**, 62, 47–62.

152 Schuster, W., Michael, J., *Z. Pflanzenzüchtg.*, **1976**, 77, 56–66.

153 Lefort-Buson, M., Dattée, Y., *Agronomie*, **1982**, 2, 315–321.

154 Schuler, T.J., Hutcheson, D.S., Downey, R.K., *Can. J. Plant Sci.*, **1992**, *72*, 127–136.

155 Lefort-Buson, M., Guillot-Lemoine, B., Dattée, Y., *Genome*, **1987**, *29*, 413–418.

156 Brandle, J.E., McVetty, P.B.E., *Can. J. Plant Sci.*, **1990**, *70*, 935–940.

157 Kott, L.S., Erickson, L.R., Beversdorf, W.D., in: *Canola and Rapeseed - Production, Chemistry, Nutrition and Processing Technology*, Shahidi, F. (ed.), Van Nostrand Reinhold, New York, **1990**, pp. 47–78.

158 Henderson, C.A.P., Pauls, K.P., *Theor. Appl. Genet.*, **1992**, *83*, 476–479.

159 Downey, R.K., in: *Proceed. 2nd International Symposium on the biosafety results of field tests of genetically modified plants and microorganisms*, May 11–14, 1992, Goslar, Germany, Casper, R., Landsmann, J. (eds.), **1992**, Biologische Bundesanstalt für Land- und Forstwirtschaft, Braunschweig, Germany, pp. 17–21.

160 Knutzon, D.S., Thompson, G.A., Radke, S.E., Johnson, W.B., Knauf, V.C., Kridl, J.C., *Proc. Natl. Acad. Sci. USA*, **1992**, *89*, 2624–2628.

161 Canvin, D.T., *Canad. J. Bot.*, **1965**, *43*, 63–69.

162 Marquard, R., Schuster, W., *Fette, Seifen, Anstrichm.*, **1981**, *83*, 99–106.

163 Grami, B., Baker, R.J., Stefansson, B.R., *Can. J. Plant Sci.*, **1977**, *57*, 937–943.

164 Shahidi, F. (ed.), *Canola and Rapeseed - Production, Chemistry, Nutrition and Processing Technology*, Van Nostrand Reinhold, New York, **1990**.

165 Thies, W., *Fat Sci. Technol.*, **1991**, *93*, 49–52.

166 Stefanson, B.R., Hougen, F.W., *Can. J. Plant Sci.*, **1964**, *44*, 359–364.

167 Downey, R.K., *Can. J. Plant Sci.*, **1964**, *44*, 295–297.

168 Röbbelen, G., *Mutation Breeding Rev.*, **1990**, *6*, 1–44.

169 Rakow, G., Stringam, G.R., McGregor, D.I., *Proceed. 7th Intern. Rapeseed Congress*, GCIRC, Poznan, Poland, **1987**, Vol. 2, pp. 27–32.

170 Pleines, S., Friedt, W., *Fat Sci. Technol.*, **1988**, *90*, 167–171.

171 Pleines, S., Friedt, W., *Theor. Appl. Genet.*, **1989**, *78*, 793–797.

172 Kräling, K., Röbbelen, G., in: *Proceed. 8th Intern. Rapeseed Congress*, GCIRC, Saskatoon, Canada, **1991**, Vol. 5, pp. 1536–1540.

173 Rakow, G., *Z. Pflanzenzüchtg.*, **1973**, *69*, 62–82.

174 Röbbelen, G., Nitsch, A., *Z. Pflanzenzüchtg.*, **1975**, *75*, 93–105.

175 Röbbelen, G., Thies, W., in: *Brassica Crops and Wild Allies*, Tsunoda, S., Hinata, K., Gómez-Campo, C. (eds.), Japan Scientific Soc. Press, Tokyo, **1980**, pp. 253–283.

176 Scarth, R., McVetty, P.B.E., Rimmer, S.R., Stefansson, B.R., *Can. J. Plant Sci.*, **1988**, *68*, 509–511.

177 Eierdanz, H., Hirsinger, F., *Fat Sci. Technol.*, **1990**, *92*, 463–467.

178 Wong, R.S.C., Beversdorf, W.D., Castagno, J.R., Grant, I., Patel, J.D., *European Patent Applicat.* 88312397.8, **1988**, Publicat. no. 0323753.

179 Wong, R., Patel, J.D., Grant, I., Parker, J., Charne, D., Elhalwagy, M., Sys, E., in: *Abstract 8th Intern. Rapeseed Congress*, GCIRC, Saskatoon, Canada, **1991** A-16.

180 Auld, D.L., Heikkinen, M.K., Erickson, D.A., Sernyk, J.L., Romero, J.E., *Crop Sci.*, **1992**, *32*, 657–662.

181 Jönsson, R., *Hereditas*, **1977**, *86*, 159–170.

182 Zhou, Y.M., Liu, H.L., *Acta Agronom. Sinica*, **1987**, *13*, 1–10.

183 Chen, J.L., Beversdorf, W.D., *Theor. Appl. Genet.*, **1990**, *80*, 465–469.

184 Chen, B.Y., Heneen, W.K., *Heredity*, **1989b**, *63*, 309–314.

185 Töregård, B., Podlaha, O., in: *Proceed. 4th Intern. Rapeseed Congress*, Giessen, Germany, Deutsche Gesell. Fettwissenschaft, **1974**, pp. 291–300.

186 Norton, G., Harris, J.F., *Phytochem.*, **1983**, *22*, 2703–2707.

187 Bernerth, R., Frentzen, M., *Plant Sci.*, **1990**, *67*, 21–28.

188 Appelqvist, L.-Å., in: *The Biology and Chemistry of the CRUCIFERAE*, Vaughan, J.G., Macleod, A.J., Jones, B.M.G (eds.), Academic Press, New York, **1976**, pp. 221–278.

189 Mahler, K.A., Auld, D.L., *Fatty acid composition of 2100 accessions of* Brassica. *Winter rapeseed breeding program*, Univ. of Idaho, Moscow, USA, **1988**.

190 Taylor, D.C., Weber, N., Hogge, L.R., Underhill, E.W., Pomeroy, M.K., *J. Am. Oil Chem. Soc.*, **1992**, *69*, 355–358.

191 Löhden, I., Frentzen, M., *Planta*, **1992**, *188*, 215–224.

192 Heiser, C.B., jr., in: *Sunflower Science and Technology*, Carter, J.F. (ed.), Agronomy Monograph 19, Am. Soc. Agron., Madison, Wisconsin, USA, **1978**, pp. 31–53.

193 Miller, J.F., in: *Principles of Cultivar Development - Crop Species*, Fehr, W.R. (ed.), Macmillan Publ. Comp., New York, **1987**, Vol. 2, pp. 626–668.

194 Fick, G.N., in: *Oil Crops of the World*, Röbbelen, G., Downey, R., Ashri, A. (eds.), McGraw-Hill Publ. Comp., New York, **1989**, pp. 301–318.

195 Friedt, W., Ganßmann, M., *Vortr. Pflanzenzüchtg.*, **1992**, *22*, 131–144.

196 Dorrell, D.G., in: *Sunflower Science and Technology*, Carter, J.F. (ed.), Agronomy Monograph 19, Am. Soc. Agron., Madison, Wisconsin, USA, **1978a**, pp. 407–440.

197 Lofgren, J.R., in: *Sunflower Science and Technology*, Carter, J.F. (ed.), Agronomy Monograph 19, Am. Soc. Agron., Madison, Wisconsin, USA, **1978**, pp. 441–456.

198 Chandler, J.M., in: *Chromosome Engineering in Plants: Genetics, Breeding, Evolution*, Tsuchiya, T., Gupta, P.K. (eds.), Part B, Elsevier, Amsterdam, **1991**, pp. 229–249.

199 Seiler, G.J., *Field Crops Research*, **1992**, *30*, 195–230.

200 Leclercq, P., *Ann. Amélior. Plantes*, **1969**, *19*, 99–106.

201 Dorrell, D.G., *Crop Sci.*, **1978b**, *18*, 667–670.

202 Robertson, J.A., Chapman, G.W., jr., Wilson, R.L., jr., *J. Am. Oil Chem. Soc.*, **1978**, *55*, 266–269.

203 Goyne, P.J., Simpson, B.W., Woodruff, D.R., Churchett, J.D., *Austral. J. Exp. Agric. Anim. Husb.*, **1979**, *19*, 82–88.

204 Zimmerman, D.C., Fick, G.N., *J. Am. Oil. Chem. Soc.*, **1973**, *50*, 273–275.

205 George, D.L., McLeod, C.M., Simpson, B.W., *Austral. J. Exp. Agric.*, **1988**, *28*, 629–633.

206 Goffner, D., Cazalis, R., Percie du Sert, C., Calmes, J., Cavalie, G., *J. Exp. Bot.*, **1988**, *39*, 1411–1420.

207 Connor, D.J., Sadras, V.O., *Field Crops Research*, **1992**, *30*, 333–389.

208 Fick, G.N., in: *Sunflower Science and Technology*, Carter, J.F. (ed.), Agronomy Monograph 19, Am. Soc. Agron., Madison, Wisconsin, USA, **1978**, pp. 279–338.

209 Robinson, R.G., in: *Sunflower Science and Technology*, Carter, J.F. (ed.), Agronomy Monograph 19, Am. Soc. Agron., Madison, Wisconsin, USA, **1978**, pp. 89–143.

210 Goyne, P.J., Schneiter, A.A., Cleary, K.C., Creelman, R.A., Stegmeier, W.D., Wooding, F.J., *Agron. J.*, **1989**, *81*, 826–831.

211 Diepenbrock, W., Pasda, G., in: *Physiological Potentials for Yield Improvement of Selected Oil and Protein Crops*, Diepenbrock, W., Becker, H.-C. (eds.), Paul Parey Scientific Publ., Berlin, **1993**, (in press).

212 Skoric, D., *Field Crops Research*, **1992**, *30*, 231–270.

213 Havekes, F.W.J., Miller, J.F., Jan, C.C., *Euphytica*, **1991**, *55*, 125–129.

214 Hahn, V., Friedt, W., *Vortr. Pflanzenzüchtg.*, **1992**, *22*, 145–151.

215 Whelan, E.D.P., in: *Sunflower Science and Technology*, Carter, J.F. (ed.), Agronomy Monograph 19, Am. Soc. Agron., Madison, Wisconsin, USA, **1978**, pp. 339–369.

216 Georgieva-Todorova, J., *Z. Pflanzenzüchtg.*, **1984**, *93*, 265–279.

217 Friedt, W., *Field Crops Research*, **1992**, *30*, 425–442.

218 Chandler, J.M., Beard, B.H., *Crop Sci.*, **1983**, *23*, 1004–1007.

219 Friedt, W., Nichterlein, K., Dahlhoff, M., Köhler, H., Gürel, A., *Fat Sci. Technol.*, **1991**, *93*, 368–374.

220 Hammann, T., Friedt, W., *Proceed. 8th EUCARPIA Congress*, Angers, France, pp. 651–652.

221 Gürel, A., Nichterlein, K., Friedt, W., *Plant Breeding*, **1991**, *106*, 68–76.

222 Schneiter, A.A., *Field Crops Research*, **1992**, *30*, 391–401.

223 Kiniry, J.R., Blanchet, R., Williams, J.R., Texier, V., Jones, C.A., Cabelguenne, , M., *Field Crops Research*, **1992**, *30*, 403–423.

224 Masirevic, S., Gulya, T.J., *Field Crops Research*, **1992**, *30*, 271–300.

225 Rogers, C.E., *Field Crops Research*, **1992**, *30*, 301–332.

226 Schuster, W., Kübler, I., Marquard, R., *Fette, Seifen, Anstrichm.*, **1980**, *81*, 443–449.

227 Urie, A.L., *Crop Sci.*, **1985**, *25*, 986–989.

228 Simpson, B.W., McLeod, C.M., George, D.L., *Austral. J. Exp. Agric.*, **1989**, *29*, 233–239.

229 Yodice, R., *Fat. Sci. Technol.*, **1990**, *92*, 121–126.

230 Soldatov, K.I., in: *Proceed. 7th Intern. Sunflower Conference*, Krasnodar, USSR, Internat. Sunflower Assoc., **1976**, pp. 352–357.

231 Miller, J.F., Zimmerman, D.C., Vick, B.A., *Crop Sci.*, **1987**, *27*, 923–926.

232 Fernández-Martínez, J., Jimenez, A., Dominguez, J., Garcia, J.M., Garcés, R., Mancha, M., *Euphytica*, **1989**, *41*, 39–51.

233 Garcés, R. and Mancha, M., in: *Lipid biochemistry, structure and utilization*, Quinn, P.J., Harwood, J.L. (eds.), Proceed. 9th Internat. Symposium Plant Lipids, Wye, Portland Press, London, **1990**, pp. 387–389.

234 Sperling, P., Hammer, U., Friedt, W., Heinz, E., *Z. Naturforsch.*, **1990**, *45c*, 166–172.

235 Lee, J.A., in: *Cotton*, Kohel, R.J., Lewis, C.F. (eds.), **1984**, Agronomy Monograph 24, Am. Soc. Agron., Madison, Wisconsin, USA, pp. 1–25.

236 Waddle, B.A., in: *Cotton*, Kohel, R.J., Lewis, C.F. (eds.), Agronomy Monograph 24, Am. Soc. Agron., Madison, Wisconsin, USA, **1984**, pp. 233–263.

237 Percival, A.E., Kohel, R.J., *Adv. Agron.*, **1990**, *44*, 225–256.

238 Fryxell, P.A., in: *Cotton*, Kohel, R.J., Lewis, C.F. (eds.), Agronomy Monograph 24, Am. Soc. Agron., Madison, Wisconsin, USA, **1984**, pp. 27–57.

239 Lee, J.A., in: *Principles of Cultivar Development - Crop Species*, Fehr, W.R. (ed.), Macmillan Publ. Comp., New York, **1987**, Vol. 2, pp. 126–160.

240 Kohel, R.J., in: *Oil Crops of the World*, Röbbelen, G., Downey, R., Ashri, A. (eds.), McGraw-Hill Publ. Comp., New York, **1989**, pp. 404–415.

241 Cherry, J.P., Leffler, H.R., in: *Cotton*, Kohel, R.J., Lewis, C.F. (eds.), Agronomy Monograph 24, Am. Soc. Agron., Madison, Wisconsin, USA, **1984**, pp. 511–569.

242 Endrizzi, J.E., Turcotte, E.L., Kohel, R.J., in: *Cotton*, Kohel, R.J., Lewis, C.F. (eds.), Agronomy Monograph 24, Am. Soc. Agron., Madison, Wisconsin, USA, **1984**, pp. 81–129.

243 Endrizzi, J.E., in: *Chromosome Engineering in Plants: Genetics, Breeding, Evolution*, Tsuchiya, T., Gupta, P.K. (eds.), Part B, Elsevier, Amsterdam, **1991**, pp. 449–469.

244 Singh, A.K., Moss, J.P., Smartt, J., *Adv. Agron.*, **1990**, *43*, 199–240.

245 Niles, G.A., Feaster, C.V., in: *Cotton*, Kohel, R.J., Lewis, C.F. (eds.), Agronomy Monograph 24, Am. Soc. Agron., Madison, Wisconsin, USA, **1984**, pp. 201–231.

246 Percy, R.G., Turcotte, E.L., *Crop Sci.*, **1991**, *32*, 1437–1441.

247 Perkins, H.H., jr., Ethridge, D.E., Bragg, C.K., in: *Cotton*, Kohel, R.J., Lewis, C.F. (eds.), Agronomy Monograph 24, Am. Soc. Agron., Madison, Wisconsin, USA, **1984**, pp. 437–509.

248 Gururajan, K.N., Henry, S., Krishnamurthy, R., *Indian J. Agric. Sci.*, **1992**, *62*, 316–318.

249 Chamkasem, N., Johnson, L.A., *J. Am. Oil Chem. Soc.*, **1988**, *65*, 1778–1780.

250 Hammons, R.O., in: *Peanut Science and Technology*, Pattee, H.E., Young, C.T. (eds.), Am. Peanut Res. and Educ. Soc. Inc., Yoakum, Texas, USA, **1982**, pp. 1–20.

251 Nigam, S.N., Dwivedi, S.L., Gibbons, R.W., *Plant Breed. Abstracts*, **1991**, *61*, 1127–1136.

252 Coffelt, T.A., in: *Oil Crops of the World*, Röbbelen, G., Downey, R., Ashri, A. (eds.), McGraw-Hill Publ. Comp., New York, **1989**, pp. 319–338.

253 Singh, B., Singh, U., *Plant Foods Human Nutrit.*, **1991**, *41*, 165–177.

254 Knauft, D.A., Norden, A.J., Gorbet, D.W., in: *Principles of Cultivar Development - Crop Species*, Fehr, W.R. (ed.), Macmillan Publ. Comp., New York, **1987**, Vol. 2, pp. 346–384.

255 Stalker, H.T., Moss, J.P., *Adv. Agron.*, **1988**, *41*, 1–40.

256 Stalker, H.T., *Am. J. Bot.*, **1991**, *78*, 630–637.

257 Singh, A.K., Stalker, H.T., Moss, J.P., in: *Chromosome Engineering in Plants: Genetics, Breeding, Evolution*, Tsuchiya, T., Gupta, P.K. (eds.), Part B, Elsevier, Amsterdam, **1991**, pp. 65–77.

258 Lu, J., Pickersgill, B., *Theor. Appl. Genet.*, **1993**, *85*, 550–560.

259 Smartt, J., Gregory, W.C., Gregory, P.M., **1978**, *27*, 665–675.

260 Guok, H.P., Wynne, J.C., Stalker, H.T., *Crop Sci.*, **1986**, *26*, 249–253.

261 Smartt, J., *Econ. Bot.*, **1979**, *33*, 329–337.

262 Wynne, J.C., Gregory, W.C., *Adv. Agron.*, **1981**, *34*, 39–72.

263 Wells, R., Bi, T., Anderson, W.F., Wynne, J.C., *Agron. J.*, **1991**, *83*, 957–961.

264 Cherry, J.P., *J. Agric. Food Chem.*, **1977**, *25*, 186–193.

265 Norden, A.J., Gorbet, D.W., Knauft, D.A., Young, C.T., *Peanut Sci.*, **1987**, *14*, 7–11.

266 Worthington, R.E., Hammons, R.O., Allison, J.R., *J. Agr. Food Chem.*, **1972**, *20*, 727–730.

267 Sanders, T.H., *Lipids*, **1979**, *14*, 630–633.

268 Sanders, T.H., *J. Am. Oil Chem. Soc.*, **1982**, *59*, 346–351.

269 Khan, A.R., Emery, D.A., Singleton, J.A., *Crop. Sci.*, **1974**, *14*, 464–468.

270 Moore, K.M., Knauft, D.A., *J. Heredity*, **1989**, *80*, 252–253.

271 Schuster, W., Marquard, R., *Fette, Seifen, Anstrichm.*, **1974**, *76*, 207–217.

272 Nichterlein, K., Marquard, R., *Agrochimica*, **1985**, *29*, 265–275.

273 Schuster, W., *Fat Sci. Technol.*, **1987**, *89*, 15–27, 47–60.

274 Lay, C.L., Dybing, C.D., in: *Oil Crops of the World*, Röbbelen, G., Downey, R., Ashri, A. (eds.), McGraw-Hill Publ. Comp., New York, **1989**, pp. 416–430.

275 McHughen, A., *Plant Breed. Abstracts*, **1992b**, *62*, 1031–1036.

276 Durrant, A., in: *Evolution of Crop Plants*, Simmonds, N.W. (ed.), Longman, London, **1976**, pp. 190–193.

277 Batta, S.K., Ahuja, K.L., Raheja, R.K., Labana, K.S., *Ann. Biol.*, **1985**, *1*, 80–85.

278 Green, A.G., *Can. J. Plant Sci.*, **1986a**, *66*, 499–503.

279 Green, A.G., *Europe Patent 431,833*; **1991**.

280 Green, A.G., *J. Am. Oil Chem. Soc.*, **1984**, *61*, 939–940.

281 Gill, K.S., Yermanos, D.M., *Crop. Sci.*, **1967a**, *7*, 623–627.

282 Gill, K.S., Yermanos, D.M., *Crop. Sci.*, **1967b**, *7*, 627–631.

283 Nichterlein, K., Nickel, M., Umbach, H., Friedt, W., *Fat Sci. Technol.*, **1989**, *91*, 272–275.

284 Pirson, H., *Züchter*, **1955**, *25*, 186–190.

284a Williams, I.H., Martin, A.P., Clark, S.J., *J. Agric. Sci.* (Cambridge), **1990**, *115*, 347–352.

285 Albrechtsen, R.S., Dybing, C.D., *Crop. Sci.*, **1973**, *13*, 277–281.

286 Diepenbrock, W., Iwersen, D., *Plant Res. Develop.*, **1989**, *30*, 104–125.

287 Sekhon, K.S., Gill, K.S., Ahuja, K.L., Sandhu, R.S., *Oléagineux*, **1973**, *28*, 525–526.

288 Green, A.G., Marshall, D.R., *Euphytica*, **1984**, *33*, 321–328.

289 Nichterlein, K., Marquard, R., Friedt, W., *Plant Breeding*, **1988**, *101*, 190–199.

290 Yermanos, D.M., Beard, B.H., Gill, K.S., Anderson, M.P., *Agron. J.*, **1966**, *58*, 30–32.

291 Nickel, M., Nichterlein, K., Friedt, W., *Vortr. Pflanzenzüchtg.*, **1989**, *15*, Poster 15–11.

292 Nickel, M., *Dissertation*, **1993**, Universität Giessen, Germany.

293 Seetharam, A., *Euphytica*, **1972**, *21*, 489–495.

294 Green, A.G., *Proceed. Austral. Plant Breed. Conference*, Adelaide, Australia, **1983**, pp. 302–304.

295 Petrova, A., *Plant Cell Reports*, **1986**, *3*, 210–211.

296 Fouilloux, G., *FLAX: Breeding and Utilisation*, Marshall, G. (ed.), Kluwer Academic Publ., Dordrecht, **1989**, pp. 14–25.

297 Nichterlein, K., Umbach, H., Friedt, W., *Euphytica*, **1991**, *58*, 157–164.

298 McHughen, A., Holm, F., *Euphytica*, **1991**, *55*, 49–56.

299 McHughen, A., Rowland, G.G., *Euphytica*, **1991**, *55*, 269–275.

300 McHughen, A., *AgBioTech News and Information*, **1992a**, *4*, 53N-56N.

301 Diepenbrock, W., Pörksen, N., *J. Agron. Crop Sci.*, **1992**, *169*, 46–60.

302 Dybing, C.D., Zimmerman, D.C., *Crop. Sci.*, **1965**, *5*, 184–187.

303 Marquard, R., Schuster, W., Iran-Nejad, H., *Fette, Seifen, Anstrichm.*, **1978**, *80*, 213–218.

304 Schuster, W., Marquard, R., Iran-Nejad, H., *Fette, Seifen, Anstrichm.*, **1978**, *80*, 173–180.

305 Rowland, G.G., Bhatty, R.S., *J. Am. Oil. Chem. Soc.*, **1990**, *67*, 213–214.

306 Tonnet, M. L., Green, A.G., *Archives Biochem. Biophys.*, **1987**, *252*, 646–654.

307 Stymne, S., Tonnet, M.L., Green, A.G., *Arch. Biochem. Biophys.*, **1992**, *294*, 557–563.

308 Green, A.G., *Theor. Appl. Genet.*, **1986b**, *72*, 654–661.

309 Rowland, G.G., *Can. J. Plant Sci.*, **1991**, *71*, 393–396.

310 Dybing, C.D., *Crop. Sci.*, **1968**, *8*, 313–316.

2 Non-Food Uses of Vegetable Oils and Fatty Acids

W. Lühs and W. Friedt

2.1 Introduction

Despite the fact that the world annual production of vegetable oils and fats, currently more than 60 MT (metric tons), is increasing more quickly than demand, there is a steady commercial interest in the improvement of existing crops and in the domestication of wild plants to produce seed oils with desirable and marketable properties. The desire for novel or modified oils is linked both to health considerations and to the need for improved raw materials for the oleochemical industry. Industrial feedstocks are considered as one potential for market expansion of vegetable oil produced in agriculture in order to develop new uses for existing cultivated land as discussed in Chapters 6 and 9.

Vegetable oils have always been used for both edible purposes (cf. Chapter 1) and for a wide range of industrial applications, such as illumination oil, soaps, cosmetics, pharmaceuticals, emulsifiers, lubricants and greases, drying and semi-drying oils in paints, varnishes and other coatings, plastics and polymers, synthetic rubber manufacture, fat liquors for the leather industry, and livestock feed. In many of these applications, fatty oils are employed due to their ability to dry or to polymerize. In the cases where a polymerization reaction is not desired or is a troublesome side-effect, they compete to some extent with mineral oils, which have the advantage of being stable towards oxidation. Besides these speciality uses, vegetable oils or their ester derivatives have shown potential as supplements or substitutes for diesel fuel, respectively.

Non-food applications of vegetable oils and/or fatty acids have not received much attention in recent years because of the overwhelming interest in nutrition. Consequently, in many cases vegetable oils have been replaced by mineral oil-based products. Until now, fatty chemicals derived from natural, i.e., renewable sources, have shared markets and have competed with petroleum-derived counterparts. However, due to their natural origin and because of the current demand for biodegradable and environmentally friendly sources of raw material and products, it is important to exploit some of the well-known uses of vegetable oils which exist now or have existed previously under different economical or ecological conditions. While the shift in favor of extensive consumption of natural oils will obviously occur as the mineral oil shortage proceeds, certain areas of application have always remained the domain of naturally derived base stocks. In Europe, emphasis on "green" products, i.e., products compatible with mankind and the environment, has encouraged increased interest in producing biodegradable lubricants and using natural fatty alcohols for detergents and personal care products. The oleochemical industry is highly sophisticated and new inventions and methods, including biotechnology in oleo- and polysaccharide chemistry, have given the opportunity for further development of products with new applications. The most frequently mentioned examples in this case are the newly developed alkylpolyglycoside surfactants.

2.2 Description of the raw materials

Due to their relatively low price and their ready availability, animal fats, such as tallow, are still the primary feedstocks, especially in Western Europe. On the other hand, the vegetable oils most widely used industrially are soybean, coconut, palm, palm kernel, tall, linseed and castor oil. Apart from the last three examples, which are considered as more or less unsuitable for edible purposes, the soap and oleochemical industry has traditionally received much of its raw materials from the by-products of the edible oil refineries, viz., soapstocks, acid oils and fatty acid distillates [1–6].

Obviously, oil and fats now available to the oleochemical industry are derived in large part from those usually used for food purposes. These are not always the best starting materials for the production of other chemicals. For sometime, representatives of the chemical industry have declared their interest in an improvement of oil crops from an industrial point-of-view [2,7,8]. Generally, the industrial value of a natural oil depends on the proportion of a specific fatty acid and the ease/expense with which that acid can be purified for further modification or chemical reactions. Since purification is in most cases coupled with material losses, environmental problems (i.e., waste) and hence additional costs, a trend turning away from low grade to better quality oleochemical feedstocks has become more and more apparent [3,8]. Major commercial fatty acids from vegetable oils consist of normal (i.e., unbranched) hydrocarbon chains with a range from 12 to 22 carbon atoms, terminating in a reactive

Table 2.1. Common and unusual fatty acids as raw materials for oleochemical industry

Trivial name	Systematic name	Principal source
Caprylic	Octanoic	*Cuphea* spp.
Capric	Decanoic	*Cuphea* spp.
Lauric	Dodecanoic	Coconut, palm kernel
Myristic	Tetradecanoic	Coconut, palm kernel
Palmitic	Hexadecanoic	Tallow, palm
Stearic	Octadecanoic	Tallow
Oleic	*cis*-9-Octadecenoic	Tallow, TOFAs*, rapeseed, palm, high-oleic sunflower, *Euphorbia lathyris*
Linoleic	*cis,cis*-9,12-Octadecadienoic	TOFAs, soybean, sunflower
Linolenic	all-*cis*-9,12,15-Octadecatrienoic	Linseed, *Perilla ocymoides*
Petroselinic	*cis*-6-Octadecenoic	*Apiaceae*
α-Eleostearic	*cis*-9-*trans,trans*-11,13-Octadecatrienoic	Tung (*Aleurites* spp.)
Calendic	*trans,trans*-8,10-*cis*-12-Octadecatrienoic	*Calendula officinalis*
Dimorphecolic	9-hydroxy-*trans,trans*-10,12-Octadecadienoic	*Dimorphotheca pluvialis*
Vernolic	*cis*-12,13-epoxy-*cis*-9-Octadecenoic	*Vernonia* spp., *Euphorbia lagascae*
Ricinoleic	12-hydroxy-*cis*-9-Octadecenoic	*Ricinus communis*
Lesquerolic	14-hydroxy-*cis*-11-Eicosenoic	*Lesquerella* spp.[#]
---	*cis*-5-Eicosenoic	*Limnanthes* spp.
Eicosenoic	*cis*-11-Eicosenoic	*Brassicaceae*, jojoba
Erucic acid	*cis*-13-Docosenoic	*Brassicaceae*, nasturtium
Nervonic	*cis*-15-Tetracosenoic	*Lunaria annua*

**Tall oil fatty acids; --- = no trivial name; [#] accompanied with minor proportions of other hydroxy acids, such as densipolic (12-hydroxy-*cis,cis*-9,15-octadecadienoic), auricolic (14-hydroxy-*cis,cis*-11,17-eicosadienoic) and ricinoleic.

carboxyl group, although the C_{16} and C_{18} fatty acids are by far the major components, viz., palmitic, stearic, oleic, linoleic and linolenic (Table 2.1). Apart from ricinoleic acid which contains a hydroxyl group, all the other fatty acids of commercial importance are saturated or unsaturated with one, two or three isolated (i.e., non-conjugated) double bonds at mid-chain and distal to the carboxyl group [2,7,9].

In addition to the studies conducted on potential "new crops", i.e., their introduction, evaluation, cultivation and commercialization [10–15], considerable effort has gone into improving existing oil crops to establish new utilities and cost-benefit advantages for vegetable fats and oils. These studies are devoted mainly towards the enhancement of yield and oil content, the modification of oil quality, the better exploitation of by-products, and the development of disease and pest resistant varieties (see also Chapters 1, 3 and 6).

2.2.1 Soybean oil

Because of its relatively low cost and dependable supply, soybean oil is one of the most important vegetable oil sources for industrial use. Products include paints based on alkyd resins, plasticizing and stabilizing agents in vinyl plastics, printing inks and other general applications such as sulfurized oil factices in rubber manufacture, core oils in metal casting, diesel fuel substitutes, and grain dust suppressants. The soapstock, an oil processing by-product, is used as a feed additive and as raw material for low-cost oleochemicals used in the petrochemical and rubber industry [3,16–20].

2.2.2 Palm oil

Before the improvements in production, quality and processing of palm oil that were achieved in recent years, it was mainly the inferior grades with higher free fatty acid content (i.e., ffa>5%) that were used for industrial purposes, e.g., the manufacture of soap. While the supply of inedible grades of tallow and lard as the main source of desirable C_{16}-C_{18} acids stagnates, the needs of the soap and oleochemical industries will increasingly be met from palm oil fractions and products. In particular, palm stearin and the refinery by-products, palm acid oil (PAO) and palm fatty acid distillate (PFAD), have become available and competitive in large quantities within the last decade. Due to different refining processes, both PAO and PFAD contain those impurities and minor compounds originally found in crude oils. Consequently, they have a wide range of quality and composition determing the degree of usability, e.g., for soap making [21–23]. About 10% of palm oil is used for non-edible applications, which include soap, candle and crayon manufacture and plating oil in the traditional hot-tinning process. During the 1980s, several Southeast Asian countries, particularly Malaysia, successfully attempted to add value to the raw palm oil by producing basic oleochemicals such as fatty acids and ester derivatives. The large multinational companies such as Unichema, Henkel and Akzo, possessing the technological know-how and the financial means, have promoted this development to take advantage of the favorable local conditions as well as the supply of cheap raw material. Future developments even suggest a potential use of palm-oil methyl esters as a fuel for combustion engines supplementary to diesel fuel [21–25].

2.2.3 Coconut oil

Coconut palm is known as a "tree of life" for many inhabitants in developing nations who depend on its versatile utility. Every part of the coconut fruit can be utilized, including the fiber for coir products, the shell for charcoal, the milk as a beverage, and the meat as food. Nevertheless, its most important product is copra, the meat after drying. Copra contains 65–70% oil which is used in a wide range of edible and industrial purposes. The type of use of coconut oil varies around half edible and half inedible depending on market conditions [23,26,27].

Despite the exceptionally high production in the 1990/91 season, the production of coconut oil has generally stagnated in recent years at a level of about 2.8 MT. The Philippines and Indonesia are the most important producers, accounting for about two-thirds of coconut oil world production. However, as opposed to the large Southeast Asian oil palm plantations, 96% of coconut oil is produced by millions of smallholders dispersed across the tropics in nearly 90 countries, and much of it is consumed locally as a domestic cooking oil. Besides this fact, coconut production is significantly affected by unforeseen events such as typhoons, droughts and irregular monsoons, all of which make estimations of production very difficult and prices for coconut oil rather unstable [27,28].

Coconut oil is distinguished from most other vegetable oils by its high content of relatively low molecular weight saturated fatty acids. The low melting point makes it very useful as a fat for synthetic creams, confectionery and biscuit fats, hard butters and similar products. Since the triacylglycerols are easily hydrolyzed, an unpleasant soapy off-flavor is sometimes produced in edible products. However, the industrial value of the oil derives from its high proportion of lauric acid ($C_{12:0}$, 45–50%), which is used for soaps, cosmetics, shampoos, lauryl alcohol-based detergents and other cleansing agents [22,28–31].

2.2.4 Tall oil

Tall oil: Crude tall oil is a by-product of the sulfate or Kraft process for pulping of pine wood. It is not a typical oil, as it is a mixture of rosin acids (40%), fatty acids (50%), and unsaponifiable compounds (10%) such as sterols, higher alcohols, waxes and hydrocarbons. Fractional distillation provides tall oil rosin and tall oil fatty acids (TOFAs) containing varying amounts of rosin and a proportion of 90–98% as the fatty acid fraction, comprising approximately 50% oleic acid, 41% linoleic acid, 7% conjugated linoleic acid, and 2% saturated fatty acids. Tall oil fatty acids are readily used for a wide range of inedible products, e.g., TOFAs have become an important raw material for protective coatings such as oil-modified alkyd resins or for dimer acid products [1,3,6,32,33].

2.2.5 Rapeseed oil

Rapeseed oil has many potential uses other than as an oil for nutrition. Historically, rapeseed oil was used mainly in industry and for domestic lighting. However, the industrial applications were limited until steam power came into use. Until the replacement of steam power by diesel engines, the use of rapeseed oil as a lubricant or additive to petroleum-based lubricants was developed because rapeseed oil clung to water-treated metal surfaces better than other lubricants [1,34,35]. Today, two different types of rapeseed oil are available, viz., the traditional high-erucic acid (HEAR) and the low-erucic acid (LEAR) or Canola type (see also Chapter 1). Rapeseed oil is currently incorporated in lubricants for two-cycle engines, e.g., in motorcycles, boats and chainsaws. It is principally suitable as a fuel for modified or specialty engines, such as the German-designed "Elsbett" motor. In Europe and North America, rapeseed oil-derived methyl esters in small-scale projects have proved useful as a diesel fuel substitute. However, this application is uneconomic at the moment due to the low price of diesel fuel [5,20,36–41]. Besides being used industrially in many applications in which almost any vegetable oil can be used, rapeseed oil with a high content of erucic acid in particular has considerable advantages in specific applications due to the properties of this long-chain fatty acid. Special attributes of HEAR oil include high smoke and flash points, oiliness and stability at high temperatures, ability to remain fluid at relatively low temperatures, and durability [34,35,42–46].

2.2.6 Linseed oil

Linseed oil is characterized by its high percentage of α-linolenic acid ($C_{18:3}$). Consequently, linseed is one of the few plants from which this fatty acid can be extracted in a state of high purity. The quality of linseed oil for technical purposes depends largely on its degree of unsaturation, indicated by the iodine value (I.V.). Due to the strong influence of climatic conditions during linseed production, this property is quite variable (I.V. = 165–204), however, for most applications a minimum iodine number of 177 is required. The crude oil contains relatively high amounts of non-oil and mucilaginous substances, which have to be removed by means of refining to reduce the appearance of break material when the oil is heated rapidly to a high temperature [47]. The susceptibility of linolenic acid to autoxidation and polymerisation gives thin, cross-linked and tough films and imparts a drying quality to the oil. This particular property is traditionally exploited in the manufacture and usage of oil paints, varnishes, exterior house paints, and other industrial coatings. It is also used as a raw material for oil clothes, core oil (for sand cores) in metal casting, printing inks, automobile brake linings, concrete protection, linoleum and binding agents. White and pastel paints based on linseed oil have the disadvantage of turning yellow with age. Evidently, the tendency for discoloration (yellowing) of the oil films is caused mainly by the high degree of unsaturation, and particularly by the high proportion of linolenic acid. In recent decades, the market for linseed oil has declined as petroleum-based compounds increasingly have replaced linseed oil in the manufacture of these products. Besides the enormous growth of synthetic polymers, e.g., latex or vinyl and acrylic resin-based emulsions for surface coating applications, floor tiles based on vinyl or phenolic resins have largely replaced linseed oil-based linoleum [1,23,33,47]. However, more recently a revival of interest in linseed oil as basic material, e.g., for paintings or floor coverings, has been observed. Paints based on linseed oil are distinguished by a high compatibility towards the environment and human health [1,48,49].

2.2.7 Other drying oils with minor importance

Perilla oil is obtained from the seeds (oil content about 38%) of *Perilla ocymoides* (syn. *frutescens*). It was used considerably in the past as a strong drying oil (I.V.= 192–208) resembling linseed oil in many characteristics. However, due to significantly higher iodine values and probably a higher proportion of linolenic acid, it dries faster than linseed oil under comparable conditions [47].

Vegetable oils with high proportions of conjugated fatty acids comprise another important group of drying oils, which are particularly suitable for manufacture of fast-drying finishes and other protective coatings. The oil of the fruit kernels (oil content about 17.5%) of the tung tree (*Aleurites* spp.), a member of the *Euphorbiaceae*, which has been grown in China for centuries, contains about 80% α-eleostearic acid. Oiticica oil, which is obtained mainly from the nuts (oil content 55–63%) of the Brasilian oiticica tree (*Licania rigida, Rosaceae*), consists largely of α-licanic acid (75–80%), a C_{18} conjugated trienoic acid identical to that from tung oil but with a keto group in the C_4-position. However, due to a shortage of supply, these natural compounds have been replaced by epoxy resins, urethane polymers, and other synthetic intermediates [33,47,50].

2.2.8 Sunflower oil

In general, sunflower oil has not yet been used for inedible purposes, because of its limited supply and higher price as opposed to soybean oil and other industrially used oils. Nevertheless, due to its naturally high proportion of linoleic acid as well as advances in drying oil technology, the oil displays good drying properties without the undesirable yellowing associated with the high linolenic acid oils such as linseed oil [51]. In addition, due to the development of high oleic cultivars, sunflower oil will become a more important feedstock for oleochemical industry, provided that the economics of using this oil instead of much cheaper raw materials, such as tallow, can be improved [8,52; see also Chapter 1].

2.2.9 Castor oil

Castor oil is obtained from the seed of the castor plant, *Ricinus communis* (oil content, about 50%). Castor is a member of the *Euphorbiaceae* family and is nowadays grown as an annual crop in tropical and subtropical countries, predominantly Brazil and India. The oil is often considered as a strategic commodity and, thus, in the 1950s efforts were made to establish castor as a domestic crop in the USA in order to overcome the dependence upon imports. However, this development was delayed because non-shattering dwarf varieties with relatively uniform seed maturity suitable for mechanized production and once-through harvest systems were not available at that time. Nevertheless, several agronomic and economic factors have changed since domestic castor production was abandoned in the early 1970s and, consequently, reestablishment of the crop has been proposed in recent years [53–56]. In addition to the tung tree, castor is often considered as a traditional source of industrial oil. Because of the presence of several limiting factors, including a complex of allergenic compounds (found in all parts of the castor plant), a highly toxic endosperm protein (ricin), and a comparatively harmless alkaloid (ricinine), the meal of the castor seed is used almost exclusively as an organic fertilizer. The use of castor oil for medicinal purposes, e.g., its use as a laxative, has declined through the years and is now of minor importance. On the other hand, castor oil and its derivatives have many non-food applications, which are mainly based on the very high content (about 85–90%) of an unusual fatty acid, ricinoleic acid, which this plant synthesizes from phosphatidylcholine-bound oleate via hydroxylation. The hydroxy group of ricinoleic acid imparts, a very high degree of viscosity and oxidative stability to the oil which makes castor oil at least four times more stable than olive oil. Unlike other oils, castor oil is miscible with alcohol, but only slightly soluble in petroleum ether at room temperature [54–58]. Since castor oil contains ricinoleic acid in such a high purity, it can be used directly in a wide range of applications, including heavy duty lubricating oils and greases, fluids for hydraulic pumps or brake systems, plasticizers, surface coatings, wetting agents for industrial disinfectants, polishes, surfactants, pharmaceuticals, cosmetics, printing inks, textile dyeing and leather manufacture [1,33,54,56,57,59].

Castor oil is useful in protective coatings as a plasticizer in alkyd and urethane resins, while the blown oil can function as a nitrocellulose plasticizer. By far the most important use in coatings is the production of dehydrated castor oil (DCO) employed in the manufacture of paints and varnishes. The oil is converted via dehydration to an effective drying oil, although, in its natural state it is completely non-drying. Chemically the dehydration of castor oil, which is carried out under heating with acidic catalysts, is concerned with the functionality of the ricinoleic acid portion of the triacylglycerols. The reaction consists of the removal of the hydroxyl group and of a hydrogen of a neighbouring carbon atom, which gives rise to the formation of a mixture of dienoic fatty acids, viz., 9,11- and 9,12-octadecadienoic acids. DCO is noted for non-yellowing and outstanding color retention properties in protective coatings [33,57]. Besides hydrogenation for the manufacture of polishes and ethoxylation yielding emulsifiers, other uses are derived by sulfation/sulfonation, which results in surface-active compounds such as the formerly known Turkey red oil which assisted in dyeing and finishing textiles [1,56,57,59]. Additional details about the special properties and industrial applications of ricinoleic acid will be described in section 2.5.4.

2.2.10 Jojoba oil

Jojoba oil is obtained from the seeds (oil content, 50–55%) of *Simmondsia californica*, which is a drought- and salinity-tolerant, evergreen perennial shrub or small tree, endemic to the southwestern USA and Mexico [54,60–62]. Chemically it is not a typical oil, since instead of glycerol esters it is composed of liquid wax esters. These are mainly C_{40} and C_{42} mono esters (about 80%) of unsaturated C_{18}-C_{24} fatty acids and alcohols [63,64]. For comparison, the wax fraction of sea animals, namely sperm whales, is similarly compounded, containing mono esters derived from a mixture of saturated and unsaturated fatty acids/alcohols in the range of C_{14} to C_{22} [47,65]. Prior to international agreements for the limitation of whale hunting, sperm whale oil was used extensively as a natural source of industrial fatty alcohols and in the manufacture of special lubricants, e.g., sulfurized oil for high-pressure/temperature conditions. Since the physico-chemical properties of jojoba oil closely resemble those of sperm whale oil, the former has been suggested as a suitable substitute for sperm whale oil, in addition to synthetic analogs. The most promising applications of jojoba oil are in cosmetics and pharmaceuticals, surface-active agents (e.g., as emulsifiers), specialized lubricants and cutting oil formulations. Jojoba oil also can be hydrogenated to produce a solid white wax, which has specific properties useful for floor finishes and all kinds of polish [20,54,60,63,64]. However, jojoba has to be considered as a rather new crop because complete domestication, development of agronomic performance and commercialization still lag behind. Consequently, jojoba oil is currently not available at a quantity and price that makes it attractive to industry for widespread use [12,20,54,61,62,66].

2.2.11 Novel oil crops as sources of unusual fatty acids

Less than 300 species of the approximately 300,000 known plant species are grown in organized agriculture throughout the world. Some uncultivated plants could become more appealing commercial crops if proper research, development, and initial financial support were adequately supplied.

In the past fifty years, chemical surveys of wild species and germplasm have revealed many plants that produce seed oils with potential for industrial exploitation. Especially noteworthy is the large screening program for new sources of industrial oils started in 1957 by the U.S. Department of Agriculture. Since then, scientists at the Northern Regional Research Center in Peoria, Illinois, have analyzed the oil and other seed components of some 8000 species of plants and more than 75 new fatty acids have been discovered. However, it was not until the 1970s and even later that interest increased in breeding and intensified agronomic research for the establishment, evaluation and production of selected plant species [9,10,50,67–79].

Generally, a seed oil that has unique properties, such as unusual carbon chain length, unsaturation different from that of common oil crops, or an oxygenated functionality such as hydroxy, epoxy, or keto groups, offers opportunities for further development (see Table 2.1). However, domestication and breeding can only be successful if considerable variation for other desirable traits such as yield-determining factors (e.g., seed retention, uniform flowering, determinate growth habit), suitable by-products (e.g., meals with reduced levels of toxic compounds), and resistance to pests and diseases is present in the species of interest. In contrast to the well-known oil crops, where large and variable germplasm collections enabled improvement through plant breeding, the availability of genetic resources of potential industrial oil crops is extremely limited. An example of this can be found in *Crambe abyssinica*, which has a seed oil rich in erucic acid, but where the genetic variation has proved to be very narrow. Certain new oil crops, such as jojoba, lesquerella and vernonia, have industrial potential and may be grown in semi-arid regions not suitable for most of the traditional crops. But the slow pace of the commercialization of these new crops and their products is the major impediment to their cultivation on a larger scale. The industrial potential of some of these minor species with seed oils containing a high proportion of specific fatty acids will be discussed in more detail in section 2.5.4 [cf. Refs. 10–15,79–87].

2.3 Non-food uses of vegetable oils and their derivatives

In contrast to the rapid growth in supply of vegetable oils, their total non-food uses have increased little since the 1960s. The competitively priced petroleum-derived chemicals have restricted the growth of industrial markets for vegetable oils, but they are not the only reason for this lack of development. Some physico-chemical properties of vegetable oils such as reactivity or viscosity display both advantages and disadvantages depending on the application. Since most vegetable oils are too viscous and too reactive towards atmospheric oxygen, their usability in cosmetics, lubricants, hydraulic working fluids and fuel is often limited. Conversely, for other applications including coatings, detergents, polymers, and certain lubricants, the oils must be made even more reactive by introducing additional functionalities. Consequently, the demand for natural fatty oils to be used directly for non-edible purposes is of comparatively minor importance. Much larger amounts are transformed by chemical processes into basic materials for important industries, including the following: saponification, for traditional soap making (cf. section 2.3.2.1); hydrogenation, in the fat hardening process; hydrolysis and alcoholysis, for the manufacture of fatty acids and fatty acid esters, respectively, as the principal building blocks in the oleochemical industry (cf. section 2.4); polymerization and copolymerization, in the paint and varnish industry; glycolysis, for preparing partial glycerol esters, e.g., as a preliminary step of alkyd resin production (cf. section 2.3.2.3); and other interestification reactions in microbial conversion processes.

2.3.1 Refining and transformation processes

Natural oils and fats used for industrial products are often of inferior quality, i.e., unsuitable for food purposes due to undesirably high amounts of impurities and minor components including moisture, free fatty acids, phosphatides, stearins, waxes, sterols, pigments, trace metals, ashes, carbohydrates, protein degradation products. These non-triacylglycerol compounds may restrict the usability of the oil because of negative effects on color, drying properties or reactivity. Consequently, in order to avoid problems of product deterioration in subsequent production steps, initial refining processes such as degumming, neutralization, bleaching and filtering are usually indispensable to remove these impurities and contaminants [21,88,89].

One of the major chemical transformations of oils and fats is alkaline saponification for soap making (cf. section 2.3.2.1). In addition to this process, refined oil for industrial applications is chemically modified by means of hydrogenation, blowing (oxidation), heat bodying (polymerization), or double processing, i.e., moderate heat bodying first and then air blowing, to impart the oils unique properties. During hardening of fats, hydrogen is added to the double bonds of the fatty acids in the presence of special metallic catalysts. Oils and fats may be fully or partially hydrogenated depending upon the final use of the end product. The overall reaction, however, is quite complex due to the degree of unsaturation of the oil/fatty acids as well as the steric environment. For instance, the process is complicated by the occurrence of both positional and geometrical isomers of the unsaturated bonds [90].

Oxidative and thermal polymerization of highly unsaturated oils such as linseed, soybean and sunflower gives high-molecular weight cross-linked products. The formulation of such gums is directly concerned with the application of these drying and semi-drying oils in paints, varnishes and other coatings. For oils which have iodine values that are too low to be used directly as a drying oil (e.g., rapeseed oil), blowing has long been practiced to increase viscosity [33–35,91]. The process consists of passing a stream of air through the oil at elevated temperatures (40–150 °C). The oxidation process causes polymerization at the double bonds with cross-linking of the fatty acids. This air treatment gives the oil desirable properties in the formulation of paints, varnishes, inks and other coatings including excellent pigment wetting and adhesion, non-penetration over porous surfaces, improved drying and glossy surfaces. Heat bodying of linseed or soybean oil at 300–320°C in closed kettles under vacuum or inert gas is actually a polymerization process leading to the formation of dimers, trimers and higher polymers accompanied by progressively higher viscosity. The uses of heat-polymerized oils are quite similar to those of blown oils; however, the latter produce less resistant films than heat-bodied oils because of their greater content of oxygen-containing groups. Bodied oils, e.g., from linseed oil, are water-resistant, durable and can therefore be used in the manufacture of exterior house paints [33].

Epoxidized oils (e.g., soybean or linseed oil) are prepared by means of oxidation utilizing hydrogen peroxide in the presence of acetic or formic acid. The peroxy acid is the actual oxidizing agent converting the double bonds to epoxide- or oxirane rings. Epoxidized vegetable oils or their fatty acid esters are excellent plasticizers/stabilizers for vinyl polymers. Further, they can be important components of refrigeration lubricating oils and of internally plasticized phenolic resins for the electronic industry [1,89,92].

In addition to esterification, i.e., recombining fatty acids with mono- or polyhydroxy alcohols (see section 2.4.2), interesterification processes have numerous commercial applications. The latter include reactions in which triacylglycerols or other fatty acid esters are caused to react either with alcohols or with other esters by the interchange of acyl moieties resulting in the formation of new esters. The reaction of fatty acid esters with an alcohol is called alcoholysis, e.g., glycerolysis for the production of mono- or diacylglycerols, or methanolysis for the preparation of fatty acid methyl esters; the reaction between the esters, e.g., in a mixture of different triacylglycerols, is termed transesterification. The latter process is used particularly in the oils and fats industry to modify important physico-chemical properties such as melting and crystallization behavior. While natural oils and fats usually display a non-random distribution of the fatty acyl residues amongst the triacylglycerols, interesterification, i.e., acyl migration promoted through the presence of chemical catalysts, usually leads to a random distribution [93,94].

Interesterification reactions catalyzed by extracellular microbial lipases possessing pronounced substrate specificities and/or selectivities result in a non-random distribution of acyl moieties, e.g., the lipase from *Geotrichum candidum* [95]. Generally, the natural function of lipases is to catalyze lipolysis, i.e., the hydrolysis of glycerides or other fatty acid esters, but this reaction is easily reversible under low-water conditions. Since several lipolytic enzymes have become available of good quality and in relatively large quantities, considerable attention has been paid to their utility as selective preparative reagents, even on a commercial scale (see Chapter 7). In comparison with chemically catalyzed processes, the reactions catalyzed by means of enzymes at moderate temperatures frequently can be carried out faster, they are selective and often also stereospecific. Several enzymatic approaches for converting natural oils and fats into value-added products have been reported in recent years, including commercial application of lipases for fat splitting [19,96–105]. Important examples of such biotransformations are the use of lipolytic enzymes for the synthesis of distinct mono-, di- and triacylglycerols or phospholipids, i.e., particularly those compounds, which are rather rare in natural sources and very difficult to prepare by commercial means, e.g., the production of cocoa butter substitutes or oleic acid-enriched vegetable oils [97,99,102,106,107]. For instance, the upgrading of high-erucic acid oils (HEAR) via synthesis of trierucin has been achieved on a laboratory scale utilizing both chemical and enzymatic interesterification [108,109]. Further examples are the preparation of monoesters, especially of those resembling natural waxes, by lipase-catalyzed reactions [103,110–112; see also section 2.4.2].

2.3.2 General applications

2.3.2.1 Soaps

Saponified fatty acids ("soaps") are the simplest surface-active agents, which have been used since ancient times for cleaning and disinfection. Soaps were traditionally home-made from animal fat and wood ashes or lye, until the early middle ages when the professional manufacture of soap began. Soap making on an industrial scale was established in the late 18th century. In particular, the process for caustic soda production made soap much cheaper and its use very common [89,113,114].

   In principle, for traditional processes of soap making, i.e., soap-boiling in kettles, the fatty acids can be obtained from the oil through alkaline saponification or from the post-refining of acids derived in the initial processing of crude oil. Preparation of fatty acids from the oil involves higher costs but yields a product of higher purity. Modern processes for soap making are based on continuous splitting of fats into fatty acids, followed by continuous distillation and neutralization of the fatty acids. This integrated process via the fat-splitting route is strictly distinguished from soap making based on commercial distilled fatty acids, which are consumed in solid and liquid speciality soaps, paste and cream soaps, waterless hand cleaners, powdered hand soaps, etc. [4,6,113]. During the last few decades, soaps as the classical surface-active agents have been gradually replaced by synthetic surfactants, except in bar soaps where fatty acid soaps have retained their important cleaning applications [113,115].

   The raw materials commonly used in the manufacture of toilet and cosmetic soaps are defined mixtures of tallow and coconut oil (e.g., a 85/15 ratio) in order to obtain an optimum balance between lathering and water solubility (C_{12}-C_{14}) as well as detergency (C_{16}-C_{18}) [22,113,116]. The stearin fraction of palm oil is competetively priced compared with tallow, but soaps based on palm stearin do not have the desirable white color that can be obtained by using tallow. However, both, palm acid oil and palm fatty acid distillate can be used directly for making low-quality laundry soaps [21,22,117].

2.3.2.2 Cosmetics and pharmaceutics

Many of the uses of fats and oils in cosmetics are the same as those of ancient times [1,118]. Today, besides coconut oil, limited amounts of peanut and castor oil are marketed for medicinal purposes and for some speciality soaps, detergents, cosmetics (e.g., face and shaving creams, hair lotions) and toilet preparations, because these oils are widely believed to be beneficial to the skin. Fatty oils and their derivatives are readily absorbed by the skin and mainly function as emollient materials in creams, ointments, lotions and other cosmetics. In pharmaceutical preparations they are also used as carriers for fat-soluble vitamins and medicinal substances [1,23,31,54,56, 59,119]. Lecithins, as by-products of vegetable oil refining, are excellent emulsifiers and liposome-forming substances. Due to their amphipathic molecular structure in particular, liposomes have gained importance in the formulation of medicaments and cosmetics [120]. On the other hand, the most profitable commercial commodities from jojoba oil are a wide range of cosmetic products such as different kinds of oil, lotion or cream formulation for the care of body, face and hair. The oil has demonstrated its usability as an effective antifoaming agent in preparing antibiotics in fermenters (e.g., penicillin and tetracycline) and also as a treatment for skin disorders [1,15,60,63,121].

2.3.2.3 Plastics and polymers

The alkyd resins (alkyds) are important polymers with a wide range of applications, which are produced by the esterification of dibasic acids (e.g., phthalic or maleic anhydride) with polyhydroxy alcohols (e.g., glycerol, pentaerythritol). The usefulness of the polyol-dibasic combination on its own is very limited, but the functionality of the polyols can be controlled by means of so-called oil modification, i.e., tying up a certain proportion of unlinked hydroxyls with monobasic acids, usually fatty acids (e.g., TOFAs). Since the degree of cross-linking is limited, oil-modified alkyds dissolve in common solvents and represent tough and flexible polymers suitable for protective coatings. Vegetable oils, such as soybean and linseed oil, are also used directly as oil modifiers, but in these cases, an alcoholysis reaction between the oil and polyol is necessary to form mixed partial esters, before the actual esterification process. The choice of fatty acids versus total oils is determined entirely on a cost/performance basis. The former have some favorable properties as compared with oils, but most of them are also more expensive [6,33].

Urethane resins (syn. polyurethanes, urethane alkyds) can be considered as alkyd resin analogs in which diisocyanates represent the dibasic acid component. Epoxy resins, formed by polymerization of an epoxide and a diphenolic compound (epichlorohydrin and bisphenol A, respectively) and usually used for thermosetting adhesives, can be used also as polyols to form varnish-like products by esterification with drying oil fatty acids (linseed, soybean, DCO, tung, oiticica, or tall oil). These epoxy resin esters are inferior to comparable alkyd resins in color and color retention, but they have better resistance to various chemicals, adhesion, flexibility and resistance towards abrasion [33].

As opposed to their comparatively limited use as extenders and modifiers of epoxy resins, epoxidized glycerides/fatty acids, e.g., soybean and linseed oil, are predominantly used as plasticizing additives in polyvinyl chloride (PVC) and other vinyl resins. Plasticizers are usually polar liquids with a high boiling point, which are incorporated into polymers to increase their flexibility, workability or extensibility. For example, PVC originally was a hard horny material that had limited utility on its own. The epoxy plasticizers promote the action of primary plasticizers such as dioctyl phthalate (DOP). They are used mainly for their stabilizing effect on PVC against deterioration caused by heat and light, i.e., they bind the released hydrogen chloride [1,17,89,92, 123,124].

2.3.2.4 Surface coatings

Drying and semi-drying oils, especially linseed and soybean oil, as well as heat-polymerized and bodied oils, are used in the manufacture of paints, enamels, varnishes and other protective and decorative coatings, including alkyd resins and printing inks [33]. Previously, the natural oils, which were used in the manufacture of coating vehicles (pigment carrier or binder) alone or in combination with natural resins, consisted almost entirely of oils with a pronounced degree of unsaturation due to high proportions of linolenic (e.g., linseed or perilla oil) or conjugated fatty acids (e.g., tung or oiticica oil). Recently, with the availability of many synthetic resins and certain oil derivatization processes (e.g., dehydration of castor oil), it has also become possible to use oils with semi- or even non-drying properties in the manufacture of alkyd resins and other synthetic coatings. In some of these cases, the function of the oil or rather of the long and relatively flexible fatty acid chains is more that of a plasticizer than of an active drying component. A detailed discussion of the wide range of surface coatings is beyond the scope of this chapter, but the reader is referred to the comprehensive review of Formo [33].

Besides the synthesis of emulsifiers and additives, the predominant function of oils/fatty acids in the production of coating is their use as educts for alkyd resins and/or reactive diluents. For the latter function in particular, there is a strong need for reactive diluents that will reduce levels of non-reactive solvents in coatings formulations for the sake of air-pollution control. Depending on chain length, double bonds and functionalities of the fatty acids (e.g., linolenate, ricinoleate) these resins are used for air-drying or stoving lacquers, which may be solvent- or water-borne systems [6,33,125].

Recently, decorative-type coatings like vegetable oil-based printing inks have achieved more and more importance. Because of the various printing processes and surface properties presented by different paper materials, plastic films or metallic foils, a great diversity of inks is required to fill the needs of the printing trade. Vegetable oils are well suited for printing inks applied in lithography (e.g., offset) and letter printing. These inks are pastes usually consisting of pigments, the pigment carrier or binding agent, and optional additives. The composition of the binding agent, including natural and/or alkyd resins as solid compounds dissolved in a hydrocarbon solvent, has considerable influence on the printing and drying process. Vegetable oils, such as soybean and linseed oil, which are usually heat-bodied or blown to give various degrees of viscosity, can be partially used as substitutes for the petroleum-based resins and the mineral oil solvents; especially for reduction of the latter because of the harmful emission of volatile organic compounds [20,33,126,127].

2.3.2.5 Oiled cloth, linoleum and rubber additives

In the manufacture of oil cloth, for example, cotton fabric is usually impregnated on one side with drying oils, such as linseed or bodied oils, as well as with paints, filler material and varnish or wax [1].

A relatively old process is the manufacture of linoleum, which has been used for floor covering and - in its simplest form - is prepared as a binder from oxidized drying oils. Inorganic or organic filler compounds consist of natural (e.g., rosin) or artificial resins (e.g., alkyds), as well as pigments and ground cork or wood. Viscosity is considerably increased mainly through oxidation but also through polymerization processes at the double bonds of the triacylglycerols, especially from linseed oil. The addition of the resin fillers yields a stable, resilient yet thermosetting product, the so-called linoleum cement. The cement is mixed with the other filler compounds, pressed on a backing of burlap or other coarse material, and then has to ripen for a certain time [1].

The production of factice (vulcanized oil), originally introduced as a rubber substitute, is a principal industrial application of vegetable oils. Compounded with rubber it is used in the rubber industry to increase stability towards ageing and changes in shape. Similar to the vulcanization process of rubber, the vegetable oil and sulfur react to form a polymer. White factice is a light-coloured, compressible material used as an extender or modifier for rubber. This material is produced by the reaction of a relatively saturated or semi-drying oil with sulfur monochloride at room temperature. Brown factice is made from drying oils after first blowing and bodying until thickened, followed by the reaction with sulfur at its melting point (about 120–175°C). The viscous dark liquid dries to form a hard solid product which finds use as a rubber extender and for the preparation of linoleum [1,34,35].

Today, in contrast to the oils, the fatty acids are used more frequently as additives in compounding of rubber, especially used for the manufacture of tires. The fatty acids act, e.g., as softeners or plasticizers so that the rubber formulation is more manageable. Further, they function as lubricants and mold-release agents to prevent the rubber sticking to the processing equipment. During vulcanization, the fatty acids (e.g., rubber-grade stearate) or rather their zinc salts accelerate the reaction by cross-linking with the rubber molecules [6,116].

2.3.2.6 Lubricants and hydraulic oils

The usability of vegetable oils and their derivatives (e.g., sulfurized jojoba oil) as lubricants for metal surfaces is generally well regarded. In comparison to mineral oils they exhibit a better adhesion to the metal surface providing a thin-film lubricating boundary. Additionally, undesirable frictional wear and temperature rise are generally lower. In the past, high-erucic acid rapeseed (HEAR) oil in particular has shown very favorable lubricating properties, e.g., for steam engines [1,34,38,63,91]. Last, but not least, a major advantage of natural oils is their biodegradability. On the other hand, vegetable oils are inherently less stable than commercial mineral base fluids used in lubricants or hydraulic fluids. Due to hydrolysis they become acidic and corrosive in use, and because of the relatively high degree of unsaturation, they are susceptible to polymerization or gum formation under high pressure and/or temperature [128,129]. Castor oil, as opposed to mineral oil and other fatty oils, has a lower pour point and a higher degree of viscosity; the latter changes only slightly with temperature. These properties make the oil suitable for heavy-duty lubricants as well as hydraulic oils and automobile brake fluids, where a certain degree of compressibility is required. Prior to introduction of cold-resistant mineral oils and ester-based synthetic lubricants, castor oil was used extensively in the lubrication of jet aircraft engines. For many years it has been an additive in special blends of lubricating oils; those marketed under the Castrol® brand for use in high-performance cars are well known [1,122].

Because of drastic changes in technology and the availability of cheaper, petroleum-based lubricants, the use of vegetable oils for these purposes has been rather limited during the last few decades. However, no pure vegetable or mineral oil possesses all the properties required by modern technology, so that blending and supplementing of additives will always be necessary. Due to increasing mineral oil prices and the increasing awareness of the deleterious effects of the mineral oil and its additives (e.g., chloroparaffins such as PCBs) on the environment, it is likely that in the foreseeable future a shift to vegetable oil-based products will take place. The present approaches, using vegetable oils in hydraulic fluids, as well as in lubricating oils for chainsaws in foresty, and last but not least the large aquatic area, have to be considered as only the beginnings of this development. With regard to the principal disadvantage of fatty oils, viz., their relatively low durability caused primarily by oxidation and hydrolysis, it is important to note, that oils from different plants display a range of stabilities; for example, jojoba oil has a better oxidative stability than rapeseed oil. Since it is usually accepted that mineral base fluids need additives for improving their lubricating properties, the requirement for certain additives (e.g., antioxidants and flow improvers) is also valid for vegetable oil-based lubricants [1,20,129–131].

The advantageous properties of vegetable oils, however, have ensured their continuous usage in specialized lubricating agents, including high-pressure additives for use in mechanical transmission and gears, e.g., sulfurized jojoba oil [63,132]. Polymerized rapeseed oil as a supplement (up to 20%) to mineral oils has improved the thermal resistance of lubricants used in machinery operating at high temperature in the presence of moisture, e.g., steam engine cylinders and valves, pneumatic tools, air compressors and internal combustion engines. Equipment subjected to steam or air containing moisture is best lubricated by means of oils, which are soluble in the mineral oil, emulsifying, and preventing spot rust. Hydrogenated fats and oils are used in blended mineral oils suitable in the manufacture of lubricating greases made from metal (sodium, calcium, lithium) soaps [1,20,34,35].

2.3.2.7 Metal treatment

Vegetable oils and their polar derivatives are used solely or in blends with mineral oil for making metalworking fluids, which require not only lubricating but also cooling and covering properties. Core oils are used as binders for sand in forming the core of the mold in which metal castings are made. The cores are baked at temperatures of about 200 to 230°C until they become hardened through heat polymerization of the oil. The oils used in core making include linseed, soybean, and rapeseed oil, which are usually bodied to different viscosities. In continuous casting processes, rapeseed oil in particular can be used for proper mold lubrication. Quenching of metals is the process of withdrawing heat from metal rapidly to obtain the required structure. Fatty oils, especially from HEAR, can be used as additives in quenching oils for increasing the cooling rate [1,34,35].

In the hot-tinning process, during and after coating, clean steel sheets with molten tin, plating oils (e.g., palm oil) are required which can withstand the high temperatures (about 240–285°C) and do not volatilize or polymerize. The oil prevents oxidation of the hot tin bath and the freshly tinned plate. Cutting oils are required for lubrication of tools for cutting, machining, stamping and drawing of metal. The fatty oils are used as such or as sulfated/sulfonated oils in emulsions with water. Compared to petroleum oil and greases they have several advantages in this area, including much greater resistance to shearing, better adherence to metal surfaces under high operation temperatures, economical degreasing in caustic bath instead of solvent treatment, and usually cleaner metal surfaces [1].

The formulation of rust-protective coatings varies with the end use of the product. Some rustproofing compounds must dry at room temperature, or have to withstand ultraviolet radiation, humidity or even salt water, but have to be easily removeable by means of cleaning with solvents [1].

2.3.2.8 Textile and leather industry

The sulfation/sulfonation through treatment of fatty oils (e.g., tallow, castor or olive oil) with concentrated sulfuric acid (or sulfur trioxide) is a relatively old process which yields surface-active compounds imparting a desirable softness and fullness to fabrics. In particular, the preparation of Turkey red oil from castor oil has historical importance in textile dyeing. The desired oils contain a sufficient quantity of oleic or ricinoleic acid, but minor amounts of saturated or polyunsaturated fatty acids, which are less suitable for this kind of derivatization. Sulfated oils were the only commercially important non-soap surfactants until they lost many of their applications starting with the advent of the synthetic detergent industry. Today, the most important applications of sulfated/sulfonated oils are as emulsifiers in the formulation of spinning oils for textiles, and oils for leather treatment [1,89,115].

Various vegetable oils and a variety of fatty derivatives are used as lubricants and softening agents imparting manageability to the yarn in combining, spinning, weaving and other operations of textile manufacture. As opposed to mineral oil-based products they are more easily removeable from the fibers of the finished textile; hence they possess a great importance in this area [1].

Vegetable oils are used in the fat liquoring of leather after chrome-tanning treatment. Fat liquors consist of raw oils, bisulfited or sulfated oils, soaps or other emulsifying agents. The oil-water emulsion is taken up by the tanned, fibrous leather matrix, and after drying, the oil film surrounds the fibers and prevents fiber cohesion, giving flexibility and firmness to the finished leather product [1,34,35].

2.3.2.9 Miscellaneous uses

Fatty oils, such as rapeseed oil, were once used extensively as burning oils for illumination, but they have been entirely replaced by cheaper petroleum products [1,34,35]. Although the manufacture of candles is based on paraffin wax and beeswax, hydrogenated oils or lower grades of stearate are suitable compounds for hardening. Most candles consist of 10–20% stearic acid, which improves the appearance, burning qualities and stiffness of the candles [1,22,133].

Linseed oil and other drying oils are used both as internal binders and for external treatment in the manufacture of presswood fiber boards. Linseed oil in emulsions or in formulations with mineral solvents has proved effective for curing concrete. An important function in this case is the protection of concrete roadways, bridges, etc., against deterioration caused by alternate freezing and thawing as well as salt treatment [1].

Emulsifiers and other surface-active agents from fats and oils are used as carriers in formulating pesticides, e.g., to stabilize fungicidal and insecticidal suspensions in paraffin oil. Contrary to commonly used petroleum oils, vegetable oils used as an additive in postemergence weed sprays have the advantage that by themselves they bear little environmental impact and hazard of crop injury or harmful residues. Additionally, such fatty materials may find application in controlled-release formulations of pesticides for delivering the bioactive compound under specific environmental conditions (e.g., weather, moisture) or during critical periods in pest or crop development [17,18,134,135]. Finally, their use as dust suppressants in grain handling may become a significant area of agriculturally utilized vegetable oils, e.g., soybean or rapeseed oil [17,18,20,136].

2.4 Oleochemistry – industrial utilization of fatty acids and their derivatives

Until now, oleochemicals derived from natural fats and oils have shared markets and have competed with petroleum-derived counterparts, e.g., fatty alcohols, fatty amines, and even synthetic fatty acids and glycerol [4,5,137–139]. Fatty acids, the basic building blocks in oleochemistry, are obtained by means of fat splitting. They are further transformed by chemical reactions involving either the carboxyl group or the hydrocarbon chain. On a commercial scale nearly all, i.e, 96% of these reactions are directed to the derivatization of the carboxyl group, only 4% deal with the modification of the hydrocarbon side chain. However, increased flexibility and new derivatives have to come from oils providing different functionalities, such as double bonds in different positions along the hydrocarbon chain or present functional groups such as hydroxy, epoxy, or keto groups [7,8,89,140].

According to a recent study by Frost & Sullivan Inc. overviewing the market of oleochemicals in Western Europe [cited in Ref. 5] the largest end use of fatty acids (24.2%) is as basic building blocks in the manufacture of important oleochemicals, including fatty alcohols, amines, esters, amides, alkanolamides and oligomer acids (Fig. 2.1). Soaps and detergents (17.2%) are the second largest application. Other applications are represented by established market segments such as plastics, personal care products, coatings/resins and rubber. In particular, the use of lubricants and specialty chemicals for the leather and textile industry are predicted to grow, according to the report of Frost & Sullivan [cited by Ref. 5].

Although the applications of oleochemicals are exceedingly diverse, they have one unifying functionality: processes at the surface or at the interface between two phases are modified in one way or another. The long hydrocarbon

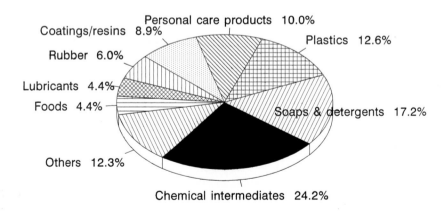

Source: Oleochemicals Market in Western Europe, Frost & Sullivan Report E1216

Figure 2.1. Proportional breakdown of fatty acid consumption by end-use applications in Western Europe in 1988 (adapted from Ref. 5; percent by volume)

Figure 2.2. Sources of vegetable oils as renewable raw material for industrial applications (clockwise from top center): seeds of soybean, rapeseed, castor, linseed, jojoba, and sunflower; the center shows the coconut palm with fruits (from Ref. 2).

chain favors the oily or hydrophobic phase, whereas the polar carboxyl group favors the aqueous or hydrophilic phase. The topography and steric environment, i.e., the shape itself as well as the space occupied by the fatty acid chains affect the interaction and packing of the chains. In turn, this strongly influences physico-chemical properties such as melting point, surface activity, foaming, micelle stability and structure, enzymatic reactions as well as membrane permeability and fluidity. Since it is predominantly the nature of the functional carboxyl group that is modified through oleochemical reactions, the resulting molecules can fulfil quite specific functions due to the properties of the hydrophobic alkyl chain. For instance, in textile production these effects of surfactant properties include enhancement of wetting, dispersion, dyeing, emulsification, foaming, lubrication, water repulsion and conduction of electrical charges. This is particularly true when fatty amide-based anti-blocking agents provide the slip of polyethylene films, when fatty amine salts act at the water-oil interface as a surfactant, or when fatty acid esters used in synthetic lubricants work at metal-to-metal interfaces [46,115,135,141,142].

2.4.1 Glycerol and derivatives

Fatty acids of storage lipids are chemically linked with glycerol or more rarely with fatty alcohols, as in the case of natural waxes (e.g., jojoba oil). Therefore, glycerol is a coupled product of fat-splitting reactions, i.e., the production of fatty acids [89,93]. The large soap and detergent producers are very interested in this lucrative by-product which has a wide range of applications. The major uses include pharmaceutics, cosmetics, food emulsifiers (e.g., bread and other bakery products), resins, cellulose, polyols, polyurethane, tobacco products, explosives, etc. [1,4,143–145]. Glycerol as such is used as humectant in various foods and tobacco. Triacetin (triacetylglycerol) is important in the production of cigarette filters. The use of glycerol in the manufacture of alkyd resins has already been mentioned (see section 2.3.2.3). Several classes of emulsifiers are based on partial fatty acid esters, with glycerol as the alcohol component, in which the free hydroxyl groups impart varying degrees of hydrophilicity. For example, glycerol monostearate is a common ingredient in several kinds of cream used in cosmetics and pharmaceutical products. Ethoxylated monoacylglycerols with different degrees of ethylene-oxide addition are particularly suitable as superfatting agents in shampoos and foam bath formulations, cosmetic cleaners and transparent creams, or as an emulsifier for extremely mild cosmetic preparations such as suntan lotions. Certain glycerol monoesters (e.g., monolaurin) function in cosmetics and food preservatives as antimicrobial agents [22,29,31,114,145–147]. Although glycerol is in relatively abundant supply at present, the prospects of increasing supplies entering the market as the industrial uses of vegetable oils expand, will make it important to enlarge further its already diverse list of uses.

2.4.2 Fatty acids and derivatives

Fatty acids (FAs) can be obtained following alkaline saponification of storage lipids by treating the resulting soaps with mineral acids. Due to economic considerations, FAs are produced today by hydrolysis based on counter-current splitting with high temperatures (up to 260°C) and pressures (up to 60 bar). Batch autoclave splitting and saponification are less important, but are still used for special oils where continuous splitting methods have their limitations [2,89,139]. In subsequent steps the crude FAs are separated from impurities or residual lipids and fractionated to achieve special cuts ($C_{8/10}$, $C_{12/14}$, $C_{16/18}$, etc.) or FAs of defined chain length. This process results in fractionation only according to the carbon number, since the vapor pressure of the FAs varies significantly with the chain length but not considerably with the degree of unsaturation. Further processing includes hydrogenation for the manufacturing of saturated FAs, or crystallization combined with several filtration steps to separate them into the solid stearin fraction and the liquid olein fraction according to their differing melting points [1,89,139,148–150].

The reactive carboxylic group of the fatty acids makes them ideal starting materials for the production of important basic oleochemicals including methyl esters, alcohols, amides, amines and quaternary ammonium compounds [1,2,6,89,116,139]. Besides the production of carboxyl group derivatives, a considerable amount of distilled fatty acids is used, as such, or in the form of alkali metal salts. The manufacture of high quality or speciality soaps and personal care products has already been mentioned (see section 2.3.2.1) in order to make a distinction with the fatty acids used in common soap making. In the so-called superfatting process, lauric-range fatty acids and stearic acid are added to some toilet soaps to correct their tendency to defat the skin due to their slightly alkaline character. In the production of mild synthetic detergent toilet soap bars (e.g., sebamed®, a product of Sebapharma, Germany; or Lever Brothers' Dove® and Caress® toilet soaps) stearic acid is used as a solvent in preparing the coconut oil-based acyl isethionates (see below), and remains in the soap as an emollient or cleansing cream [6,29,113,115,151]. Further important applications are in polishes and buffing compounds, household and industrial cleaning, or in the compounding of alkyd resins, high grade candles, and rubbers for manufacturing tires [1,6,139,144]. Heavy metal salts of fatty acids, commonly called metallic stearates, function as mould release agents in the manufacture of plastics or as a lubricant for coated papers. In food and pharmaceuticals, they are used to coat tablets or spices to increase flow and prevent sticking. They find manifold applications in cosmetics, rubber, petroleum, greases, varnishes, etc. [1,6,29,116,123].

Fatty acid esters represent another class of derivatives that are used extensively. Simple monobasic acid esters of short-chain alcohols, such as the *n*-propyl, isopropyl, *n*-butyl, isobutyl, or hexyl esters, are used for all kinds of lubricants in the plastics industry (e.g., internal and external lubricants), metal treatment (e.g., drilling, cutting or rolling oils) and textile manufacture. In addition, many of these fatty acid esters (e.g., isopropyl myristate) are consumed as emollients in cosmetics and other personal care products [1,6,29,31,123,131,147].

Synthetic wax esters can be prepared by means of both esterification of long-chain fatty alcohols with fatty acids or alcoholysis of triacylglycerols. The products possess hydrophobic, oily properties similar to natural waxes (sperm whale oil, jojoba oil), and applications are found in lubricants, plasticizers, cosmetics, polishes and buffing compounds, e.g., for conservation in floor finishes [89,111,112,119,131,152].

Esterification with polyfunctional alcohols (e.g., ethylene and propylene glycol, polyethylene glycol, pentaerythritol, glycerol, sorbitol, sucrose) leads to another important group of fatty acid esters, which are widely used as emulsifiers, wetting agents, protective coatings, synthetic lubricants, detergents, auxiliaries in textile and plastics industry, agrichemicals, and antifoaming agents. Many of them are mild and have zero or very low levels of toxicity, which makes them suitable for use in food, pharmaceuticals, and cosmetics. For instance, the sorbitan esters, e.g., the well-known Spans® and Tweens®, are used in cosmetics but also as emulsifiers for polymer production. The use and advantages of partial glycerol esters has been described earlier (see section 2.4.1) [1,6,22,29,33, 115,139,146,147,153].

Synthetic lubricants, or synlubes, are suitable in depressing the pour point of lubricant formulations as they possess high thermal and oxidative stability as well as retaining their lubricity and liquidity under a wide range of conditions. Important components of synlubes are polyol esters, which are the monobasic acid esters of branched polyols (e.g., pentaerythritol, trimethylolpropane, neopentyl glycol). Besides the petroleum-based polyhydroxy alcohols as a backbone, these polyol esters contain short-chain monofunctional acids available from the oleochemical industry; examples are heptanoic acid ($C_{7:0}$) and pelargonic acid ($C_{9:0}$), derived from ricinoleic and oleic/erucic acid, respectively, or the C_8/C_{10} fraction of coconut fatty acids. The uses of synlubes include jet-engine lubricants, crank-case oils for automobiles, hydraulic fluids, metalworking lubricants, industrial gear lubricants, and long-life automotive engine oils [1,6,131,147, 154,155].

Fatty acids, mainly of the lauric range, are used to prepare a variety of anionic speciality surfactants, including acyl isethionates, acyl taurates and acyl sarcosinates. The former two represent surface-active sulfonates with an intermediate ester or amide group, respectively, between the hydrophobic acyl chain and the sulfonate group. These compounds were first produced in the 1930s in Germany under the Ipegon® trade name. The acyl isethionates (i.e., Ipegon A type surfactants) are derived from sodium isethionate ($HOCH_2CH_2SO_3Na$), which is also used for the manufacture of *N*-methyltaurine, the precursor for the production of acyl taurates (i.e., Ipegon T). Both types of surfactants have excellent detergency, lime soap dispersancy, and wetting ability, as they represent early attempts to overcome the well-known disadvantages of fatty acids, e.g., hard water sensitivity and acid insolubility. The acyl isethionates occupy the largest volume in this group due to their use in synthetic detergent soap bars, which has already been mentioned. Acyl taurates are used as foaming agents and conditioners in shampoos. Further, both have been used in textile operations, ore flotation, and agricultural formulations. In contrast, acyl sarcosinates are anionic surfactants containing no sulfate or sulfonate group. They represent condensation products of fatty acids (or rather their chlorides) and sarcosine (*N*-methylglycine), a synthetic amino acid. The *N*-lauroyl derivative has been used primarily in toothpaste formulations and other oral care products; other cosmetic applications include shampoo formulations and hand cleaners. They are also suitable for usage in carpet cleaner products, ore flotation, and as additives for corrosion inhibitors [6,29,115,151].

2.4.3 Fatty acid methyl esters and derivatives

Fatty acid methyl esters (FAMEs) are produced either by esterification of fatty acids with methanol or more frequently by the means of alcoholysis, i.e., interesterification of triacylglycerols with methanol. Essential advantages of methyl esters as compared to their fatty acids are that they can be transported, stored and processed through fractional distillation more easily due to their lower melting/boiling point and their lower corrosive activity. Furthermore, production costs for several fatty acid derivatives, such as fatty alcohols or certain fatty esters, can be reduced via the methyl ester route [1,4,89]. Consequently, FAMEs predominantly serve as important intermediates, since about 80% of methyl esters in Western Europe are used for the conversion to fatty alcohols (see section 2.4.4). Other areas of application include the following: the production of alkanolamides, which yields high-purity products, the so-called superamides; the reaction with isopropyl alcohol is the preferred method for producing isopropyl esters; and sucrose esters formed by reaction of methyl esters with sucrose, in which one to six or seven of the eight available hydroxyl groups are esterified with fatty acid molecules. The advantage of these processes is due to the volatility of methanol, which strongly favors the interesterification, especially in the formation of new fatty acid esters [2,4,116,139,147,153].

Oleochemistry succeeded in developing interesting surfactants from methyl esters by direct sulfonation at the α-position of the carbon side chain, the so-called methyl ester sulfonates (MES, α-sulfo-fatty acid methyl esters, α-ester sulfonates), which perform well in detergents and are environmentally acceptable because of their biodegradability [156–158]. In comparison to fatty alcohol-based surfactants, MES are obviously of great advantage, particularly on a cost basis. Since fatty alcohols are inevitably derived from fatty acid methyl esters, it seems more efficient in the case of MES to add the hydrophilic group directly to the methyl ester rather than to the alcohol [156,158].

2.4.4 Fatty alcohols and derivatives

In oleochemistry, fatty or higher alcohols are produced by high pressure hydrogenation of fatty acids, or more frequently of FAMEs [2,89,139]. The high volume natural alcohols are made from distinct oils and fats: saturated $C_{12/14}$ alcohols either come from coconut or palm kernel oil, while for the manufacture of $C_{16/18}$ alcohols, tallow is predominantly used, which, however, is principally interchangeable with palm oil. On the other hand, higher alcohols -

especially those of C_8-C_{14} chain lengths - can also be synthesized out of ethylene or *n*-paraffins as basic petroleum building blocks. Worldwide, the petrochemical industry still accounts for about 60% of the total production of fatty alcohols. In Western Europe, in contrast, the majority of fatty alcohols is based on natural fats and oils; and this situation is additionally favored through consumers emphasis on products derived from natural/renewable raw materials [4–6,139,144,159].

The fatty alcohols either are used as such, e.g., as plasticizers, lubricant and textile auxiliaries and in cosmetics (e.g., lauryl alcohol), or they are further processed to a large variety of surface-active agents of different types: the nonionic fatty alcohol ethoxylates (AE) - made from fatty alcohols through reaction with ethylene oxide (EO); the anionic fatty alcohol ether sulfates (AES) - derived by additional sulfation of the former; and the anionic fatty alcohol or alkyl sulfates (AS) via sulfation of fatty alcohols [2,30,115,139,144,157,159–161]. It is important to mention that in the detergent end-market there is an additional competition between alcohol-based surfactants and those which are derived exclusively from petroleum feedstocks such as the anionic linear alkylbenzene sulfonates (LAS). Besides the fact that alcohol derivatives have to some degree shown a better biodegradability than LAS, the good performance of AEs in lower temperature washing as well as their very good water solubility and reduced impairment of washing performance due to water hardness, are factors which favor the use of nonionic alcohol-based surfactants. However, the use of these nonionics in formulations of heavy-duty laundry detergents is not equally pronounced in Western Europe compared with the USA [144,157,160]. Aside from the dominance of lauric-range detergent feedstocks, fatty alcohol sulfates of the $C_{16/18}$ range derived from tallow or palm oil are well suited as LAS substitutes in powder detergents, since they combine similar washing properties with better biodegradability, e.g., the new surfactant product line Plantaren® by Henkel, Germany [156,157]. Recently, a new group of nonionic surfactants has been derived from natural-based fatty alcohols and starch. The so-called alkyl polyglycosides display excellent lathering properties for the use as rinsing and cleansing agents as well as in cosmetics (e.g., hair shampoos) and food technology [156,157].

2.4.5 Fatty amines and derivatives

The fatty amines are by far the most important nitrogen-containing oleochemicals derived from fatty acids. Raw materials for fatty amines and their derivatives are commonly of natural origin, since they are usually manufactured either from fatty acids ($C_{12/14}$, $C_{16/18}$) via the conversion to intermediary nitriles followed by hydrogenation with a nickel catalyst, or due to a single-step reaction of fatty acids/FAMEs with ammonia under special hydrogenating conditions. However, less commonly, they also can be referred to petrochemical feedstocks, since production from fatty alcohols is possible. Fatty amines are classified as primary, secondary and tertiary amines, corresponding to the number of alkyl groups attached to the nitrogen atom. Quaternary ammonium compounds (quaternaries, quats) are tetrasubstituted ammonium salts, which can be prepared by the reaction of tertiary amines with various alkylating agents and by several other routes [2,89,116,139,162]. Fatty amines and their derivatives are very reactive chemicals in all kinds of physico-chemical process related to surfaces. They possess a wide range of diverse applications including antimicrobial disinfectants (bacteriostats), textile and paper softeners, laundry detergents and aftertreatment aids, textile specialties, emulsifiers (e.g., for asphalt, cosmetics), organophilic clays (organobentonites, Bentone®) used as viscosifiers in formulations of paints and greases, additives to lubricants, corrosion inhibitors, and antistatic agents in formulations of plastics [1,4,6,115,116,123,162–165].

In all areas of the petroleum industry (exploration, production, refining) the uses of so-called oleoamines are manifold, e.g., for well-drilling auxiliaries, as emulsifying, deemulsifying or anticorrosive agents, etc. [166]. Tertiary amines and quats, based on the C_{8-10} fraction of coconut fatty acids, find a wide use as flotation agents in hydrometallurgy, i.e., metals recovery processing based on solvent extraction [167].

One important product type in the whole group of fatty amine derivatives is made up of distearyl dimethyl ammonium chloride (DSDMAC) or methyl sulfate, which is a quat produced from hydrogenated tallow fatty acids. DSDMAC has been the leading household fabric softener in the world, although recently products containing this very effective agent are under attack in Western Europe because of concerns about its limited level of biodegradability [5,6,115,162,163].

Some amines are used for further reaction with acrylonitrile followed by hydrogenation to give fatty alkyl diamines. Because of their enhanced functionality, these products possess excellent bonding ability on surfaces such as metals, textiles, plastics and minerals. Besides the applications typical for amines, the diamines can be used in the preparation of asphalt, as they assist in emulsifying the heavy crude oil or tar fractions by binding the hydrocarbons to the siliceous aggregate rock used for roadbeds [6,116]. Primary as well as secondary amines can also react with ethylene oxide to produce various ethoxylated amines, which have a great variety of end uses including internal antistatic agents in plastic formulations, corrosion inhibitors, and levelling agents for dyes and lubricants in textile industry [6,116,123].

Fatty amine oxides, produced by oxidizing tertiary amines with hydrogen peroxide, represent detergents and shampoo ingredients (e.g., lauramine oxide) of outstanding cleaning and foaming properties. Not only do they clean the hair but they also give it body and sheen, as well as making it more manageable. Furthermore, amine oxides have the advantages of being antistatic, antiseptic, mild to the skin, biodegradable and of low oral toxicity [29,89,115, 116,153].

Amphoterics with a betaine structure containing both an anionic and a cationic moiety in the same molecule, comprise another class of specialty surfactants that are prepared from fatty acids, especially from the lauric range. Betaines can be prepared by reacting fatty tertiary amines (e.g., lauryl dimethylamine) with sodium chloroacetate. However, relatively high costs have so far restricted their use to specialty applications such as shampoos and industrial cleaners where their mildness, low skin and eye irritation as well as their stability over a broad pH range are important attributes [29,89,115,116,153,162].

2.4.6 Fatty amides

Fatty amides are nitrogen derivatives, which are usually prepared industrially by the reaction of fatty acids with anhydrous ammonia or amines. Primary amides ($RCONH_2$) of the C_{18}-C_{22} range (e.g., erucamide) are important slip- or antiblocking agents in the production of poly olefine-films. Because of the lower friction between two polyethylene (or polypropylene) sheets, the force necessary to pull them apart is reduced. The fatty amides function also as internal or external lubricating additives for a variety of polymers, i.e., in the latter case they effect the release of the hot resin melts from the metal surfaces of the processing equipment. Monosubstituted amides (R′CONHR), i.e., secondary amides produced by reacting fatty acids with a primary amine, perform in the same way. The diamide *N,N*-ethylene bis(stearamide) (EBSA), composed of ethylenediamine and two molecules of stearate, accounts for the largest proportion of all fatty amides and is used as internal lubricant. In this case the function is to reduce the cohesive forces of the polymer chains and improve the flow of the polymer during processing [6,46,116,123,168,169].

Alkanolamides, i.e., the mono- and diethanolamides of fatty acids are the simplest, but by far the most important members of the class of surface-active alkylamides, which are used in all kind of shampoo, cleanser and other light-duty liquid detergents. Long-chain monoethanolamides, which are important nonionic surfactants, can be readily sulfated to the corresponding anionic surfactants. The latter have good foaming and dispersing properties as recommended for use in shampoos and bar soaps. Other useful ingredients for cosmetic products are obtained from esterification of fatty acid monoethanolamides ($RCONHC_2H_4OH$) with maleic acid, followed by sulfonating the product to give the corresponding sulfosuccinate monoesters. Because of their high foaming power and foam stability, these ester products are used for both hair shampoos and carpet cleaners. Alkanolamides also possess very good rust prevention properties. They are important constituents of anticorrosion additive formulations (e.g., oleic acid diethanolamide). Furthermore, they stabilize the emulsion and increase the lubricating properties of cooling lubricants such as drilling or cutting liquids [1,115,116,151,153].

2.4.7 Modifications of fatty acid chain length and functionality

The introduction of additional functionalities such as another carboxyl group or the oxirane ring formation into fatty acids, which are usually monofunctional, gives rise to polymer chemistry. Important chemical conversions with potential uses in for the oleochemical industry are epoxidation, distinct polymerization reactions, oxidative cleavage, hydroformylation, and metathesis. Also, the microbial conversion of fatty acids will probably increase in importance in the near future [2,8,19,110,140,170,171].

2.4.7.1 Epoxidation

This is a versatile reaction suitable for all kinds of unsaturated fatty material, not only for triacylglycerols, which has already been discussed with regard to epoxidized oils. The epoxide-or oxirane ring is formed by organic peroxides (e.g., performic or peracetic acid), which attach active oxygen atoms to the double bonds of the fatty acid. Following the splitting of the oxirane ring by nucleophilic substances, trifunctional compounds can be obtained. However, the potential applications of these products have not been investigated in detail, so far [2,3,8,89,92].

2.4.7.2 Dimerization

The preparation of dimer acids is conducted by heating unsaturated C_{18} fatty acids in the presence of special clay catalysts. The reaction, commercially applied either to oleate or more frequently to tall oil and soybean fatty acids, is actually a polymerization resulting in a variety of products. The complex mixture includes about two third of the desirable dibasic acids with 36 carbon atoms, but also monomeric acids and a smaller amount of higher polymers (e.g., trimer acids). Besides unreacted starting fatty acid(s) the monomer fraction contains branched-chain and cyclized isomers as products of rearrangements [2,3,6,172–175].

The dimer acids (C_{36}) are used in the production of polyamide resins via their reaction with ethylenediamine or other polyamines. Dimer-based polyamides are highly flexible, film-forming, tough, adhesive as well as water- and corrosion-resistant. These products are used for both reactive or non-reactive type resin formulations, e.g., as curing agents for epoxy resins as well as hot-melt adhesives and specialty inks. Other applications of dimer acids include lubricants, greases, cutting oils and polyester hot-melt adhesives [6,89,170]. Monomeric acids can be considered as an important by-product of commercial dimer acid production. By means of solvent crystallization a liquid fraction termed "isostearic acid" is obtained, which contains a mixture of saturated C_{18} branched-chain fatty acids [174,176]. Since "isostearic acid" combines both the liquid property of oleic acid as well as thermal and oxidative stability of stearic acid, it has a wide range of uses. These include, as an emollient in cosmetic products, textile softeners and antistats, coupling agents and emulsifiers, textile lubricants, greases and synlubes [1,6,19,31,176,177].

2.4.7.3 Diels-Alder reaction

Addition reactions of isolated or conjugated systems of double bonds can be used for introducing ring formations or alkyl branching groups into fatty acids. For instance, Diels-Alder addition of acrylic acid to "conjugated" linoleic acid in the presence of a isomerization catalyst leads to a C_{21} dicarboxylic acid. This cyclic acid (Diacid®) has value in heavy-duty industrial detergents such as specialty soaps for use in lubricants and latex emulsions. Selected ester derivatives have potentially attractive properties as synthetic lubricants [1–3,170,178,179].

2.4.7.4 Ozonolysis

Other dibasic acids produced commercially from fatty oils are sebacic and azelaic acid. The former is a C_{10} dicarboxylic acid obtained by high-temperature alkaline cleavage of castor oil (see section 2.5.4), while the latter is a product of oxidative cleavage of oleic acid with ozone. Since production of ozone is a problematical point in this process, efforts are being made to find oxidative procedures that are cheaper and more environmentally preferable to ozonolysis [3,170,179–181].

Azelaic acid (nonanedioic acid) is important in the production of various esters used as plasticizers and in diester-based synthetic lubricants (e.g., dioctyl azelate). In addition, it can be used as a monomer in various specialty polymers, i.e., in a similar way to its petroleum-derived counterpart, namely adipic acid (a C_6 dicarboxylic acid). The second product of oleate ozonolysis is pelargonic (nonanoic) acid, which is used for polyol ester-containing synthetic lubricants as well as in compounding of alkyd resins, flavors and cosmetics [3,6,154,170,179].

2.4.7.5 Hydroformylation

Unsaturated fatty acids or their esters can be converted by reacting either with carbon monoxide and hydrogen or with formaldehyde under defined conditions, e.g., at elevated temperature and in the presence of specific homogeneous catalyst systems, into interesting formyl derivatives. For instance, methyl oleate gives an equimolar mixture of methyl-9 and methyl-10 formylstearate. From a starting material containing linoleate or linolenate, the corresponding diformyl and triformyl stearates are formed. In general, the introduced aldehyde group is well known as a versatile functional group able to undergo a wide range of reactions, including oxidation to the corresponding acid, acetalation, reduction to hydroxy compounds, and reductive amination to amino compounds. Several of the ester and acetal derivatives have proved useful as plasticizers with low volatility characteristics suitable for poly vinyl resins. Other formyl derivatives can find application in rigid urethane foams, urethane-modified coatings, polyamides, ester lubricants and wool finishing [3,124,171,182–184].

2.4.7.6 Metathesis

Metathesis is a disproportionation reaction, which is conducted on a commercial scale with petrochemical olefins in order to modify their chain length [185]. One of the most promising applications of this reaction in oleochemistry is to functionalize fatty material. For instance, unsaturated fatty esters (e.g., methyl oleate or erucate) can be transformed to corresponding bifunctional hydrocarbon derivatives. Thus, catalytic metathesis can be considered as a further option for preparation of dibasic acids. Warwel *et al.* [186] have proposed two-step syntheses with ethylene cleavage as the preliminary reaction, which results in a higher conversion rate than the single-step procedures without ethylene assistance. Co-metathesis reactions with short-chain olefins such as ethylene or *n*-hexene offer a method for the transformation of distinct long-chain fatty acid derivatives (e.g., oleate, erucate) into those with the more important C_{10}-C_{14} chain lengths. In the past few years the development of suitable catalyst systems has been the main problem, and the translation into industrial applications still largely depends on further improvements in this area [140,186–188].

2.5 Fatty acids – sources and applications

2.5.1 Medium-chain fatty acids (laurics)

2.5.1.1 Sources

In principle, medium-chain fatty derivatives, e.g., lauryl alcohol (dodecanol), for industrial applications can be obtained either from lauric oils, rich in medium-chain fatty acids (C_{12-14}), or from the petrochemical industry, so that the two sources are in competition. Due to dramatic oscillations in the price for coconut oil in the 1980s, most of these important basic chemicals were produced synthetically from petroleum feedstocks, especially in the USA. However, in recent years there is a trend back to lauric-based oleochemicals [3,16,27,28,161].

The demand for lauric oils is met almost entirely by coconut and palm kernel oil. Although both oils usually are interchangeable within all applications, there are minor differences in oil composition due to coconut oil having higher proportions of C_8 and C_{10} fatty acids as compared to a higher level of oleic acid and a higher iodine value in palm kernel oil. This may lead to different processing steps to yield desired products for specific end uses. In the future, however, palm kernel oil will increase its market share of laurics because it can undercut coconut oil in price due to its abundant supply [22,27,28,150,161; see also Chapters 1 and 6].

In the past few years, several projects concerned with increasing the availability of laurics and other fatty acids of relatively low molecular weight have been initiated. In particular, attention has been focused on the genus *Cuphea* in the family *Lythraceae* [13,28,86,189–193; see also Chapters 3 and 6]. The seeds of this herbaceous species most commonly have 30–33% oil, which is a rich source of C_8 to C_{14} fatty acids. In addition, one important advantage of *Cuphea* seed oils as compared to traditional sources of lauric oils is that high concentrations (up to 95%) of specific single fatty acids such as C_8 or C_{10} are possible [64,194–198]. Because of the unfavorable agronomic characteristics of *Cuphea*, including indeterminate flowering, seed shattering, plant and seed stickiness, and seed dormancy, progress to date has been rather slow, although recent new developments in *Cuphea* breeding show considerable promise for the future [13,189,199–208]. Meanwhile, scientists at Calgene Inc., Davis, California, pursued another route by showing that rapeseed can be genetically engineered to produce medium-chain fatty acids. Because rapeseed is an annual crop growing in temperate climates, it would be easier to adjust supply to demand which could stabilize the price of lauric oils for end users [27,209,210; see also Chapter 6].

2.5.1.2 Applications

Most of the supply of lauric oils is used in the manufacture of soaps, shampoos and cosmetic products, detergents and other surface-active agents, including industrial lubricants, coatings, plastics and other specialty products, e.g., oilfield and mining chemicals. The detergent industry dominates the increasing consumption of medium-chain length surfactants, including coconut-based anionics (alkyl sulfates) and nonionics (fatty alcohol ethoxylates) for both detergent powders and liquids, or lauric soaps in liquid formulations of domestic laundry detergents, e.g., Persil flüssig®, a product of Henkel, Germany. Besides good degradability lauryl alcohol sulfates have excellent foaming ability, so they are widely used as detergents or dispersing/emulsifying agents in all kinds of cosmetics. The anionic fatty alcohol ether sulfates, especially of the $C_{12/14}$ range, are suitable in light-duty liquids used in dishwashing detergents and hair shampoos; since they combine desirable properties: good foaming ability and superior cleaning action in synergistic mixtures, low irritation to the skin or eyes, and good biodegradability. Recently introduced alkyl polyglycosides based on starch and fatty alcohols from lauric oils have been shown to be starting points for fairly new surface-active compounds. Generally, there is a growing trend in personal care products towards the use of "natural" raw materials, i.e., petroleum-derived constituents are replaced by fatty derivatives from lauric oils. For instance, shampoos and hair care formulations often contain lauryl foaming agents prepared from naturally derived alcohols [5,6,27,29,30,144,161].

Recently, the short-and medium-chain fatty acids present in lauric oils have found a special medicinal application. Unlike the long-chain saturated fats they are readily digested and not stored in the body as fat. Consequently, triacylglycerols of caprylic ($C_{8:0}$) and capric ($C_{10:0}$) fatty acids are used in specialized clinical diets to improve nutrition of newborns or to alleviate certain digestive disorders caused by metabolism of long-chain fatty acids, e.g., for weight control of obese patients [27,211,212].

2.5.2 Stearic and oleic acid

2.5.2.1 Sources

Due to their great abundance, tallow, tall oil fatty acids, and recently also palm oil are the prime raw materials for the preparation of stearic and/or oleic acid in the oleochemical industry [4–6; see also section 2.2]. Depending on the distillation as well as separation technique used, e.g., mechanical pressing, solvent separation or Henkel hydrophilization process, various degrees of hardness/cloudiness, un-/saturation (I.V.) and purity are obtained. It is noteworthy that commercial stearic and oleic acid achieved their growth in recent decades as relatively impure fatty acids. For instance, commercial "oleic acid" usually contains not more than 70% pure oleic acid, i.e., it is accompanied by saturated C_{16}/C_{18} and polyunsaturated C_{18} fatty acids in approximately equal proportions. Further, it has the tendency to develop a dark color during heating due to oxidation and minor unsaponifiable constituents [1,89,149].

Oleyl alcohol is the most widely applicated liquid fatty alcohol. In nature, it is the most abundant of the unsaturated alcohols and occurs widely in fish and marine mammal oils, i.e., prior to banning of whale hunting it was extensively extracted from sperm whale oil (about 70% oleyl). Today, oleyl alcohol is produced on large commercial scale from other natural fatty materials (e.g., tallow), usually from the methyl esters via catalytic high-pressure hydrogenation providing complete preservation of the double bond. By this process, for instance, tallow alcohol is obtained consisting of 40–45% oleyl and about 45% saturated C_{16}-C_{18} fatty alcohols [89,213].

High-oleic acid oils: Some chemical reactions, e.g., ozonolysis or dimerization of oleic acid, could be more interesting if higher-purity materials of known and constant composition were available (see also section 2.4.7). However, increasing industrial demands for oleic acid with higher purity can only be accomplished either by costly processing and upgrading steps or, more suitably, by employment of oils already containing a high level of oleic acid (about 85%), e.g., from high-oleic sunflower cultivars (see also Chapter 1) or novel oil crops like *Euphorbia lathyris* (caper spurge) [52]. The latter is a Mediterranean herb with considerable adaptability to arid environments. Besides its former application as source of oil for illumination it is nowadays still useful as an officinal herb and in gardens for the control of moles and voles. The oil content of the seeds is about 50%, and seed production can reach 3.0 metric tons per hectare. Because of some major drawbacks, however, domestication and cultivation still lag behind. Like most members of the spurge family (*Euphorbiaceae*), it contains a cocarcinogenic milky sap which can cause irritation and injury to eyes and skin upon exposure. Further, the plant shows some more typical wild-type characteristics such as the widely distributed biennial life-cycle, indeterminate flowering, lengthy maturation period, dehiscent fruits, and pronounced seed dormancy, all of which impede its large-scale production [13,214,215].

2.5.2.2 Applications

Stearic acid has always been one of the most important fatty acids. Its low price and good properties cannot easily be matched by other fatty acids in many applications. Commercial grades of "stearic acid", actually a mixture often containing more palmitic acid than stearic, are extracted mainly from hydrogenated tallow or palm oil fatty acids. These lower grades are used as such or as metal salts in several kinds of technical applications such as the manufacture of plastics, coated paper, rubber, coatings, candles, crayons, waxes, lubricating greases, polishes and buffing compounds. Cosmetic grade stearic acid is used in a wide range of personal care products including viscosifying, pearlizing and superfatting agents in speciality soaps, creams, lotions, shampoos, shaving creams, tanning lotions and rubbing oils [1,6,22,25,29; see also section 2.4.2).

Among the more common uses of oleic acid are the manufacture of various liquid soaps and detergents, alkyds and other surface coatings, plasticizers, cosmetics and toiletries, pharmaceutical preparations and textile chemicals (see also section 2.4.2). Egan *et al* [213] have comprehensively described the properties and uses of oleyl alcohol and its derivatives. The refined oleyl alcohols as such have advantages in cosmetic and pharmaceutical formulations because of their light color, fluidity and low odor. The technical grades are more recommended as intermediates in the preparation of surface-active compounds (e.g., ethoxylates, sulfates, esters) for manifold use in cosmetics and detergents. The industrial grades of oleyl alcohol and its derivatives are mainly used in lubricants, metalworking fluids, plasticizers, textile auxiliaries and petroleum additives.

Recently, natural mixtures of C_{16}-C_{18} fatty alcohols, both saturated and unsaturated (mainly oleyl alcohol), from tallow, palm oil or even canola, have become increasingly regarded as suitable and competitive feedstocks for the production of alcohol ethoxylates/alkyl sulfates as wash-active substitutes for the less degradable linear alkylbenzene sulfonates based on petroleum [213,156,157,161; see also section 2.4.4].

In addition, nylon-9 is an example of a product that can best be made from high-oleic oils via azelaic acid, the ozonolysis product of oleic acid (see section 2.4.7). Because of its low moisture absorption, this special type of polyamide is well suited for engineering polymers used in electric and electronic industry. Along with other vegetable oil-based nylons with even longer repeating units in the polymer, e.g., nylon-610 and nylon-11 from castor as well as nylon-1313 from high-erucic acid oil (the numbers refer to the length of the repeating units in the polymer), it has better dimensional stability and dielectric properties in moist environments than those derived from petroleum feedstocks and used for synthetic fibers (e.g., nylon-6 and nylon-66). Furthermore, all these polyamides made from vegetable oils-based monomers are expected to have continuing uses in the replacement of metals in the automotive area and/or other materials in building construction [3,216; see also below].

2.5.3 Erucic acids

2.5.3.1 Sources

Among the very long-chain fatty acids, erucic acid has the greatest importance. It is present in the seed oil of the *Brassicaceae* (*Cruciferae*) and *Tropaeolaceae*, in which it is a major component. The seed oil of members of the former botanical family, such as *Brassica* spp., *Crambe abyssinica*, *Sinapis alba*, *Lunaria annua*, usually contains 40–60% erucic acid [47,73,76,217–220]. On the other hand, the seed oil of nasturtium (*Tropaeolum majus*) contains as much as 80% erucic acid, which is extraordinary in the plant kingdom

[77,220–222]. Several other species of rather distantly related botanical families contain minor amounts of erucic acid, including jojoba (*Simmondsia californica*, ca. 15%), *Limnanthes* spp. (12–14%), *Borago* spp. (2–3%) and white lupin (*Lupinus albus*, up to 3%) [77,85,223–226].

Erucic acid as an important commercial long-chain fatty acid has usually been derived from rapeseed. With the advent of low-erucic acid rapeseed types, production of traditional high-erucic acid rapeseed (HEAR) has waned rapidly worldwide (see Chapter 1). Consequently, during the past 20 years the development of alternative crops for production of erucic acid has been strongly promoted [10,44,217,227,228]. Among potential candidates in the *Cruciferae* family, crambe (*Crambe abyssinica*) has received most attention in this regard. Crambe is a herbaceous annual of Mediterranean origin, which has high levels of erucic acid, i.e., 55–60% as compared to about 50% in currently available HEAR. Since the fiber-rich hull accounts for 20 to 40% of the pod weight and usually sticks to the seed, the harvested product contains only some 30% oil. Furthermore, the limited genetic diversity in available germplasm of *C. abyssinica* has hindered any real progress in the improvement of yield, resistance to insect pests and diseases (e.g., *Alternaria brassicicola*), as well as meal quality. The latter is mainly concerned with the presence of glucosinolates, which diminishes the value of the meal as stock feed [13,44,54,87,228–235]. In Sweden and Germany, white or yellow mustard (*Sinapis alba*), usually used as spice or condiment crop, has been proposed as another summer-annual cruciferous crop suitable for the production of erucic acid. *S. alba* is easily grown, resistant to insects, diseases and seed shattering, and requires only a short growing season; but both oil content (30–35%) and seed yield are comparatively low [236–239]. Although both crambe and white mustard have the advantage that they would not interfere with domestic production of low-erucic rapeseed in Europe, the development of industrial rapeseed with erucic acid contents as high as 60% or more is still considered to be the most promising route due to its higher yield and oil content [240; see also Chapters 1 and 6). In the USA, the production of both HEAR winter types and crambe is favored for technical purposes, since that country produces its edible oil almost exclusively from other oil crops such as soybean [15,20,43, 44,241–244].

Money plant or Honesty (*Lunaria annua*), native to northern Europe, is well-known as an ornamental cruciferous plant. Its seed (30–40% oil) has a unique oil composition containing more than 90% (n-9)-*cis*-monounsaturated fatty acids. In particular, erucic (42%) and nervonic acid ($C_{24:1}$, 21%) are present; and it has even been possible to select single seed lots with a total of about 72% of these two long-chain acids. This is probably an attractive combination for industrial use where such fatty acids are required. *L. annua* is normally grown as a biennial, but annual lines are also available. However, extensive breeding and agronomy research programs are needed before commercial production will be feasible [73,83,84, 245–248].

2.5.3.2 Applications

Their long-chain fatty acid compositions make high-erucic acid oils unique and valuable for industrial uses. Many applications have been produced on occasion in pilot quantities, but often, promising results suggest commercial possibilities only if erucic acid were to be available on a larger scale and at a lower price (see Chapter 6). Consequently, current demand for high-erucic acid oil is based mainly on established markets for erucamide. This nitrogen derivative is used as an anti-block and slip-promoting agent in polyolefine films to modify the physical interface properties of the film layers. The additives are required to improve handling of polyolefine film and bags for both manual and automatic packaging. For this special application, the properties of the erucic acid derivative are highly valued, even though it costs roughly twice as much as oleamide. Erucamide is especially recommended where high temperatures are involved because its higher molecular weight imparts slower migration to the surface, lower votality and, consequently, better surface lubrication. An additionally desirable characteristic of erucamide is its antistatic property, for which its oleic acid-derived counterpart is the major competitor [6,46,168,169; see also section 2.4.6]. Erucic acid can be cleaved oxidatively to brassylic (1,13-tridecanedioic) acid by similar methods used to oxidize oleic acid to azelaic acid, e.g., ozonolysis. The co-product pelargonic acid, which is already commercially available from oleic acid, has well established uses as a component of plasticizers and synthethic lubricants (cf. section 2.4.7).

Like other dibasic acids (e.g., azelaic or sebacic acid) brassylic acid has the following principal applications: (1) diesters and linear polyesters as plasticizers for polyvinyl chloride (PVC), (2) diesters as synthetic engine lubricants, and (3) polymer intermediates in polyamide, polyester and polyurethane coatings, fibers, adhesives and resins. For instance, diesters of brassylate with a wide range of alkyl moieties have shown satisfactory functions as primary or secondary plasticizers of PVC. Apart from lower volatility, dicyclohexyl brassylate behaves similar to bis(2-ethylhexyl) phthalate (DOP), a standard plasticizer for PVC. On the other hand, ethylene brassylate, a cyclic diester prepared from brassylate and ethylene glycol, is used in the perfumery industry as a perfume fixative and as a synthetic musk fragrance [42]. Polyamides derived from brassylate, such as nylons-613, -13 and -1313, are not yet produced commercially, but their properties have been extensively investigated. They would have significant potential for use in manufacturing molded and extruded items (especially for the automotive area), electrical insulation materials, coatings and adhesives [3,42,43,249–251].

It is clear that the splitting of erucic acid into brassylic and pelargonic acid will find no adequate commercial application as long as oxidative cleavage routes alternative to ozonolysis are unavailable [cf. Ref. 15]. Nonetheless, Sonntag [45] predicts that "the acceptance or rejection of nylons-13 and -1313 as engineering plastics in the early 21st century may determine, to a large degree, the extent of commercial development of C_{22} oleochemicals".

In addition, Sonntag [45] has described a wide range of new patented applications of erucate and behenate (docosanoic acid, i.e., the hydrogenation product of erucic acid) including surfactants and detergents, plastics and plastic additives, photographic and recording materials, food additives, cosmetics and personal care products, pharmaceuticals, textiles, inks and paper, lubricants and fuel additives, and a lot of other uses. A critical reading of this valuable summary of patent literature reveals that the applications mentioned for erucic/behenic acid and their derivatives often represent the intention to translate the slightly better efficiency and usefulness of higher molecular weight materials onto established applications of C_{18} oleochemicals, including the acids, metal soaps, alcohols, esters, partial glycerol esters and nitrogen derivatives [see also Refs. 1,34,35,42,45,46,252].

2.5.4 Fatty acids with unique functionalities

Fatty acids with unusual specific properties and unsuspected possibilities are present in numerous "minor", i.e., less known seed oils as potential raw materials for industrial applications (see also Table 2.1). Among them are: fatty acids with double bond position different from those of common unsaturated acids (*Apiaceae*, *Limnanthes* ssp.), hydroxy acids (castor, *Lesquerella* spp.), epoxy acids (*Vernonia* spp., *Euphorbia lagascae*), conjugated polyene acids (*Calendula officinalis*, *Dimorphotheca pluvialis*), acetylenic acid (*Crepis alpina*), and cyclopropene acids (*Malvales*, *Sterculia* spp.). Apart from castor, as an established source of hydroxy acids, none of these oils has yet attained commercial importance and availability [9,10,13,47,50,64,77–79,84,85,87, 253–255].

2.5.4.1 Sources

Apiaceae (*Umbelliferae*): Some effort has also been expended in production research for several species of this family, such as coriander (*Coriandrum sativum*), fennel (*Foeniculum vulgare*) and dill (*Anethum graveolens*), which are usually grown for condiment, essential oils and spices. Their fatty oil is of interest because it contains high levels (70–80%) of petroselinic acid. This oleic acid-like isomer opens up another potential approach to the manufacture of medium-chain acids, since it can be split at the double bond into lauric and adipic acid by oxidative cleavage. However, oil percentages in the range of 30 to 40% are needed for an economically viable oil crop; they seem to be improbable in the species mentioned above [64,83,84,206,256]. Another approach is to transfer the gene(s) responsible for petroselinic acid formation (a Δ_6 stearate desaturase) from one of the above species, such as coriander, to an established oilseed crop, such as rapeseed [257; see Chapter 6].

Meadowfoam (*Limnanthes* spp.) is a wild herbaceous genus, which is native to the Northamerican Pacific coast and is especially abundant in northern California and southern Oregon. The most promising species is *L. alba* followed by *L. douglasii*; both are highly allogamous as they have a pronounced requirement for insect pollination. Further, they show much variation in yield and morphological characters, inbreeding depression and heterosis. Primary breeding and agronomic evaluation had to emphasize selection for higher seed yield and oil content, erect and taller growth habit, non-shattering of seed, and large seed size. A higher degree of self-pollination might also increase seed set by removing the dependence of the crop on foraging honey bees. In order to increase the economic potential of *Limnanthes* another goal is improving meal quality by reducing the amount of undesirable glucosinolates as toxic constituents [258–262]. However, the status of *Limnanthes* domestication and commercialization is still in the experimental and developmental stage [13,15,206]. The seed oil of meadowfoam has attracted attention because of its fatty acid composition, which is unique in several ways: (1) more than 95% of its component fatty acids are monoenes or dienes of C_{20} and C_{22} chain length, (2) about 90% of these fatty acids carry double bonds in the Δ_5 position, and (3) an almost complete lack of polyunsaturates with reduced oxidative stability, such as linoleic and linolenic acid [64]. Previous studies concerned with genetic variation among *Limnanthes* species/varieties revealed that the seeds contain about 20–33% oil, while the total fatty acids consist of 52 to 77% *cis*-5-eicosenoic acid, 8–29% C_{22} monoenes (i.e., the sum of *cis*-5-docosenoic and erucic acid) and 7–20% *cis,cis*-5,13-docosadienoic acid [75,260]. Recent analyses of *L. alba* seed oil gave the following result: 63% *cis*-5-eicosenoic, 10–12% erucic, 2–4% *cis*-5-docosenoic and 18% *cis,cis*-5,13-docosadienoic acid [87,263].

Since stereospecific analyses have shown that erucoyl is esterified in the *sn-2* position of the triacylglycerol [87,264], *Limnanthes* is also regarded as a model system for fatty acid/lipid metabolism [222,265–269]. Isolation of the enzyme(s) involved in the erucic acid pathway in meadowfoam, viz., *sn-2* acyltransferase, is a goal of biotechnology research seeking to transfer the corresponding gene(s) to rapeseed [270–273; see also Chapter 6].

Castor (*Ricinus communis*) is one of the few naturally occurring oils that is very nearly a pure compound, since it is the only commercially important oil containing major amounts of a hydroxy acid, i.e., about 85–90% ricinoleic acid. The crop and the use of its oil as such has been described earlier (see section 2.2).

Lesquerella spp., native to arid regions of North America, belong to the *Brassicaceae* family. Like castor they produce several unsaturated hydroxy fatty acids with contents ranging from 50 to 75%, which are similar to ricinoleic acid, viz., lesquerolic, densipolic, auricolic (for systematic names see Table 2.1). Small amounts of ricinoleic acid also occur in *Lesquerella* seed oils. The seeds of *L. fendleri* (bladder pod), one of the most promising species currently under breeding and agronomic development in the USA, contain about 25% oil, of which 55–60% is lesquerolic acid. Whereas castor seed meal is toxic and essentially useless, lesquerella seed meals - although containing glucosinolates - are relatively high in lysine and have good potential as protein supplements [10,13,15,62,64,74,83,274–276].

Stokes aster (*Stokesia laevis*) and *Vernonia* (iron weed) spp. are members of the *Compositae* family, and contain about 75% vernolic (*cis*-12,13-epoxy-*cis*-9-octadecenoic) acid in their seed oil. The former is a rather underexploited perennial aster that is indigenous to the USA and has been used ornamentally. With regard to the latter genus, *V. galamensis* (syn. *V. pauciflora*) has been proposed as being more commercially attractive than *V. anthelmintica* due to its better seed retention, comparatively high oil content (40%) and epoxy oleic acid proportion (ca. 80%). *Vernonia* spp. are annual herbaceous short-day plants native to Tropical Africa and India and, consequently, they are of particular interest as potential industrial oilseed crops for developing countries, e.g., Zimbabwe, Kenya and Pakistan. However, germplasms of *V. galamensis* have been selected that are neutral to day length and, therefore, suitable for the development of a temperate industrial crop [10,15,20,64,80,83,206, 277–282].

Euphorbia lagascae, a wild spurge native to Spain, is another interesting source of vernolic acid (58–62%). The plant has a relatively high seed oil content (42–50%), but it bears the same problematic characteristics as *E. lathyris* (see section 2.5.2.1). Because of limited natural variability, mutagenesis breeding research is being undertaken in order to eliminate wild-type properties such as seed shattering [64,283].

Calendula officinalis (marigold) contains about 20% oil in its fruits (achenes), of which about 60% is calendic acid, a conjugated fatty acid similar to that present in tung oil. This plant belongs to the *Compositae* and currently is grown as an ornamental or medicinal plant, although breeding is underway to develop genotypes for oil production [80,84,87]. Due to a high percentage (about 60%) of dimorphecolic acid, the seed oil (24% on a dry fruit weight basis) of *Dimorphotheca pluvialis* (cape marigold) is unique in its physico-chemical properties, in particular with respect to a high reactivity accompanied with low oxidative and thermal stability, as well as very high viscosity. The hydroxy group can be dehydrated, as is often done with castor oil, to give a conjugated triene formation [80,84,87].

2.5.4.2 Applications

Petroselinic acid: The double bond of this fatty acid can be cleaved by ozonolysis to give lauric and adipic acids as co-products. The latter is an important dibasic acid used in the manufacture of certain nylon polymers. Whereas conventional petroleum-based production of adipic acid causes release of harmful nitrogen monoxide (N_2O), the route from petroselinic acid would avoid such environmental pollution. This beneficial side-effect could be taken in account as soon as ozone-based cleavage is adequately replaceable by another method. However, the abundant and inexpensive supply of both lauric and adipic acid does not favor the commercial development of oilseed crops bearing petroselinic acid at the present time [64,206,284,285].

Meadowfoam oil: Most of the *Limnanthes* C_{20-22} fatty acids are monoenoic; and even the principal dienoic ($C_{22:2}$) has unsaturated bonds far removed from each other, thus making this acid react more like a monoenoic fatty acid in terms of its oxidative stability. Furthermore, Δ_5 bonds have shown to be more stable than olefins with double bonds located nearer the center of the hydrocarbon chain. Consequently, *Limnanthes* oils possess high oxidative stability [64,83,87,263,286]. Besides this valuable property, cosmetic preparations from meadowfoam oil can also benefit from its low general toxity and skin irritation [287]. Additionally, it can be converted to satisfactory lubricant additives by sulfurization in the presence of methyl lardate [132]. Esterification of meadowfoam fatty acids with corresponding fatty alcohols results in liquid wax esters of the range C_{40} and C_{42}, which have properties very similar to those of jojoba and sperm whale oil [258,288,289]. For some reactions of *cis*-5-eicosenoic acid as major component of *Limnanthes* oil, the position of the double bond with respect to the carboxyl group is probably advantageous, e.g., for lactonization, estolide formation or dimerization [64,290–292]. Currently, the oil and fatty acid derivatives are being tested for several technical applications such as factices, polymer additives and lubricants [64,83,263,291,293,294].

Ricinoleic acid: Besides the traditional uses of castor oil-based products, such as lubricants, coatings, surfactants (see section 2.2), ricinoleic derivatives also possess a wide range of applications in polyesters, nylon-type polyamides and urethane polymers, plasiticizers, cosmetics, flavorings, fungistats and high-performing greases, e.g., those of lithium 12-hydroxystearate [1,3,33, 56,57,59].

Ricinoleic acid is used on a commercial scale for the production of a particular type of polyamide, viz., nylon-11. Castor oil is transesterified with methanol to obtain methyl ricinoleate, which is subsequently cracked at high temperatures (about 500 °C) to methyl 10-undecylenate and heptaldehyde. Hydrolysis of the methyl ester gives 10-undecylenic acid, which is further converted to 11-aminoundecanoic acid, the bifunctional monomer for production of nylon-11 (Rilsan®). Because of its vibration and shock resistance this polyamide is important in automotive, truck, aviation and similar high-performance plastics applications. Heptaldehyde is used as such in preparing synthetic flavors and fragrances, or it is oxidized to heptanoic acid ($C_{7:0}$) suitable in the production of polyol ester-based synthetic lubricants [2,6,56,57].

Following heating at 250–275°C in the presence of excess alkali and catalyst, ricinoleic acid is split into sebacate (decanedioic acid) and 2-octanol as a co-product. The former is an important dibasic C_{10} intermediate for plasticizers in polyvinyl chloride films, diester-based synthetic lubricants, polyesters and polyamides, while 2-octanol can be used to prepare heptanoic acid by caustic oxidation [2,57,170,179,295].

Sebacic acid: This is used in the manufacture of nylon-610 via reaction with hexamethylenediamine. This polyamide finds major applications in moldings, extrusions, signal wire coatings, bristles and synthetic fibers [57,170,179]. The usual variety of alcohols (e.g., 2-ethylhexanol) can be used to produce the corresponding diesters from sebacic acid. These diesters are used either as plastic additives to contribute to low temperature flexibility and permanency [123,124,170,179], or in formulations of synthetic lubricants. With regard to the latter, for instance, diesters such as bis(2-ethylhexyl) sebacate are suitable in lubricants that combine good oxidative and high-temperature stability with fluidity at low temperatures. Their uses include lubricants for turbo and jet aircraft engines, compressors, gas turbines, hydraulic fluids and instruments [1,57,131,154,155,170,179].

Recently, ricinoleic acid derivatives such as simple esters, mono-and diacylglycerols, glycol esters and their hydroxystearate analogs have gained importance in cosmetic preparations, since they impart oxidative stability, mildness to the skin, emolliency, ease of emulsification, pigment and dye-carrying capacity, as well as a wide range of viscosity [56,59].

Other hydroxy fatty acids: The presence of large amounts of hydroxy fatty acids in the seed oils of *Lesquerella* species makes them attractive candidates for castor oil replacements and other industrial uses including formulations of lithium-type greases and lubricants, protective coatings and plastics, e.g., polyamides such as nylon-13 and nylon-1212. However, the current raw material for dodecanedioic acid (C_{12}) as an intermediate in the production of the latter polyamide is petroleum [15,62,64,179].

Vernolic acid: Vernonia oil is a natural epoxy oil that is chemically distinct from epoxidized soybean and linseed oil due to its much lower viscosity. Because of their lower oxirane content vernonia oils and/or derivatives of the oil are less suitable as plasticizers/stabilizers in the manufacture of polyvinyl chloride (cf. section 2.3.2.3); but they are more likely to attain a role as reactive diluents for epoxy- and alkyd resins in the manufacture of baked coatings (e.g., on steel) and adhesives, where there is urgent need to reduce the volatile organic compounds given off in conventional processes. In contrast, the specific properties of *Euphorbia lagascae* seed oil in industrial applications are less well evaluated [20,64,83,87,278].

Calendic acid: The seed oil of *Calendula officinalis* has properties similar to tung oil, i.e., a relatively high viscosity, and because of its very low oxidative stability, a rapid tendency to air-drying that makes it suitable for application as a binder in paints and coatings [87].

2.6 Conclusions

Vegetable oils have many actual and potential uses other than as an oil for nutrition. Although in many cases fatty oils and their derivatives have been replaced by mineral oil-based products, the past few years have witnessed increasing interest in renewable, biodegradable and environmentally friendly sources of raw material and products.

Besides a wide range of applications in which almost any natural fat or oil can be used, different oils are favored in specialized areas, including the following: soybean and linseed oil are suitable in the manufacture of paints, enamels, varnishes and other protective and decorative coatings, while in their epoxidized form these oils are predominantly used as plasticizing and/or stabilizing additives in vinyl plastics; lauric oils and tallow are important feedstocks for the production of soaps and detergents; in addition to jojoba and other specialty oils, lauric oil derivatives are well regarded in the manufacture of a wide range of cosmetic products and toiletries; and tall oil fatty acids and castor oil are traditional raw materials for many industrial applications. Fatty acids, the basic building blocks in oleochemistry, are used as such or they are further transformed to important oleochemicals, including fatty alcohols, esters, amines, quaternary ammonium compounds, amides, alkanolamides and oligomer acids. The largest areas of applications are represented by established markets such as soaps and detergents, plastics, personal care products, coatings/resins, rubber, lubricants and specialty chemicals, e.g., for the leather and textile industry.

Fatty acid methyl esters predominantly serve as important intermediates, especially in the production of fatty alcohols. The latter are used as such or they are further processed to a large variety of surface-active agents, including fatty alcohol ethoxylates, fatty alcohol ether sulfates, alkyl sulfates and the recently introduced alkyl polyglycosides.

The fatty amines, by far the most important nitrogen-containing oleochemicals derived from fatty acids, possess a wide range of diverse applications, e.g., disinfectants, fabric softeners, laundry detergents, textile auxiliaries, emulsifiers, lubricants additives, corrosion inhibitors, antistatic agents in plastics, flotation agents and oil field chemicals.

The oil and fats available to the oleochemical industry are derived in the large part from those usually used for food purposes, i.e., the C_{16} and C_{18} fatty acids are by far the major components. On the other hand, extensive chemical surveys have revealed a variety of minor plant species, e.g., *Cuphea*, *Limnanthes*, *Lesquerella*, *Vernonia*, bearing seed oils with fatty acids that display unusual specific properties and unsuspected possibilities with regard to industrial exploitation. Since the slow pace of commercialization of these new crops and their products is still impeding the cultivation on larger scale, considerable effort in research and development, financial support and close contact between growers, plant breeders, chemists and industry is indispensable.

Acknowledgements: We wish to thank Mrs Ingeborg Scholz and Dr. F. Ordon for help in preparing the manuscript. Own scientific work reported herein was supported in part by research grants from *Bundesministerium für Forschung und Technologie*, Bonn, *Bundesministerium für Ernährung, Landwirtschaft und Forsten*, Bonn, *Gemeinschaft zur Förderung der privaten deutschen Pflanzenzüchtung*, Bonn, and *Deutsche Forschungsgemeinschaft*, Bonn.

References

1 Formo, M.W., in: *Bailey's Industrial Oil and Fat Products*, Swern, D. (ed.), John Wiley & Sons, New York, **1982**, Vol. 2, pp. 343–405.

2 Baumann, H., Bühler, M., Fochem, H., Hirsinger, F., Zoebelein, H., Falbe, J., *Angew. Chem. Int. Ed. Engl.*, **1988**, *27*, 41–62.

3 Pryde, E.H., Rothfus, J.A., in: *Oil Crops of the World*, Röbbelen, G., Downey, R., Ashri, A. (eds.), McGraw-Hill Publ. Comp., New York, **1989**, pp. 87–117.

4 Kaufman, J., Ruebusch, R.J., *INFORM*, **1990**, *1*, 1034–1048.

5 Haumann, B.F., *INFORM*, **1991a**, *2*, 438–447.

6 Modler, R.F., Janshekar, H., Yoshida, Y., *CEH Marketing Research Report - Natural Fatty Acids*, Chemical Economics Handbook, SRI International, Palo Alto, California, USA, June **1992**.

7 Stein, W., in: *Improvement of Oil-Seeds and Industrial Crops by Induced Mutations*, Proceed. Advisory Group Meeting, Intern. Atomic Energy Agency (IAEA), Vienna, Austria, **1982b**, pp. 233–242.

8 Zoebelein, H., *INFORM*, **1992**, *3*, 721–725.

9 Schmid, R.D., in: *Proceedings of the World Conference on Biotechnology for the Fats and Oils Industry*, Applewhite, T.H. (ed.), Am. Oil Chem. Soc., Champaign, Illinois, **1988** pp. 169–172.

10 Princen, L.H., *Econ. Bot.*, **1983**, *37*, 478–492.

11 Princen, L.H., in: *Fats for the Future*, Cambie, R.C. (ed.), Ellis Horwood Publ., Chichester, UK, **1989**, pp. 205–216.

12 Gillis, A., *J. Am. Oil Chem. Soc.*, **1988**, *65*, 6–20.

13 Hirsinger, F., in: *Oil Crops of the World*, Röbbelen, G., Downey, R., Ashri, A. (eds.), McGraw-Hill Publ. Comp., New York, **1989**, pp. 518–532.

14 Janick, J., Simon, J.E. (eds.), *Advances in New Crops - Proceedings of the First National Symposium New Crops: Research, Development, Economics*, Timber Press, Portland, Oregon, USA. **1990**.

15 Haumann, B.F., *INFORM*, **1991b**, *2*, 678–692.

16 Pryde, E.H., *Econ. Bot.*, **1983**, *37*, 459–477.

17 Sonntag, N.O.V., *J. Am. Oil Chem. Soc.*, **1985**, *62*, 929–933.

18 Mounts, T.L, Wolf, W.J., Martinez, W.H., in: *Soybeans: Improvement, Production, and Uses*, Wilcox, J.R. (ed.), Agronomy Monograph 16, Americ. Soc. Agron., Madison, Wisconsin, USA, **1987**, 2nd edition, pp. 819–866.

19 Bagby, M.O., Carlson, K.D., in: *Fats for the Future*, Cambie, R.C. (ed.), Ellis Horwood Publ., Chichester, UK, **1989**, pp. 301–317.

20 Latta, S., *INFORM*, **1990**, *1*, 434–443.

21 Wong, M.H., *J. Am. Oil Chem. Soc.*, **1983**, *60*, 268A-273A.

22 Berger, K.G., Ong, S.H., *Oléagineux*, **1985**, *40*, 613–621.

23 Salunkhe, D.K., Chavan, J.K., Adsule, R.N., Kadam, S.S., *World Oilseeds: Chemistry, Technology, and Utilization*, Van Nostrand Reinhold, New York, **1992**.

24 De Vries, R.J., *J. Am. Oil Chem. Soc.*, **1984**, *61*, 404–407.

25 Ong, A.S.H., in: *Fats for the Future*, Cambie, R.C. (ed.), Ellis Horwood Publ., Chichester, UK, **1989**, pp. 285–300.

26 Satyabalan, K., in: *Oil Crops of the World*, Röbbelen, G., Downey, R., Ashri, A. (eds.), McGraw-Hill Publ. Comp., New York, **1989**, pp. 494–504.

27 Haumann, B.F., *INFORM*, **1992**, *3*, 1080–1093.

28 Arkcoll, D., *Econ. Bot.*, **1988**, *42*, 195–205.

29 Johnson, D.H., *J. Am. Oil Chem. Soc.*, **1978**, *55*, 438–443.

30 Takei, K.,Tsuto, K., Miyamoto, S.,Wakatsuki, J.A., *J. Am. Oil Chem. Soc.*, **1985**, *62*, 341–347.

31 Kalustian, P., *J. Am. Oil Chem. Soc.*, **1985**, *62*, 431–433.

32 Logan, R.L., *J. Am. Oil Chem. Soc.*, **1979**, *56*, 777–779.

33 Formo, M.W., in: *Bailey's Industrial Oil and Fat Products*, Swern, D. (ed.), John Wiley & Sons, New York, **1979** Vol. 1, 4th edition, pp. 687–816.

34 Ohlson, R., in: *Rapeseed - Cultivation, Composition, Processing and Utilization*, Appelqvist, L.-Å., Ohlson, R. (eds.), Elsevier Publ. Comp., Amsterdam, **1972**, pp. 274–287.

35 Niewiadomski, H., *Rapeseed - Chemistry and Technology*, Elsevier, Amsterdam, **1990**, pp. 332–363.

36 Peterson, C.L., Auld, D.L., Korus, R.A., *J. Am. Oil Chem. Soc.*, **1983**, *60*, 1579–1593.

37 Strayer, R.C., Blake, J.A., Craig, W.K., *J. Am. Oil Chem. Soc.*, **1983**, *60*, 1587–1592.

38 Tavenius, M.T., *G.C.I.R.C. Bulletin*, **1986**, *3*, 49–50.

39 Stöver, H.-M., Münch, E.-W., Sitzmann, W., *Fat Sci. Technol.*, **1988**, *90*, 547–550.

40 Anon., *INFORM*, **1992a**, *3*, 169.

41 Koonen, B., *Fat Sci. Technol.*, **1992**, *94*, 359–365.

42 Nieschlag, H.J.,Wolff, I.A., *J. Am. Oil Chem. Soc.*, **1971**, *48*, 723–727.

43 United States Department of Agriculture, *High Erucic Acid Oil: From Farm to Factory*, Coop. State Res. Serv., USDA,Washington, DC, USA, **1989**.

44 Erickson, D.B., Bassin, P., *Rapeseed and Crambe: Alternative Crops with Potential Industrial Uses*, Agric. Exp. Station Kansas State Univ., Manhattan, USA, Bulletin No. 656, **1990**.

45 Sonntag, N.V.O., *INFORM*, **1991**, *2*, 449–463.

46 Leonard, E.C., in: *Proceedings World Conference and Exhibition on Oilseed Technology & Utilization*, Applewhite,T. (ed.), Am. Oil Chem. Soc., Champaign, Illinois, **1993**, (in press).

47 Sonntag, N.O.V., in: *Bailey's Industrial Oil and Fat Products*, Swern, D. (ed.), John Wiley & Sons, New York, Vol. 1, **1979**, pp. 289–477.

48 McHughen, A., *Plant Breed. Abstracts*, **1992**, *62*, 1031–1036.

49 Eggers, U., *Fat Sci. Technol.*, **1992**, *94*, 279–280.

50 Smith, C.R., Jr., *Progress in the Chemistry of Fats and other Lipids*, **1970**, *11*(1), 137–177.

51 Dorrell, D.G., in: *Sunflower Science and Technology*, Carter, J.F. (ed.), Agronomy Monograph 19, Americ. Soc. Agron., Madison, Wisconsin, USA, **1978**, pp. 407–440.

52 Eierdanz, H., Hirsinger, F., *Fat Sci. Technol.*, **1990**, *92*, 463–467.

53 Zimmerman, L.H., *Adv. Agron.*, **1958**, *10*, 257–288.

54 Weiss, E.A., *Oilseed Crops*, **1983**, Longman, London.

55 Atsmon, D., in: *Oil Crops of the World*, Röbbelen, G., Downey, R., Ashri, A. (eds.), **1989**, McGraw-Hill Publ. Comp., New York, pp. 438–447.

56 Vignolo, R., Naughton, F., *INFORM*, **1991**, *2*, 692–699.

57 Naughton, F.C., *J. Am. Oil Chem. Soc.*, **1974**, *51*, 65–71.

58 Bafor, M., Smith, M.A., Jonsson, L., Stobart, K., Stymne, S., *Biochem. J.*, **1991**, *280*, 507–514.

59 Hein, H., *Fette, Seifen, Anstrichm.*, **1985**, *87*, 283–289.

60 Benzioni, A., Forti, M., in: *Oil Crops of the World*, Röbbelen, G., Downey, R., Ashri, A. (eds.), McGraw-Hill Publ. Comp., New York, **1989**, pp. 448–461.

61 Naqvi, H.H., Ting, I.P., in: *Advances in New Crops - Proceedings of the First National Symposium New Crops: Research, Development, Economics*, Janick, J., Simon, J.E. (eds.), Timber Press, Portland, Oregon, USA, **1990**, pp. 247–251.

62 Thompson, A.E., in: *Advances in New Crops - Proceedings of the First National Symposium New Crops: Research, Development, Economics*, Janick, J., Simon, J.E. (eds.), Timber Press, Portland, Oregon, USA, **1990**, pp. 232–241.

63 Miwa, T. K., *J. Am. Oil Chem. Soc.*, **1984**, *61*, 407–410.

64 Kleiman, R., in: *Advances in New Crops - Proceedings of the First National Symposium New Crops: Research, Development, Economics*, Janick, J., Simon, J.E. (eds.), Timber Press, Portland, Oregon, USA, **1990**, pp. 196–203.

65 Hamilton, R.T., Long, M., Raie, M.Y., *J. Am. Oil Chem. Soc.*, **1972**, *49*, 307–310.

66 Milthorpe, P.L., Dunstone, R.L., *Austr. J. Experim. Agric.*, **1989**, *29*, 383–395.

67 Regel, C., *Angew. Botanik*, **1940**, *22*, 400–413.

68 Regel, C., von, *Pflanzen in Europa liefern Rohstoffe*, E. Schweizerbart'sche Verlagsbuchhandl., Stuttgart, Germany, **1945**.

69 Hackbarth, J., *Die Ölpflanzen Mitteleuropas*, Monographien aus dem Gebiete der Fettchemie, Bd. 15, Wissenschaftl. Verlags-Gesell., Stuttgart, Germany, **1944**,

70 Boguslawski, E., von, *Fette, Seifen, Anstrichm.*, **1952**, *54*, 737–743.

71 Wolff, I.A., Jones, Q., *Chemurgic Digest*, **1958**, *17*(9), 4–8.

72 Earle, F.R., Melvin, E.H., Mason, L.H., van Etten, C.H., Wolff, I.A., *J. Am. Oil Chem. Soc.*, **1959**, *36*, 304–307.

73 Mikolajczak, K.L., Miwa, T.K., Earle, F.R., Wolff, I.A., *J. Am. Oil Chem. Soc.*, **1961**, *38*, 678–681.

74 Mikolajczak, K.L., Earle, F.R., Wolff, I.A., *J. Am. Oil Chem. Soc.*, **1962**, *39*, 78–80.

75 Miller, R.W., Daxenbichler, M.E., Earle, F.R., *J. Am. Oil Chem. Soc.*, **1964**, *41*, 167–169.

76 Miller, R.W., Earle, F.R., Wolff, I.A., *J. Am. Oil Chem. Soc.*, **1965**, *42*, 817–821.

77 Hilditch,T.P.,Williams, P.N., *The Chemical Constitution of Natural Fats*, Chapman and Hall, London, 4th edition, **1964**.

78 Hitchcock, C., Nichols, B.W., *Plant Lipid Biochemistry*, Academic Press, New York, **1971**.

79 Radatz, W., Hondelmann, W., *Landbauforsch. Völkenrode*, **1981**, *31*, 227–240.

80 Princen, L.H., *J. Am. Oil Chem. Soc.*, **1979**, *56*, 845–849.

81 Hondelmann, W., Radatz, W., *Fette, Seifen, Anstrichm.*, **1982**, *84*, 73–75.

82 Hondelmann, W., Radatz, W., *Landbauforsch. Völkenrode*, **1984**, *34*, 145–154.

83 Princen, L.H., Rothfus, J.A., *J. Am. Oil Chem. Soc.*, **1984**, *61*, 281–289.

84 Meier zu Beerentrup, H., Röbbelen, 1987, *Angew. Botanik*, **1987**, *61*, 287–303.

85 Kleiman, R., in: *Proceedings of the World Conference on Biotechnology for the Fats and Oils Industry*, Applewhite, T.H. (ed.), Am. Oil Chem. Soc., Champaign, Illinois, **1988**, pp. 73–77.

86 Röbbelen, G., in: *Proceedings of the World Conference on Biotechnology for the Fats and Oils Industry*, Applewhite, T.H. (ed.), Am. Oil Chem. Soc., Champaign, Illinois, **1988**, pp. 78–86.

87 Muuse, B.G., Cuperus, F.P., Derksen, J.T.P., *Industrial Crops and Products*, **1992**, *1*, 57–65.

88 Norris, F.A., in: *Bailey's Industrial Oil and Fat Products*, Swern, D. (ed.), John Wiley & Sons, New York, **1982**, Vol. 2, 4th edition, pp. 253–314.

89 Dieckelmann, G., Heinz, H.J., *The Basics of Industrial Oleochemistry*, Peter Pomp GmbH, Essen, Germany, **1988**.

90 Allen, R.R., in: *Bailey's Industrial Oil and Fat Products*, Swern, D. (ed.), John Wiley & Sons, New York, **1982**, Vol. 2, 4th edition, pp. 1–95.

91 Korus, R.A., Mousetis, T.L., *J. Am. Oil Chem. Soc.*, **1984**, *61*, 537–540.

92 Carlson, K.D., Chang, S.P., *J. Am. Oil Chem. Soc.*, **1985**, *62*, 934–939.

93 Sonntag, N.O.V., in: *Bailey's Industrial Oil and Fat Products*, Swern, D. (ed.), John Wiley & Sons, New York, **1982**, Vol. 2, 4th edition, pp. 97–173.

94 Rozenaal, A., *INFORM*, **1992**, *3*, 1232–1237.

95 Baillargeon, M.W., *Lipids*, **1990**, *25*, 841–848.

96 Werdelmann, B.W., Schmid, R.D., *Fette, Seifen, Anstrichm.*, **1982**, *84*, 436–443.

97 Macrae, A.R., *J. Am. Oil Chem. Soc.*, **1983**, *60*, 291–294.

98 Linfield,W.M., Barauskas, R.A., Sivieri, L., Serota, S., Stevenson, Sr. R.T., *J. Am. Oil Chem. Soc.*, **1984**, *61*, 191–195.

99 Ratledge, C., *Fette, Seifen, Anstrichm.*, **1984a**, *86*, 379–389.

100 Falbe, J., Schmid, R.D., *Fette, Seifen, Anstrichm.*, **1986**, *88*, 203–212.

101 Schmid, R.D., *J. Am. Oil Chem. Soc.*, **1987**, *64*, 563–570.

102 Mukherjee, K.D., *Biocatalysis*, **1990**, *3*, 277–293.

103 Casey, J., Macrae, A.R., *INFORM*, **1992**, *3*, 203–207.

104 Bühler, M.,Wandrey, C., *Fat Sci. Technol.*, **1987**, *89*, 156–164.

105 Bühler, M.,Wandrey, C., *Fat Sci. Technol.*, **1992**, *94*, 82–94.

106 Sridhar, R., Lakshminarayana, G., Kaimal, T.N.B., *J. Agric. Food Chem.*, **1991**, *39*, 2069–2071.

107 Mukherjee, K.D., *Fat Sci. Technol.*, **1992**, *94*, 542–546.

108 Carlson, K.D., Bagby, M.O., *J. Am. Oil Chem. Soc.*, **1989**, *66*, 486.

109 Trani, M., Lortie, R., Ergan, F., *INFORM*, **1992**, *3*, 482.

110 Ratledge, C., *J. Am. Oil Chem. Soc.*, **1984b**, *61*, 447–453.

111 Mukherjee, K.D., Kiewitt, I., *J. Agric. Food Chem.*, **1988**, *36*, 1333–1336.

112 Trani, M., Ergan, F., André, G., *J. Am. Oil Chem. Soc.*, **1991**, *68*, 20–22.

113 Jungermann, E., in: *Bailey's Industrial Oil and Fat Products*, Swern, D. (ed.), John Wiley & Sons, New York, **1979a**, Vol. 1, pp. 511–585.

114 Kabara, J.J., *J. Am. Oil Chem. Soc.*, **1984**, *61*, 397–403.

115 Jungermann, E., in: *Bailey's Industrial Oil and Fat Products*, Swern, D. (ed.), John Wiley & Sons, New York, **1979b**, Vol. 1, pp. 587–686.

116 Reck, R.A., *J. Am. Oil Chem. Soc.*, **1985**, *62*, 355–365.

117 Bhattacharyya, D.K., Chatterjee, B., *J. Am. Oil Chem. Soc.*, **1984**, *61*, 417–419.

118 DeNavarre, M.G., *J. Am. Oil Chem. Soc.*, **1978**, *55*, 435–437.

119 Kroke, H.P., *J. Am. Oil Chem. Soc.*, **1978**, *55*, 444–446.

120 Schneider, M., *Fat Sci. Technol.*, **1992**, *94*, 524–533.

121 Haumann, B.F., *J. Am. Oil Chem. Soc.*, **1983**, *60*, 44–58.

122 Macfarlane, N., *Trop. Sci.*, **1975**, *17*(4), 217–228.

123 Reck, R.A., *J. Am. Oil Chem. Soc.*, **1984**, *61*, 187–190.

124 Smith, H.V., *J. Am. Oil Chem. Soc.*, **1985**, *62*, 351–355.

125 Hintze-Brüning, H., *Industrial Crops and Products*, **1993**, (in press).

126 Erhan, S.Z., Bagby, M.O., *J. Am. Oil Chem. Soc.*, **1991**, *68*, 635–639.

127 Erhan, S.Z., Bagby, M.O., Cunningham, H.W., *J. Am. Oil Chem. Soc.*, **1992**, *69*, 251–256.

128 Fabig, W., Hund, K., Groß, K.J., *Fat Sci. Technol.*, **1989**, *91*, 357–360.

129 Busch, C., *Raps*, **1992**, *10*, 165–169.

130 Mang, T., *Raps*, **1991**, *9*, 4–7.

131 Schmid, K.-H., *Fat Sci. Technol.*, **1987**, *89*, 237–248.

132 Kammann jr., K.P., Phillips, A.I., *J. Am. Oil Chem. Soc.*, **1985**, *62*, 917–923.

133 Weiss, M., Rosberg, R., Sonntag, N.O.V., *J. Am. Oil Chem. Soc.*, **1979**, *56*, 849–853.

134 Kapusta, G., *J. Am. Oil Chem. Soc.*, **1985**, *62*, 923–927.

135 Quinn, P.J., in: *Biological Role of Plant Lipids*, Biacs, P.A., Gruiz, K., Kremmer, T. (eds.), Plenum Press, New York, **1989**, pp. 443–453.

136 Hsieh, F.-H., Daun, J.K., Tipples, K.H., *J. Am. Oil Chem. Soc.*, **1982**, *59*, 11–15.

137 Stein, W., *Fette, Seifen, Anstrichm.*, **1982a**, *84*, 45–54.

138 Haupt, D.E., Drinkard, G., Pierce, H.F., *J. Am. Oil Chem. Soc.*, **1984**, *61*, 276–281.

139 Eierdanz, H., in: *Proceedings World Conference and Exhibition on Oilseed Technology & Utilization*, Applewhite, T. (ed.), Am. Oil Chem. Soc., Champaign, Illinois, **1993**, (in press).

140 Warwel, S., *Fat Sci. Technol.*, **1992**, *94*, 512–523.

141 Duncan, D.P., *J. Am. Oil Chem. Soc.*, **1984**, *61*, 233–241.

142 Proffitt, T.J., Jr., Patterson, H.T., *J. Am. Oil Chem. Soc.*, **1988**, *65*, 1682–1694.

143 Van Haften, J.L., *J. Am. Oil Chem. Soc.*, **1979**, *56*, 831–835.

144 Richtler, H.J., Knaut, J., *J. Am. Oil Chem. Soc.*, **1984**, *61*, 160–175.
145 Steinberner, U., Preuss, W., *Fat Sci. Technol.*, **1987**, *89*, 297–303.
146 Kabara, J.J., *J. Am. Oil Chem. Soc.*, **1979**, *56*, 760–767.
147 Meffert, A., *J. Am. Oil Chem. Soc.*, **1984**, *61*, 255–267.
148 Stage, H., *J. Am. Oil Chem. Soc.*, **1984**, *61*, 204–214.
149 Haraldsson, G., *J. Am. Oil Chem. Soc.*, **1984**, *61*, 219–222.
150 Combs, D.L., *J. Am. Oil Chem. Soc.*, **1985**, *62*, 327–330.
151 Cahn, A., *J. Am. Oil Chem. Soc.*, **1979**, *56*, 809–811.

152 Sánchez, N., Martinez, M., Aracil, J., Corma, A., *J. Am. Oil Chem. Soc.*, **1992**, *69*, 1150–1153.

153 Maag, H., *J. Am. Oil Chem. Soc.*, **1984**, *61*, 259–267.
154 Matthews, D.M., *J. Am. Oil Chem. Soc.*, **1979**, *56*, 841–844.
155 Struyck, W.J.A., *J. Am. Oil Chem. Soc.*, **1984**, *61*, 395–397.
156 Baumann, H., *Fat Sci. Technol.*, **1990**, *92*, 49–56.
157 Fabry, B., *Fat Sci. Technol.*, **1990**, *92*, 287–291.
158 Satsuki, T., *INFORM*, **1992**, *3*, 1099–1108.
159 Kreutzer, U.R., *J. Am. Oil Chem. Soc.*, **1984**, *61*, 343.
160 Houston, C.A., *J. Am. Oil Chem. Soc.*, **1984**, *61*, 179–184.
161 Knaut, J., Richtler, H.J., *J. Am. Oil Chem. Soc.*, **1985**, *62*, 317–327.
162 Billenstein, S., Blaschke, G., *J. Am. Oil Chem. Soc.*, **1984**, *61*, 353–357.
163 Puchta, R., *J. Am. Oil Chem. Soc.*, **1984**, *61*, 367–377.
164 Schaeufele, P.J., *J. Am. Oil Chem. Soc.*, **1984**, *61*, 387–389.
165 Mardis, W.S., *J. Am. Oil Chem. Soc.*, **1984**, *61*, 383–387.
166 LaSusa, C.D., *J. Am. Oil Chem. Soc.*, **1984**, *61*, 185–187.
167 House, J.E., *J. Am. Oil Chem. Soc.*, **1984**, *61*, 357–362.
168 Molnar, N.M., *J. Am. Oil Chem. Soc.*, **1974**, *51*, 84–87.
169 Flider, F., *Oils & Fats Intern.*, **1989**, 5(2), 29–31.
170 Johnson, R.W., *J. Am. Oil Chem. Soc.*, **1984**, *61*, 241–246.
171 Pryde, E.H., *J. Am. Oil Chem. Soc.*, **1984**, *61*, 419–425.
172 Leonard, E.C., *J. Am. Oil Chem. Soc.*, **1979**, *56*, 782A–785A.

173 Nakano, Y., Foglia, T.A., Kohashi, H., Perlstein, T., Serota, S., *J. Am. Oil Chem. Soc.*, **1985**, *62*, 888–891.

174 Link, W., Spiteller, G., *Fat Sci. Technol.*, **1990a**, *92*, 19–25.
175 Link, W., Spiteller, G., *Fat Sci. Technol.*, **1990b**, *92*, 135–138.
176 Kinsman, D.V., *J. Am. Oil Chem. Soc.*, **1979**, *56*, 823–827.
177 Haase, K.D., Taylor, G., Smith, P.A., *Seifen-Öle-Fette-Wachse*, **1988**, *114*, 231.

178 Ward, B.F., Jr., Force, C.G., Bills, A.M., *J. Am. Oil Chem. Soc.*, **1975**, *52*, 219–224.

179 Kadesch, R.G., *J. Am. Oil Chem. Soc.*, **1979**, *56*, 845A–849A.
180 Klein, H.-P., *J. Am. Oil Chem. Soc.*, **1984**, *61*, 306–312.
181 Hinze, A.G., *Fat Sci. Technol.*, **1987**, *89*, 339–342.
182 Frankel, E.N., Pryde, E.H., *J. Am. Oil Chem. Soc.*, **1977**, *54*, 873–881.

183 Kohlhase, W.L., Frankel, E.N., Pryde, E.H., *J. Am. Oil Chem. Soc.*, **1977**, *54*, 506–510.

184 Behr, A., Handwerk, H.-P., *Fat. Sci. Technol.*, **1992**, *94*, 443–447.

185 Wagner, P.H., *Chem. & Ind.*, **1992**, 330–333.

186 Warwel, S., Jägers, H.-G., Thomas, S., *Fat Sci. Technol.*, **1992**, *94*, 323–328.

187 Boelhouwer, G., Mol, J.C., *J. Am. Oil Chem. Soc.*, **1984**, *61*, 425–430.

188 Warwel, S., Döring, N., Deckers, A., *Fat Sci. Technol.*, **1988**, *90*, 125–129.

189 Graham, S.A., *CRC Crit. Rev. Food Sci. Nutri.*, **1989**, *28*, 139–173.

190 Klein, B., Pawlowski, K., Höricke-Grandpierre, C., Schell, J., Töpfer, R., *Mol. Gen. Genet.*, **1992**, *233*, 122–128.

191 Spener, F., Schuch, R., in: *Bericht Vortragstagung "Nachwachsende Rohstoffe: Industrieraps - Biotechnologie"*, Forschungszentrum Jülich GmbH/BMFT, Ref. 323 (ed.), Jülich, Germany, **1993**, pp. 134–153.

192 Töpfer, R., in: *Bericht Vortragstagung "Nachwachsende Rohstoffe: Industrieraps - Biotechnologie"*, Forschungszentrum Jülich GmbH/BMFT, Ref. 323 (ed.), Jülich, Germany, **1993**, pp. 154–165.

193 Dörmann, P., Spener, F., Ohlrogge, J.B., *Planta*, **1993**, *189*, 425–432.

194 Graham, S.A., Hirsinger, F., Röbbelen, G., *Amer. J. Bot.*, **1981**, *68*, 908–917.

195 Wolf, R.B., Graham, S.A., Kleiman, R., *J. Am. Oil Chem. Soc.*, **1983**, *60*, 27–28.

196 Thompson, A.E., Kleiman, R., *J. Am. Oil Chem. Soc.*, **1988**, *65*, 139–146.

197 Thompson, A.E., Dierig, D.A., Knapp, S.J., Kleiman, R., *J. Am. Oil Chem. Soc.*, **1990**, *67*, 611–617.

198 Graham, S.A., Kleiman, R., *Industrial Crops and Products*, **1992**, *1*, 31–34.

199 Hirsinger, F., *Angew. Botanik*, **1980a**, *54*, 157–177.

200 Hirsinger, F., *Z. Pflanzenzüchtg.*, **1980b**, *85*, 157–169.

201 Hirsinger, F., *J. Am. Oil Chem. Soc.*, **1985**, *62*, 76–80.

202 Hirsinger, F., Röbbelen, G., *Z. Pflanzenzüchtg.*, **1980**, *85*, 275–286.

203 Hirsinger, F., Knowles, P.F., *Econ. Bot.*, **1984**, *38*, 439–451.

204 Thompson, A.E., *HortSci.*, **1984**, *19*, 352–354.

205 Röbbelen, G., Witzke, S., von, in: *Plant Domestication by Induced Mutation*, Proceed. FAO/IAEA Advisory Group Meeting, Series STI/PUB/793, Intern. Atomic Energy Agency (IAEA), Vienna, Austria, **1989**, pp. 101–119.

206 Knapp, S.J., in: *Advances in New Crops - Proceedings of the First National Symposium New Crops: Research, Development, Economics*, Janick, J., Simon, J.E. (eds.), Timber Press, Portland, Oregon, USA, **1990**, pp. 203–210.

207 Knapp, S.J., in: *Advances in New Crops II*, Janick, J., Simon, J.E. (eds.), Timber Press, Portland, Oregon, USA, **1992**.

208 Roath, W.W., Widrlechner, M.P., Kleiman, R., *Industrial Crops and Products*, **1992**, *1*, 5–10.

209 Voelker, T.A., Worrell, A.C., Anderson, L., Bleibaum, J., Fan, C., Hawkins, D.J., Radke, S.E., Davies, H.M., *Science*, **1992**, *257*, 72–74.

210 Anon., *INFORM*, **1992b**, *3*, 1027.

211 Babayan, V.K., *J. Am. Oil Chem. Soc.*, **1981**, *58*, 49A-51A.

212 Babayan, V.K., *Lipids*, **1987**, *22*, 417–420.

213 Egan, R.R., Earl, G.W., Ackerman, J., *J. Am. Oil Chem. Soc.*, **1984**, *61*, 324–329.

214 Hondelmann, W., Radatz, W., *Angew. Botanik*, **1983**, *57*, 349–362.

215 Hecker, E., Sosath, S., *Fat Sci. Technol.*, **1989**, *91*, 468–478.

216 Perkins, R.B., Roden, J.J., Pryde, E.H., *J. Am. Oil Chem. Soc.*, **1975**, *52*, 473–477.

217 Appelqvist, L.-A., Jönsson, R., *Z. Pflanzenzüchtg.*, **1970**, *64*, 340–356.

218 Appelqvist, L.-Å., in: *The Biology and Chemistry of the CRUCIFERAE*, Vaughan, J.G., Macleod, A.J., Jones, B.M.G (eds.), Academic Press, New York, **1976**, pp. 221–278.

219 Mahler, K.A., Auld, D.L., *Fatty acid composition of 2100 accessions of* Brassica. *Winter rapeseed breeding program*, Univ. of Idaho, Moscow, USA, **1988**.

220 Lühs, W., Ecke, W., Friedt, W., in: *Bericht Vortragstagung "Nachwachsende Rohstoffe: Industrieraps - Biotechnologie"*, Forschungszentrum Jülich GmbH/BMFT, Ref. 323 (ed.), Jülich, Germany, **1993**, pp. 81–105.

221 Harlow, R.D., Litchfield, C., Reiser, R., *Lipids*, **1966**, *1*, 216–220.

222 Löhden, I., Frentzen, M., *Planta*, **1992**, *188*, 215–224.

223 Kleiman, R., Earle, F.R., Wolff, I.A., *J. Am. Oil Chem. Soc.*, **1964**, *41*, 459–460.

224 Gross, R., Baer, E., von, Rohrmoser, K., *J. Agronomy & Crop Science*, **1983**, *152*, 19–31.

225 Ackman, R.G., in: *High and Low Erucic Acid Rapeseed Oils*, Kramer, J.K.G., Sauer, F.D., Pigden, W.J. (eds.), Academic Press, New York, **1983**, pp. 85–129.

226 Muuse, B.G., Essers, M.L., Soest, L.J.M., van, *Neth. J. Agric. Sci.*, **1988**, *36*, 357–363.

227 Tallent, W.H., *J. Am. Oil Chem. Soc.*, **1972**, *49*, 15–19.

228 Seehuber, R., *Fat Sci. Technol.*, **1987**, *89*, 263–268.

229 Earle, F.R., Peters, J.E., Wolff, I.A., *J. Am. Oil. Chem. Soc.*, **1966**, *43*, 330–333.

230 Leppik, E.E., White, G.A., *Euphytica*, **1975**, *24*, 681–689.

231 Lessman, K.J., Meier, V.D., *Crop. Sci.*, **1972**, *12*, 224–227.

232 Lessman, K.J., *J. Am. Oil Chem. Soc.*, **1975**, *52*, 386–389.

233 Lessman, K.J., in: *Advances in New Crops - Proceedings of the First National Symposium New Crops: Research, Development, Economics*, Janick, J., Simon, J.E. (eds.), Timber Press, Portland, Oregon, USA, **1990**, pp. 217–222.

234 Carlson, K.D., Tookey, H.L., *J. Am. Oil Chem. Soc.*, **1983**, *60*, 1979–1985.

235 Hamid, S., Salma, Sabir, A.W., Khan, S.A., *Pak. J. Sci. Ind. Res.*, **1987**, *30*, 33–35.

236 Olsson, G., *Sver. Utsädesfören. Tidskrift*, **1984**, *94*, 26–29.

237 Brune-Pleines, U., *Dissertation*, Universität Giessen, Germany.

238 Brune, U., Gröne, I., Marquard, R., Friedt, W., *Bericht 38. Arbeitstagung Arbeitsgemein. Saatzuchtleiter*, Gumpenstein, Austria, **1989**, pp. 275–286.

239 Jönsson, R., *G.C.I.R.C. Bulletin*, **1985**, *2*, 41–45.

240 Lühs, W., Friedt, W., *G.C.I.R.C. Bulletin*, **1993**, (in press).

241 Calhoun, W., Crane, J.M., Stamp, D.L., *J. Am. Oil Chem. Soc.*, **1975**, *52*, 363–365.

242 Calhoun, W., Jolliff, G.D., Crane, J.M., *Crop Sci.*, **1983**, *23*, 184–185.

243 Auld, D.L., Mahler, K.A., Bettis, B.L., Crock, J.C., *Crop Sci.*, **1987**, *27*, 1310.

244 Anon., *INFORM*, **1992c**, *3*, 1027–1028.

245 Wilson, T.L., Smith, C.R., Wolff, I.A., *J. Am. Oil Chem. Soc.*, **1962**, *39*, 104–105.

246 Mukherjee, K.D., Kiewitt, I., *Phytochem.*, **1986**, *25*, 401–404.

247 Fehling, E., Mukherjee, K.D., *Phytochem.*, **1990**, *29*, 1525–1527.

248 Röbbelen, G., Möllers, C., in: *Bericht Vortragstagung "Nachwachsende Rohstoffe: Industrieraps - Biotechnologie"*, Forschungszentrum Jülich GmbH/BMFT, Ref. 323 (ed.), Jülich, Germany, **1993**, pp. 53–80.

249 Perkins, R.B., Roden, J.J., Tanquary, A.C., Wolff, I.A., *Mod. Plast.*, **1969**, *46*, 136–142.

250 Carlson, K.D., Sohns, V.E., Perkins, R.B., Jr., Huffman, E.L., *Ind. Eng. Chem., Prod. Res. Dev.*, **1977**, *16*, 95–101.

251 Nieschlag, H.J., Rothfus, J.A., Sohns, V.E., Perkins, R.B., Jr., *Ind. Eng. Chem., Prod. Res. Dev.*, **1977a**, *16*, 101–107.

252 Jönsson, A., Bokström, J., Malmvik, A.C., Wärnheim, T., *J. Am. Oil Chem. Soc.*, **1990**, *67*, 733–738.

253 Pohl, P., Wagner, H., *Fette, Seifen, Anstrichm.*, **1972a**, *74*, 424–435.

254 Pohl, P., Wagner, H., *Fette, Seifen, Anstrichm.*, **1972b**, *74*, 541–550.

255 Padley, F.B., Gunstone, F.D., Harwood, J.L., in: *The Lipid Handbook*, Gunstone, F.D., Harwood, J.L., Padley, F.B. (eds.), Chapman and Hall, London, **1986**, pp. 49–112.

256 Kleiman, R., Spencer, G.F., *J. Am. Oil Chem. Soc.*, **1982**, *59*, 29–38.

257 Murphy, D.J., *Trends in Biotechnol.*, **1992**, *10*, 84–87.

258 Gentry, H.S., Miller, R.W., *Econ. Bot.*, **1965**, *19*, 25–32.

259 Higgins, J.J., Calhoun, W., Willingham, B.C., Dinkel, D.H., Raisler, W.L., White, G.A., *Econ. Bot.*, **1971**, *25*, 44–54.

260 Pierce, R.O., Jain, S.K., *Crop. Sci.*, **1977**, *17*, 521–526.

261 Brown, C.R., Hauptli, H., Jain, S.K., *Econ. Bot.*, **1979**, *33*, 267–274.

262 Jolliff, G.D., in: *New Sources of Fats and Oils*, Pryde, E.H., Princen, L.H., Mukherjee, K.D. (eds.), Monograph No. 9, Am. Oil Chem. Soc., Champaign, Illinois, USA, **1981**, pp. 269–285.

263 Purdy, R.H., Craig, C.D., *J. Am. Oil Chem. Soc.*, **1987**, *64*, 1493–1498.

264 Phillips, B.E., Smith, C.R., Jr., Tallent, W.H., *Lipids*, **1971**, *6*, 93–99.

265 Pollard, M.R., Stumpf, P.K., *Plant Physiol.*, **1980**, *66*, 649–655.

266 Cao, Y.Z., Oo, K.C., Huang, A.H.C., *Plant Physiol.*, **1990**, *94*, 1199–1206.

267 Lardans, A., Trémeliéres, A., *Phytochem.*, **1991**, *30*, 3955–3961.

268 Lardans, A., Trémeliéres, A., *Phytochem.*, **1992**, *31*, 121–127.

269 Laurent, P., Huang, A.H.C., *Plant Physiol.*, **1992**, *99*, 1711–1715.

270 Taylor, D.C., Thomson, L.W., MacKenzie, S.L., Pomeroy, M.K., Weselake, R.J., in: *Proceed. 6th Crucifer Genetics Workshop*, McFerson, J. R., Kresovich, S., Dwyer, S.G. (eds.), USDA-ARS Plant Genetic Resources Unit, Cornell University, Geneva, NY, **1990**, pp. 38–39.

271 Wolter, F.P., Bernerth, R., Löhden, I., Schmidt,V., Peterek, G., Frentzen, M., *Fat Sci. Technol.*, **1991**, *93*, 288–290.

272 Frentzen, M., in: *Bericht Vortragstagung "Nachwachsende Rohstoffe: Industrieraps - Biotechnologie"*, Forschungszentrum Jülich GmbH/BMFT, Ref. 323 (ed.), Jülich, Germany, **1993**, pp. 106–120.

273 Thierfelder, A., Lühs, W., Friedt, W., *Industrial Crops and Products*, **1993**, *1*, 261–271.

274 Kleiman, R., Spencer, G.F., Earle, F.R., Nieschlag, H.J., *Lipids*, **1972**, *7*, 660–665.

275 Carlson, K.D., Chaudhry, A., Bagby, M.O., *J. Am. Oil Chem. Soc.*, **1990a**, *67*, 438–442.

276 Carlson, K.D., Chaudhry, A., Peterson, R.E., Bagby, M.O., *J. Am. Oil Chem. Soc.*, **1990b**, *67*, 495–498.

277 Campbell, T.A., in: *New Sources of Fats and Oils*, Pryde, E.H., Princen, L.H., Mukherjee, K.D. (eds.), Monograph No. 9, Am. Oil Chem. Soc., Champaign, Illinois, USA, **1981**, pp. 287–296.

278 Carlson, K.D., Schneider,W.J., Chang, S.P., Princen, L.H., in: *New Sources of Fats and Oils*, Pryde, E.H., Princen, L.H., Mukherjee, K.D. (eds.), Monograph No. 9, Am. Oil Chem. Soc., Champaign, Illinois, USA, **1981**, pp. 297–318.

279 Raie, M.Y., Zaka, S., Khan, S., Khan, S.A., *Fat. Sci. Technol.*, **1985**, *87*, 324–326.

280 Perdue, R.E., Jr., Carlson, K.D., Gilbert, M.G., *Econ. Bot.*, **1986**, *40*, 54–68.

281 Ayorinde, F.O., Butler, B.D., Clayton, M.T., *J. Am. Oil Chem. Soc.*, **1990a**, *67*, 844–845.

282 Ayorinde, F.O., Carlson, K.D., Pavlik, R.P., McVety, J., *J. Am. Oil Chem. Soc.*, **1990b**, *67*, 512–518.

283 Pascual, M.J., Correal, E., *Crop Sci.*, **1992**, *32*, 95–98.

284 Thiemens, M.H., Trogler, W.C., *Science*, **1991**, *251*, 932–934.

285 Thies, W., *Fat Sci. Technol.*, **1993**, *95*, 20–23.

286 Kaneniwa, M., Myashita, K., and Takagi, T., *J. Am. Oil Chem. Soc.*, **1988**, *65*, 1470–1474.

287 Emken, E.A., Adlof, R.O., Abraham, S., *Lipids*, **1991**, *26*, 736–741.

288 Miwa, T.K., Wolff, I.A., *J. Am. Oil Chem. Soc.*, **1962**, *39*, 320–322.

289 Nieschlag, H.J., Spencer, G.F., Madrigal, R.V., Rothfus, J.A., *Ind. Eng. Chem., Prod. Res. Dev.*, **1977b**, *16*, 202–207.

290 Fore, S.P., Sumrell, G., *J. Am. Oil Chem. Soc.*, **1966**, *43*, 581–584.

291 Burg, D.A., Kleiman, R., *J. Am. Oil Chem. Soc.*, **1991b**, *68*, 600–603.

292 Isbell, T.A., Kleiman, R., Erhan, S.M., *J. Am. Oil Chem. Soc.*, **1992**, *69*, 1177–1183.

293 Erhan, S.M., Kleiman, R., *J. Am. Oil Chem. Soc.*, **1990**, *67*, 670–674.

294 Burg, D.A., Kleiman, R., *J. Am. Oil Chem. Soc.*, **1991a**, *68*, 190–192.

295 Lawson, N.E., Farina, T.E., *J. Am. Oil Chem. Soc.*, **1988**, *65*, 1824–1827.

3 Breeding Oil Crops

A. E. Arthur

3.1 Introduction

The oil crops form an extremely diverse group of species from a breeding point of view. It includes, for example, inbreeding annuals like soybean, peanut, safflower and linseed, outbreeding species including sunflower, niger and maize, species which tolerate both inbreeding and outcrossing such as the allopolyploid Brassicas, sesame, poppy and cotton and highly complex long-lived perennials such as oil palm, coconut and olive.

The oil crops are also very different in their evolutionary and crop development. Some have been grown for oil for many centuries, like sesame which was a highly prized crop in Babylon and Assyria 4000 years ago or even earlier [1]. Others have been adopted as oil crops relatively recently, like safflower, which was not investigated seriously as an oil crop until the 1930s, and jojoba, which was practically unknown to agriculture until some 10–15 years ago. Others, like the monoecious maize and tetraploid cottons have been grown primarily for their non-oil products and, although well established crops for other uses, have been developed as oil crops only comparatively recently.

Although the oil crops as a group include a wide range of plant and reproductive types, with diverse histories and evolution during the development of the crops, most of the procedures and techniques used during crop improvement are reasonably standard and details of these can be found in almost any textbook of plant breeding [eg 2, 3, 4] or in books describing the oil crops species by species [eg 5]. It is the intention here to adopt a different approach and look at the oil crops grouped according to their reproductive systems and the methods available for crop improvement which are, to a great extent, determined and limited by these systems.

3.2 Breeding objectives

Although the diversity found amongst the oil crops requires, and offers, opportunities for very different approaches during breeding, the objectives set for crop improvement are often broadly similar. A list of some of these is given in Tables 3.1 and 3.2.

3.2.1 Oil and agronomy

Of primary interest in most oil crops, over and above most other factors, is the yield and quality of oil produced. The relative importance of these two factors depends on the type of oil and its end-use. For major crops such as oilseed rape or sunflower, the main consideration is yield of oil per unit area, with certain minimum standards set for oil (and meal) quality depending on its destination and use. With sesame, however, because a very high quality oil is required and demanded, oil content and profile are equally important considerations in improving this crop. To a large extent, oil yield, through seed yield, is determined by the agronomic characteristics, suitability and adaptability of the crop, especially with relatively new or undeveloped crops. Indeed, one of the main limitations placed on the use of certain species containing very desirable oils is their agronomic unsuitability. One example of such a species with large potential because of its unique, high quality oil is *Cuphea*. This is proving difficult to establish by conventional methods, which result in low yields and great loss of potential. Similar problems arise in sesame, where seed loss through pod shattering before harvesting, and poor and unreliable crop yields, greatly limit the crop's performance and acceptability. In *Limnanthes*, poor seed set in the field under certain conditions is a major limitation for the acceptance of this species as a commercial oilseed crop. The plants are allogamous, requiring insects, especially honeybees (*Apis mellifera*) [6], to pollinate the flowers to obtain seed set and, where conditions are unfavourable to such insect activity, crops will fail to reach anything like their full potential yield of seed and hence oil. Much effort is thus being put into the selection and development of autogamy in this species as a means of improvement.

Breeding objectives, as listed in Tables 3.1 and 3.2, other than oil quality and quantity which are common to most of the oil crops, include requirements for improved agronomic performance such as resistance to lodging (the bending or flattening of the crop) and improved seed retention, early flowering and maturity and generally improved adaptability to the environment and to local

Table 3.1. Objectives of general interest in each of the major oilseed crops.

GENERAL OBJECTIVES (CROP¹)	A	B	C	D	E	F	G	H	I	J	K	L	M	N	O	P	Q	R	S	T
Seed yield	A²											LL	M			P			S	
Resistance to diseases	A	BB	C		EE	FF	GG	x	II	JJ		L		N	O	P		R	S	
Resistance to pests	AA		C	D	E	FF		x	II	JJ		L		N						
Resistance to pod shattering	AA			D	EE	F	G	H												
Time to maturity, earliness	A	B	C	D		F			I	J	K	L		N		P	Q	R	S	
Oil content	A	B		D	EE	F	G	H	I	J		x								
Resistance to lodging	AA			D	E	F	G	H	I											
Seed and oil quality	A	x	C	D	E	F	G						M		O					
Antinutritional factors	A		C	D					I											
Resistance to frost/cold		B	C	D								L	M	N	O		Q	R	S	
Plant/tree height		B					G	H				L			O	P	Q			
Resistance to drought		B	C																	
Improved agronomic characters									I								Q		S	
Seed/fruit weight/size	A	B											M	N						
Suitability for mechanisation			C										M							
Quality of meal				D																
Higher self-fertility		B																	S	T
Improved competition with weeds							G	H												
Determinate growth habit								H										R		
Reduction in hull thickness								H												T
Improved adaptation														N	O					
Simultaneous fruit ripening														N	O	P				
Oil yield per area															O					

¹ A=Soybean B=Sunflower C=Peanut D=Brassicas E=Safflower F=Sesame G=Poppy H=Niger I=Cotton J=Linseed K=Maize L=Castor M=Jojoba N=Olive O=Oil Palm P=Coconut Q=Cuphea R=Euphorbia S=Limnanthes T=Crambe

² Double letters indicate that the objective is of particular interest
x indicates that the objective has been achieved or is already adequate in the crop

Table 3.2. Objectives of special interest in particular oilseed crops.

SPECIFIC OBJECTIVES	CROP¹	A	B	C	D	E	F	G	H	I	J	K	L	M	N	O	P	Q	R	S	T
Mineral deficiencies /toxicities		A																			
Head diameter			B																		
Stem diameter			B																		
Leaf Area per plant			B																		
Stability over environment				C																	
Fresh seed dormancy				C																	
Suitability for processing				C																	
Reduction of spines						E															
Daylength neutral							F														
Seed color								G													
Improved Harvest Index									H												
Improved production efficiency										I											
Increased rate of flowering										I											
Improved fibre content											J										
More vigorous plants														L							
Higher proportion ♀ to ♂ flowers														L							
Vegetative propagation															M						
Good response to irrigation/fertilisers															M						
Uniformity of production through year																	O				
Canopy establishment																				R		
Elimination of sap																				R		

¹ A=Soybean B=Sunflower C=Peanut D=Brassicas E=Safflower F=Sesame G=Poppy H=Niger I=Cotton J=Linseed K=Maize L=Castor M=Jojoba N=Olive O=Oil Palm P=Coconut Q=Cuphea R=Euphorbia S=Limnanthes T=Crambe

conditions. This latter requirement is especially important in crops such as peanut in which resistance to drought is becoming increasingly important as water supplies become limited, and in coconut, where local conditions can demand the ability to withstand cyclones and hurricanes as well as drought.

3.2.2 Disease and pest resistance

Of overwhelming importance in nearly all crops is resistance to diseases and pests; the only exception is likely to be niger, in which it is stated that these factors are not usually serious enough to require attention at the moment [7]. In all the other oil crops, breeding for resistance is a prime objective. The range of diseases and pests against which resistance is sought is very wide and is determined largely by the nature of each specific crop and the local environments in which it is grown. Details can be found elsewhere [eg 5]. From a breeding point of view, the effort required to produce resistant forms will be determined by the nature of the genetic control. For at least some of the diseases, major gene resistance has been found, often with a single gene being responsible. This is sometimes called 'vertical' [8, 9] or specific resistance. Where such resistance is available, breeding new resistant cultivars has been successful and relatively easy to achieve, unless variation for pathogenicity in the pathogen is able to overcome the host's resistance, leading to the 'boom and bust' cycles familiar in the large scale cereal growing areas of the USA. More persistent, though often very much more difficult to achieve and manipulate, is polygenic, 'horizontal' [8, 9] or general resistance. Here a number of genes acting together in a positive cumulative way are required to achieve sufficient levels of resistance but, once obtained, tend to be longer lasting in their effects and provide more durable forms of resistance.

Major gene or specific resistances have been found in a number of the oil crops. Efforts to find such resistance have been particularly effective in soybean and sunflower, where specific resistance based on one or two genes in each case has been found to Phytophthora rot and downy mildew (caused by *Peronospora manshuriea* (Naum) Syd. ex Gaun) in soybean and to rust, mildew and Verticillium wilt in sunflower. Single gene resistance to bacterial blight has been found in cotton [10]. Breeding for resistance to a number of important diseases in linseed is based on the 'gene for gene' concept put forward by Flor [11] and thoroughly reviewed by Day [12] in which resistance in the host occurs only when a pathogen contains the corresponding virulence gene and vice versa.

General or horizontal resistance, that based on polygenic systems, tends to be much more complex and therefore difficult to manipulate by breeding. Even if its performance gives something less than full resistance as is sometimes the case, it can offer useful forms of tolerance which are often sufficient to offset full scale effects of the disease or pest and tip the balance in favour of the host crop. General resistance can give a broader and more persistent form of resistance or tolerance in cases where single gene resistance is quickly overcome by changes in the pathogen, offering a valuable 'background' of resistance. This form of resistance or tolerance has been found in soybean, providing useful barriers to root knot nematode (*Meloidogyne* spp), brown stem rot caused by *Phialphora gregata* (Anington & Chambers) W. Gams and Phytophthora rot.

In cases where selection for resistance to a particular disease or pest is difficult, perhaps because assessment is complex or environmental effects mask the genetic expression of resistance, it may be possible to use indirect selection to achieve the objectives, by selecting for an easily recognised and assessed character or genetic or molecular marker which is highly correlated with resistance, indicating linkage between the genes or sets of genes, or pleiotropic effects. Either way, this can provide a convenient method for handling what can otherwise be difficult traits requiring lengthy and expensive breeding and assessment programs. An example of this approach can be found in olives, where breeding programs are inevitably long term and expensive. Some resistance to *Cycloconium oleagisum* has been found to be correlated with very thin septa in leaf tissue and it may be possible to use this to select for resistance amongst seedlings in early growth stages of the trees. Such approaches may also have great potential, particularly in long term crops, with the development of molecular markers such as isozyme patterns and restriction fragment length polymorphisms (RFLPs), as discussed in Chapter 6. Limited populations are screened for linkage between the desired trait and a suitable marker and individuals then selected from amongst segregating seedling populations for the particular marker(s). The resulting selections must still be field tested, but the method saves greatly on time and cost within the program. An example is root knot nematode in soybean which can cause significant economic losses in crops grown in Florida and against which genetic resistance is being sought amongst new cultivars. Klein-Lankhorst *et al* [13] have reported finding linkage between certain RFLP markers and the gene *Mi* in tomato which confers resistance to the nematode, offering the possibilities of such indirect selection in tomato and the hope of finding a similar linkage in soybean.

3.2.3 Genetic variability

Nearly all programs of crop improvement, whatever the objectives, benefit from the genetic variability available amongst related wild species. In some cases, large collections are maintained purposely to provide such a diverse gene pool for breeding programs, although, with increasing economic pressures and costs relating to the maintenance of such collections, these are not always as complete or comprehensive as they could be or, indeed, used to be. Efforts over the past 10 years to conserve and preserve germplasm for use in breeding programs have increased, with the realisation that, prior to this time, much valuable variability was being lost, popularly referred to as 'gene erosion'. Many of the genes used to improve traits associated with the objectives of crop improvement have been introduced from wild relatives of the crop. Some examples include resistance to rust, Fusarium and Verticillium wilts, Alternaria leaf blight and Phytophthora root rot, all introduced into safflower from the USDA world collection. Sesame species provide such a rich source of variation for breeders that FAO in 1981 and 1985 recommended that sesame germplasm collection and exchange be increased and improved [14, 15]. This is particularly true in many of the less developed oil crops which require agronomic improvement, such as resistance to pod shattering in *Cuphea* and *Limnanthes*, as discussed in Chapter 1. Similarly, many of the more established crops have benefited greatly from interspecific hybridisation and wide crossing to enhance the gene pool available or introduce desired traits not already present in the crop species.

Crosses amongst *Papaver* spp; *P.bacterium*, *P.orientale*, *P.setigerum* and the crop species P. *somniferum* have been used to extend further the genetic variation available to the breeder. Interspecific hybridisation between *Elaeis guineensis*, oil palm, and its wild American relative *E.oleifera* has interesting possibilities [16]. The F_1 hybrids are fertile and, although the yield of oil is usually depressed, secondary characters often show improvement, such as lower height increment, higher unsaturation of the oil and resistance to some diseases. The cross can also produce individual palms that may be of considerable interest for vegetative reproduction. In studies on inheritance in interspecific hybrids between *Carthamus flavescens* and *C.tinctorius*, Imrie and Knowles [17] found that differences between levels of shattering in safflower and closely related species appeared to be due to a change in one gene. They concluded that, because of its close relationship to the cultivated safflower, the wild *C.flavescens* must be regarded as a potential source of usable genetic variability. They also found that self-incompatibility in this species was controlled by alleles at one locus [18]. In some interspecific crosses, especially those amongst less closely related species, problems often arise during development of the hybrid embryo, resulting in the seed aborting before it matures, sometimes within days of fertilisation. Tissue culture techniques, such as embryo rescue and ovary culture, have helped to overcome these barriers to interspecific hybridisation and have greatly facilitated gene exchange amongst species and the transfer of desirable traits from wild relatives to the crop species in many cases.

Further efforts are required in some species, however. In soybean, for example, in which the wild perennial *Glycine* species encompass a broad range of variability, representing a potentially valuable resource for breeders, the variation has become available only through the techniques of embryo and ovule culture, and then only as tetraploids [19] since the diploid hybrids were sterile. Grant [20] reports limited success in hybridising *Glycine* relatives but suggests that accessibility of the useful agronomic traits from the wild soybean species will be limited until fertility problems are overcome.

3.2.4 Meal quality

An objective of major concern in some oil crops relates to the quality of the meal remaining after the extraction of oil. In many cases, the value of the crop is directly dependent on the suitability of the meal as a product in its own right, as an animal feedstuff or for human consumption. In the *Cruciferae*, for example, toxic compounds called glucosinolates are present in the meal. These compounds are appetite suppressors, causing weight loss in the livestock to which the meal is fed, and are involved in diseases like goitre. They are toxic in very high concentrations but are rarely fatal. Their presence in the meal greatly restricts its end use and thus decreases the overall economic value of the crop. Over the last twenty years, breeders have put considerable effort into successfully reducing the levels of these detrimental compounds in oilseed rape to enable the meal to be more widely acceptable as a protein rich source for animal feed [eg see 21, 22, 23]. Although the genetic control of these compounds is complex and levels vary within the plant and seed during development, selection for depressed levels in the seed has been adequate for commercial acceptance. Other oilseed members of the *Cruciferae*, notably *Crambe*, and species of *Limnanthes* and *Tropaeolum* (nasturtium) also possess glucosinolates which, if the meal remaining after oil extraction is to have any reasonable economic value, will require attention during breeding and selection programs to improve these crops.

Antinutritional factors also occur in some of the other oil crops. Until the discovery of glandless forms of cotton, the use of the meal was severely limited by the presence of the alkaloid gossypol; two major genes, in the recessive state gl_2 and gl_3, remove the pigment glands which produce the alkaloid from the seed and all aerial parts of the plant [24]. Soybean seeds contain enzymes which reduce the utilisation and digestion of the meal protein so that the meal has to be heat-treated to inactivate the enzymes. Research in progress to find genotypes lacking these enzymes has met with some success and a major gene has been identified which eliminates the Kunitz trypsin inhibitor [25], offering increased potential for the use of the protein- rich meal. Jojoba meal contains similar degrading toxic compounds called simmondsins which make its direct use undesirable; methods of detoxification are discussed by Verbiscar and Banigan [26].

3.2.5 Defining objectives

The list of objectives presented in Tables 3.1 and 3.2 is by no means exhaustive, but it is sufficiently comprehensive to illustrate the wide range of traits which the breeder must consider in the development and improvement of an oil crop. Not all these traits are required in each and every crop, but as can be seen from the table, many are requirements common to a large number of the oil crop species. It is interesting to consider who sets these objectives - the breeder must weigh up the relative importance of the objectives, how easily and quickly they might be achieved and at what cost. When the objectives are clearly defined and the traits required are available in the gene pool and can be readily assessed, the ease with which they might be transferred to the crop is likely to result in a relatively rapid and inexpensive improvement. This improvement will also depend on previous knowledge of the desired traits, particularly genetic behaviour and expression in the target background, whether simple, single genes or more complex quantitative genes are involved, and the effects of the environment and any genotype-environment interaction which may result in expression of the trait depending to a large extent on the genotype and environment in which it is being expressed. This latter problem can make selection for desired plant types particularly difficult, requiring testing or trialling in a range of likely environments and possibly restricting the growing of certain cultivars to specific areas. This is the case with peanut, where stability of lines or cultivars over a wide range of environments is required or, in contrast, the development of new lines which are particularly well suited to differing local conditions, pests and diseases.

The setting of objectives can be market-led, demands for certain quality standards or certain products being made by the consumer. A recent example of this relates to the increased awareness of problems associated with cholesterol and saturated fatty acids in human diets. Demand for "healthier polyunsaturates" has resulted in considerable breeding efforts being directed towards altering oil profiles of some of the major oil crops, particularly oilseed rape in which, along with the elimination of erucic acid, increased linoleic acid and decreased linolenic acid is required to meet these demands. There has been a similar trend towards the use of gamma-linolenic acid (GLA) as part of healthier diets which, together with its use in the treatment of premenstrual tension (PMT) and certain skin diseases [see eg 27, 28], has resulted in increased interest in species which can provide, or be persuaded to provide, this product. This raised profile of GLA has resulted in a substantial increase in the acreage of *Oenothera* (Evening Primrose) being grown in the EEC for example. Although it only contains approximately 10% GLA, the oil of this plant has a value some four or five times that of rapeseed oil, and breeding work on this and other GLA producing species such as Borage has increased significantly in recent years.

With the overproduction in Europe of certain crops such as the cereals, interest has turned to alternatives. Among those likely to be acceptable are some of the oilseeds, especially those like *Oenothera* and Borage producing high value edible oils, and others like *Crambe* and *Limnanthes* producing industrial oils, all providing products much in demand. Thus, the overall requirements to provide improvements in such species to make them agronomically and economically feasible as agricultural crops are, at least initially, largely political. The detailed objectives which follow to make the crop agronomically acceptable to growers and suitable for processing etc will be drawn up by those having to handle the crop, its seed and its end-use products. Political standards, such as those set by the EEC for maximum and minimum levels of erucic acid and glucosinolates in rapeseed oil and meal, have to be met by the breeders, by whatever means at their disposal; these means will be very dependant on the particular crop, its agronomic and botanical attributes, previous knowledge, germplasm available *and* the considered political importance of the crop and its possible roles. The end use requirement for edible and non-edible oils will also play its part in the setting out of the objectives and, again, be controlled to a large extent by political motives.

3.3 Breeding strategies

The extreme diversity of plant and reproductive type, as well as history and crop development in the oilseeds, necessitates the application of a wide range of breeding procedures and techniques. A summary of the salient features relevant to breeding of each of the major oilseed crops is presented in Table 3.3. Broadly speaking, the crops can be divided into groups according to their botanical reproductive systems and these are summarised in Table 3.4. It is these which largely determine the strategies adopted during crop improvement, along with additional features such as genetic behaviour and heritability of important traits and the availability of cytoplasmic male sterility for the production of hybrids where appropriate and, in more recent years, the availability of techniques emerging from developing biotechnologies and molecular biology (see Chapter 6).

The breeding strategies adopted to achieve the objectives are thus dependant on many factors, including the biology of the species concerned, prior knowledge of the relevant factors such as the heritability and genetic behaviour of traits and the nature of the objectives themselves, whether they concern overcoming problems associated with the growing of the crop, the harvesting and processing of the seed and oil, or the end-use products.

3.3.1 Reproductive type

The reproductive system of the species, whether inbreeding, outbreeding or both, with or without self-incompatibility, whether the plants are hermaphrodite, monoecious or dioecious, not only determines the strategy adopted for improvement, but also dictates the ease or otherwise of the breeding procedures and the timescales involved. Conventional annual inbreeding oilseed species, like soybean, peanut and safflower, are the easiest to handle, outbreeders like sunflower, and the diploid Brassicas are less easy. Monoecious plants like maize and castor, and dioecious plants like jojoba can be very difficult and progress with these can be slow. Progress is also restricted when dealing with the tree crops such as olive, coconut and oil palm because of the time required between germination and fruiting, often many years for these crops. Here, assessment, selection and adjustments to the breeding strategy are all necessarily delayed and the timescale for improvement can be many times that of crops with a more rapid life cycle, especially the inbreeding annuals.

Table 3.3 Botanical features of the major oilseed crops.

CROP	BOTANICAL NAME	FAMILY	Longevity	Flower Type	Reproductive Type	Propagative Type	Chromosome Number and Ploidy	ADDITIONAL COMMENTS
Soybean	Glycine max	Leguminosae	A	H	I	S	$2n=2x=40$	Highly self pollinated
Sunflower	Helianthus annuus	Compositae	A	H	O	S	$2n=2x=34$	Highly outcrossed. CMS available
Peanut	Arachis hypogaea	Leguminosae	A	H	I(O)	S	$2n=4x=40$	Allotetraploid. Self pollinating with limited outcrossing.
Oilseed Brassicas	Brassica rapa	Cruciferae	A	H	O	S	$2n=2x=20$	Some CMS available + sporophytic self incompatibility in diploids.
	Brassica nigra		A	H	O	S	$2n=2x=16$	
	Brassica napus		A/B	H	IO	S	$2n=4x=38$	
	Brassica juncea		A	H	IO	S	$2n=4x=36$	
	Brassica carinata		A	H	IO	S	$2n=4x=34$	
Safflower	Carthamus tinctorius	Compositae	A	H	I	S	$2n=2x=24$	Highly self pollinating with limited outcrossing. Structural male sterility, genic and CMS.
Sesame	Sesamum indicum	Pedaliaceae	A	H	IO	S	$2n=2x=26$	Self pollinating with up to 60% outcrossing. Male sterility available.
Poppy	Papaver somniferum	Papaveraceae	A	H	IO	S	$2n=2x=22$	Self pollinating with about 30% outcrossing. Autogamous. No CMS or SI available.
Niger	Guizotia abyssinica	Compositae	A	H	O(I)	S	$2n=2x=30$	Mostly outcrossing but will self pollinate.
Cotton	Gossypium hirsutum (90%) + G. barbadense (10%)	Malvaceae	A(P)	H	IO	S	$2n=4x=52$	Allotetraploid. Mostly self pollinated but insects can cross pollinate. CMS available.
Linseed	Linum usitatissimum	Linaceae	A	H	I	S	$2n=2x=30$	Highly self pollinated. CMS available.
Maize	Zea mays	Graminae	A	M	O	S	$2n=2x=20$	Highly cross pollinated with selfing <5%.
Castor	Ricinus communis	Euphorbiaceae	A	M	O	S	$2n=2x=20$	Mostly cross pollinated but will self. Male sterility available.

CROP	BOTANICAL NAME	FAMILY	Longevity	Flower Type	Reproductive Type	Propagative Type	Chromosome Number and Ploidy	ADDITIONAL COMMENTS
Jojoba	Simmondsia chinensis	Simmondsiaceae	P	D(H)	O	C(S)	2n=2x=52	Rare hermaphrodite flower, mostly vegetatively propagated.
Olive	Olea europa (+ O. salvia + O. sylvestris + O. laperrini + O. ferrugina)	Oleaceae	P	H	I(O)	SC	2n=2x=46	Mostly self pollinating but some cross pollinating. Highly heterozygous. Some vegetative propagation.
Oil Palm	Elaeis guineensis	Palmae	P	M(H)	O	S(C)	2n=2x=32	Strictly allogamous. Cross and self pollinates easily.
Coconut	Cocos nucifera	Palmae	P	M	O	S	2n=2x=32	Tall trees (typica) - mostly cross pollinated. Dwarf trees (nana) - mostly self pollinated. Can be unisexual.

KEY:

Longevity: A = Annual; B = Biennial; P = Perennial

Flower Type: H = Hermaphrodite; M = Monoecious; D = Dioecious

Reproductive Type: I = Inbreeding; O = Outbreeding

Table 3.4 Reproductive features of the major oilseed crops.

SELF-POLLINATING	HERMAPHRODITE					MON-OECIOUS	DI-OECIOUS
	CROSS-POLLINATING						
	SELF COMPATIBLE		EITHER/BOTH	SELF INCOMPATIBLE			
	WIND	INSECTS		SPOROPHYTIC	GAMETOPHYTIC		
Soybean Peanut Brassicas 4x * Safflower Sesame * Poppy Linseed Cotton * Olive * Cuphea * Crambe * (Jojoba) *	Olive * (Sunflower) *	Sesame * Sunflower * (Peanut) * (Brassica 4x) * (Poppy) * (Niger) * Cotton * Euphorbia Limnanthes	Cuphea * ? Crambe * ? Lesquerella	Brassica 2x Niger *		Maize Castor Oil Palm Coconut	Jojoba *

* indicates appearance under more than one heading.
( ) indicates secondary or less important state.
2x = diploid
4x = amphidiploid

The reproductive system also determines, to a large extent, the amount of variability available to the breeder and this is often the converse of the ease of handling. Obligate outbreeders, those plants with systems which promote outcrossing, whether physical as in the monoecious and dioecious plants, or genetical as in the self- incompatibility systems of the diploid Brassicas and niger, will tend to have considerably more heterogeneity in their populations. This provides the breeders with a readily available and varied gene pool, but such variation, because of the problems associated with breeding outbreeders, is usually more difficult to handle. Inbreeders on the other hand, though much easier to breed, often have a more limited gene pool available. Breeders have attempted to extend this, with significant success, by interspecific hybridisation, often using very wide crosses, and in programs of mutation breeding. In cases where traits are controlled simply, by one or two genes, the transfer of the appropriate gene(s) is easier in inbreeders than outbreeders and very much more difficult in both when the traits are complex and quantitative.

As part of the reproductive system, floral morphology and ease of pollination by hand must be considered. In species with standard flowering morphology, such as soybean, the Brassicas and sesame, the procedures for self- and cross-pollinating the flowers are straightforward and rapid, but for species with more complex flowers, such as the composites, sunflower, safflower and niger, modifications of the standard procedures may be required, which sometimes result in restrictions being imposed on the ease and efficiency of the process. Details of these procedures for each crop are given in numerous texts [see eg 3, 5].

3.3.2 Longevity and propagation

Breeding strategies are also determined to some extent by whether the crop is annual, biennial or perennial. Techniques for the latter are sometimes modifications of those applied to annuals, with adjustments to account for the extended timescale and for the fact that once the crop is established, the plants are fixed for the duration of the plantation. This is particularly relevant in the case of long term perennials, such as olive, oil palm and coconut, where there is little, if anything, that can be done to improve the crop genetically once it is planted, short of a costly grafting or, more rarely, replanting with lines with improved performance. Biennial crops also need a special mention in that many are required to over- winter in their vegetative state and thus need sufficient levels of winter hardiness, frost or cold tolerance to survive. Of the oil crops mentioned here, only the Brassicas, in particular *B.napus*, winter oilseed rape, fit into this category, although other crops may require attention in this respect for other reasons. It is usually possible to persuade biennials to flower within a shorter time period than the naturally required eighteen months or two years by subjecting them to periods of cold, usually <10 °C, and short days to encourage the development of leaf primordia before becoming initiated to flower - the process of vernalisation.

Despite the many advantages of working with annual species, one of the biggest disadvantages can be the perpetuation of particular genotypes, especially in species where inbreeding is difficult or prevented, or where the individuals are highly heterozygous. For these, and for the long term perennials like oil palm and olive, the ability to propagate the plants vegetatively can be a tremendous asset to the breeder and can be very influential in determining the breeding strategy. In recent years, with the advent and wide-spread application of tissue culture and, more specifically, micropropagation, many of the problems in this connection have been alleviated. Indeed, for oil palm and coconut which have no natural means of vegetative propagation, it has completely changed breeding approaches and facilitated studies on genotypic and environmental effects and the all important interaction between the two [16, 29]. It is now feasible to set out homogeneous plantations of these crops with the best high performing homozygous or heterozygous genotypes raised through micropropagation rather than relying on the heterogeneous populations resulting from seed-raised individuals. In oil palm, for example, which is naturally highly heterozygotic, the ability to produce large numbers of genetically identical individuals is likely to lead to a 30% increase in oil yield by simply enabling the establishment of homogeneous plantations from the best trees [30]. Equally important, but on a smaller scale, is the ability to propagate, and maintain vegetatively, superior parents such as those with good genetic combining ability. At the other end of the time scale, a number of the seed-raised annuals are relatively short lived and unless homozygous inbred lines of the best parents are available, there can be real dangers of losing important parental genotypes between generations; the ability to propagate these vegetatively overcomes many of the problems and greatly facilitates the breeding and assessment of new lines and can be an important consideration in determining breeding strategies.

Clonally propagated material can also have its problems, however, especially when comparatively large portions of the parent plant are propagated vegetatively in the conventional way. Such cloned units can contain a variety of diseases, the most serious and most difficult to deal with being the viruses, which can be passed on from one cloned generation to the next. Tissue culture methods now available have greatly facilitated the cleaning up of such clones to provide 'virus free' or 'virus tested' material, often resulting in a restoration of plant vigor and associated advantages. In the case of viruses, meristem tip culture, in which just the meristematic dome is excised and grown up into a small plantlet under aseptic conditions in culture, has proved a very successful procedure in helping to produce and maintain clean stocks of valuable cloned lines [eg 31, 32, 29].

In seed-raised species, there is sometimes a conflict of requirement with regard to seed dormancy. For most such crops, it is obviously necessary for the sown seed to germinate readily and for the stand to become established as soon as possible; any delay in germination resulting from seed dormancy will reduce the efficiency with which the crop becomes established and this may lead ultimately to lower yields and unfulfilled potential. However, in peanut, for example, fresh seed dormancy is a positive objective in breeding for quality, since the tendency for the seed to germinate prematurely in the pod greatly reduces the quality of the seed. This problem can also arise in oilseed rape, especially in wet seasons during harvesting.

As already mentioned in this section, the breeding strategy adopted for any particular crop will be determined by many factors, including the reproductive system and whether the plants are annual, biennial or perennial. Similarly, the size of the plants and their longevity and the method of cropping also play a large part in determining strategic approaches in breeding. The oil crops as a group cover a wide range of plant size, from the small annuals like linseed and sesame, through to olive, coconut, and the oil palms. The annuals can flower within some two to four months of sowing, whereas coconut can take five to eight years to reach initial flowering. Techniques and procedures have been developed to cope with these long time scales and with the space requirements of the large trees. Progeny assessment of these can be very time consuming and expensive and, because of the often highly heterozygous nature of the plants themselves, very difficult and somewhat unreliable. Techniques such as marker aided selection and looking for seedling traits which are correlated with mature characters associated with crop performance have greatly improved the efficiency of such programs.

The form and size of the seed or fruit and its suitability for agronomic practices, harvesting and processing are also important considerations. Other than the enormous size range covered by the oil crops, from sesame and poppy at the small end of the range to coconut at the other, seed and fruit size and form have other implications for the breeder. The all-too-common conflicts of thicker versus thinner seed coats, larger versus smaller seeds etc. will largely depend on the purpose being considered. In peanut and rapeseed, for example, the seed coat has to be durable enough to withstand harvesting but not so thick as to impede processing activities or contribute significantly to the meal where it dilutes the proportion of valuable protein. Efforts are being made to reduce seed coat thickness in some crops. In the diploid Brassica progenitors of rapeseed, there appears to be tight linkage between thinner testa and yellow testa color but it is proving difficult to select stable forms of these traits in rapeseed, or to transfer the traits to resynthesised *B.napus* from what are apparently suitable diploid parents [33]. Some success in reducing hull size in safflower has been reported by Urie [34,35], the recessive partial-hull gene responsible may be useful in safflower breeding programs as a means of increasing both oil and protein percentage of the seed.

3.3.3 Interspecific hybridisation

Mention has already been made of the importance of interspecific hybridisation in some crops as a means of widening the range of variability available to the breeder by extending the gene pool, and as a means of importing specific traits into the crop species. Strategies adopted to perform these transfers and exploit the newly acquired variability will depend almost entirely on the ease with which they can be achieved. Such approaches have been important and largely successful in crops like soybean, peanut, sunflower and poppy but barriers to interspecific hybridisation can and have frustrated attempts in these and other oil crops. Some of these can be overcome with tissue culture methods such as embryo rescue and ovary culture but where chromosome pairing is very disrupted it may not be possible to obtain useful hybrids at all. It is also often the case that recovering the agronomic and reproductive qualities required in the crop from interspecific hybrids can be a slow and time consuming task of backcrossing and selecting the plant types suitable for growing as a crop; selection for the transferred trait must be combined with selection amongst the segregating progeny of the interspecific cross for all the other traits required in the crop. Marker-aided selection and the use of RFLP and other molecular markers can aid this procedure by providing reliable information about the amounts of the recipient (crop plant species) and donor (wild species) genomes present in individuals in the segregating populations. Though costly and highly specialised at present, it is likely that these techniques will become less expensive and more readily available to breeders for these purposes. Other techniques used in gene transfer such as transformation and protoplast fusion also offer great scope for extending the gene pool available to breeders and others, where this is seriously limiting crop improvement.

Most of the techniques have been successfully employed in the Brassicas, both to import specific traits or genes into crop plants from related species, or to provide new combinations of genes as in the resynthesising of the amphidiploid species *B.napus* and *B.juncea* by crossing the diploid parents, rescuing and growing up the newly synthesised hybrids. Some aspects of the potential of resynthesised *B.napus* in breeding and genetic analysis are reviewed by Chen and Heneen [36]. In this way, completely new combinations of genes can give rise to greatly extended variability, some of which will be of great value to breeders. More distant relatives of *Brassica napus* such as *Moricandia* species can also be combined using such techniques which gives rise to even greater possibilities - in the case of *Moricandia*, some of the species exhibit some of the enhanced photosynthetic activity and water use efficiency of C4 plants and herein lies the possibility of transferring these highly desirable traits to the crop Brassicas [37]. Once the desired traits have been transferred, by whatever means, and the recipient plants have made any necessary recovery from the transfer process, the breeder can usually employ conventional breeding procedures to fully exploit the newly developed genetic material.

3.3.4 Pests and diseases

The importance of pests and diseases in the various crops has already been considered in the section 3.2. Breeding strategies for these will depend largely on the mode of inheritance of any resistance or tolerance available and the breeding nature of the crop itself. However, if such traits are considered crucial to the success of the crop, additional effort may be required and justified to find suitable forms and levels of resistance, such as those just described using techniques of biotechnology and genetic engineering.

3.3.5 Heritabilities

Breeding strategies will also depend on the extent and nature of the genetic control and heritability for the characters in question. Highly heritable, major gene characters with little or no influence from the environment are relatively easy to deal with in most crops but particularly in crops where mass selection is the most commonly practised method of improvement. This is the case with oil palm where good progress has been made in selecting for characters with high heritability and additive genetic variation, but improvement of yield by this method has been disappointing, largely because yield is a complex character with low heritability and mainly non-additive genetic variation [38]. In circumstances such as these it is necessary to choose traits with high heritabilities and, if possible, simple additive genetic variation, in parents with good genetic combining ability, making selections on the basis of progeny testing. To achieve this, it may be necessary to work with individual components of complex characters such as yield, since these are often more simply inherited. An example is given by Webb and Knapp [39] who estimated heritabilities and devised a selection scheme for the improvement of oil yield in a population of *Cuphea lanceolata*. They found heritabilities for seed and oil yield to be relatively low at 20 and 24% respectively, but higher values for oil content and seed weight at 46 and 58% respectively, and a high genetic correlation between oil yield and the other traits. They suggest that, in view of this genetic correlation and the higher heritabilities for oil content and seed weight, indirect selection for oil yield based on these characters may be advantageous. Statistical techniques such as path coefficients, multiple and partial regressions and various forms of mulitvariate analysis can often provide valuable information about the relative importance of individual components in complex traits during such an approach.

The effects of the environment on characters during selection are, in themselves, not usually a problem for the breeder, provided the effects are the same, or at least consistent, for all genotypes. Where this is not the case, and genotype-environment interaction is in evidence, the performance of the genotypes under investigation depends on, and varies with, the particular environment in which they are being grown. In such cases, little reliable information about the genotypes can be obtained without reference to the particular environment in which they have been trialled. Breeding plans have been devised to accommodate and assess these interactive effects.

3.3.6 Heterosis

In many oil crops, it may be possible to exploit the heterotic effects which can result in progeny from particular crosses exceeding the performance of the best parent. Indeed, for some crops hybrid breeding to make use of this hybrid vigor is a prime objective and strategies to exploit heterosis play an important role in crop development. Hybrid vigor is usually less in self-pollinated species, where additive gene action predominates, than in outbreeders where dominance gene action is more important. This is an important consideration in deciding breeding strategy since the additional cost and effort associated with a hybrid production program has to be covered by the gains from improved performance and returns of the hybrid crop. In order to be commercially viable, it is necessary to have some relatively easy way of ensuring the production of hybrids as opposed to self-pollinations in the crossing block during seed production. In crops where hybrids have been successfully introduced, as in maize and sunflower, this has been achieved with some form of male sterility. In other crops, such as oilseed rape, safflower, sesame, poppy, cotton and, to a limited extent in castor, hybrid vigor has been demonstrated but attempts to reliably produce the hybrid seed on a large enough scale have been frustrated by the lack or unreliability of suitable parental material with restricted selfing. The approach to exploit hybrid vigor does, however, offer great potential for improvement in these crops and many breeding programs include hybrid development as part of their strategy. In linseed, however, although material with restricted selfing is available, the flowers of the male-sterile plants do not open sufficiently to allow cross pollination and the petals are not retained long enough to attract the natural pollinators needed for pollination [40]. Peanut and castor are also exceptions to this; the flower morphology and low amount of pollen produced on a given day probably means that hybrid production is not practically feasible in peanut [41], and in castor the expression of hybrid vigor is rare [42], although the main yield advantage of hybrids is their strong female tendency which increases the number of capsules per inflorescence.

Heterosis is related to genetic distance - the further apart the two parents are genetically, the more chance there is of the progeny exhibiting hybrid vigor [43]. But increased genetic distance is no guarantee of heterotic effects and parental combinations have to be tested for their ability to combine and produce exceptional progeny in progeny testing trials.

An interesting heterotic effect has been obtained in China during grafting of one species of olive, *Olea ferruginae*, on to *O.europea* root stocks in which the resulting trees are apparently smaller and more productive. Such interspecific grafting methods could have potential for developing olive groves which are homogeneous, well adapted to difficult soil conditions and resistant to certain diseases [44].

3.3.7 Trait assessment

Breeding strategy is also highly dependent on the ability to easily assess the traits of interest. In many cases selections are made amongst thousands of individuals in segregating populations, as in the F_2 generation in pure line or pedigree breeding. A rapid assessment of the characters is therefore required to reduce time, space and costs, or the use of indirect selection by choosing an easily assessed character which is closely linked to, or pleiotropic with, the desired trait. In peanut, Coffelt and Hammons [45] reported on correlations between a number of traits in an F_2 population of a cross between Argentine (Spanish type) and Early Runner (Virginia type). They found highly significant positive correlations between number of pods and pod weight, number of seeds and seed weight, pod weight and number of seeds, and pod weight and seed weight. They suggested that selection for any of the four characters, number of pods, pod weight, number of seeds or seed weight, should result in a corresponding increase in the other correlated traits. Jain and Abuelgaism [46] reported that early flowering populations of *Limnanthes alba* and *L. douglasii* also had the highest yields. Such relationships between characters, provided they are consistently good and reliable, can be of considerable value to the breeder looking for improvements in complex characters like yield.

Recent developments in nuclear magnetic resonance (NMR) technology have provided breeders with opportunities to readily screen large populations non-destructively for seed quality traits, thereby facilitating a more direct approach to improvement of some important characters such as oil profiles and protein content. Technical developments in other physical and chemical assays have similarly altered breeding strategies adopted by breeders and greatly improved the efficiency of selection for improvement. During efforts to select for reduced levels of seed glucosinolates in oilseed rape, rapid methods of detecting the glucose released on the breakdown of the glucosinolate compounds were developed. These methods provided breeders with a rapid and effective means of screening large numbers of individuals for those with lowered levels of the compounds which could be confirmed later on the selected smaller numbers by more accurate laboratory-based methods.

Another approach to trait assessment is through the development of cytogenetic stocks. The creation of aneuploids, monosomics and substitution lines in a crop species can greatly facilitate the understanding of the genetic control and behaviour of traits [eg see 47, 48]. By including or excluding individual chromosomes from the normal genome, genes responsible for recognisable effects can be assigned to particular chromosomes; subsequent genetic studies, including linkage, can then provide sufficient information to map the genes in detail. This approach has been pursued in safflower where Estilai and Knowles [49] suggested that the production of aneuploids is feasible and could thus provide valuable material for detailed genetic studies in this crop. In soybean, Sadanaga and Grindeland [50] used a homozygous translocation line with cytologically recognisable interchange chromosomes to facilitate the mapping of the *w1* (white flowers) locus.

All these methods, which allow the breeder to select the desired characters more easily and rapidly, will have an effect on the strategy adopted in breeding for crop improvement. Other techniques, including those of biotechnology and molecular biology, some of which have already been mentioned, will also have considerable impact on the approach a breeder will take in breeding for particular improvements in specific crops as discussed in Chapter 6.

3.4 Breeding plans

Plant breeding is the manipulation and management of genetic variability to meet certain desired objectives, creating new genetic combinations to satisfy those requirements in the development of new cultivars. Genetic variability is the essential resource of all this work; a struggle to create in some species and a problem to control in others.

Although the genetic principles can be universally applied, the breeding problems and techniques to tackle them can be very different from one crop to another. Much effort has to be put into generating initial genetic variation in self-pollinating small seeded annual crops like safflower and soybean, after which self-pollination in successive generations relatively easily provides the breeder with opportunities to select homozygous or inbred lines. In contrast, this initial variability is already present in outbreeding species such as maize, castor and sunflower and the breeder has to work hard in successive generations to produce true breeding inbred lines. Either way, genetically variable populations are derived and desired phenotypes selected from these. Breeders must decide how to generate and manipulate this variability and to be aware of agronomic and production practices, marketing and consumer needs, climatic influences on the crop, problems of pests and diseases, local, regional, national and international considerations, world market trends, government policies, population growth trends. Knowledge of heritabilities, genetic variances and the breeding system of the crop are also required to maximise efficiency in any effective breeding program. Thus, the approaches to creating and controlling this variability are determined by many factors.

Although the reasons for wanting to create and control the variability are many and varied, the approaches to breeding for crop improvement can, essentially, be summarised in four basic schemes, as described in detail by Simmonds [2] and referred to in one form or another in most text books on plant breeding [eg see 4]. The four schemes are distinguished by the crop's reproductive system, whether predominantly inbreeding or outbreeding, and the longevity or life cycle, whether annual, biennial or perennial, and the propagative type, whether mainly seed-raised or cloned. These features give rise to the four main breeding groups - Inbred Pure Lines, concerning inbred seed-propagated crops; Open Pollinated Populations and Hybrids, involving outbreeding seed-raised crops; and Clones, concerning outbred perennials. A summary of these features for the major oil crops is given in Tables 3.3 and 3.4.

3.4.1 Inbred pure lines

Inbred Pure Lines are crosses made between parental inbreds initially and selection is practised over subsequent generations of selfing. The objective is to isolate one or more transgressive segregants from amongst the progeny. There are three options available to the breeder: selection can be applied from the F_2 generation (Pedigree method), or deferred to later generations (Bulk method); if deferred, techniques can be used to speed progress through the generations (Single Seed Descent (SSD)).

Choice of approach depends to a large extent on whether early selection for the traits of interest will be effective. Selection in early generations, as in the Pedigree method, for characters of low heritability such as yield is generally ineffective but this method significantly reduces the number of lines the breeder has to deal with more quickly, and can be very efficient if the traits have sufficiently high heritabilities. This used to be the favoured method of breeding inbreeding crops but it has been replaced by the Bulk method in recent years. Here, evaluation of all but the most easily assessed characters is delayed until F_5 to F_8 when selections are made between the emerging inbred families in replicated trials. Single Seed Descent is an extreme form of this in which no selection at all is practised in the early generations and the objective is to reach generation F_5 as quickly as possible by taking two or even three generations a year. All three schemes end up with intensive evaluation and selection between F_5 and F_8. A form of the Bulk method, known as 'long term bulks', is used in some crops, for example soybean, in which mixtures of pure lines are grown as the crop. A similar approach has been used in peanut to establish 'composite' varieties comprising phenotypically similar but genotypically dissimilar sister lines [51].

3.4.2 Open pollinated populations

In considering Open Pollinated Populations it is necessary to distinguish between Population Improvement in which cultivars are indefinitely propagated as closed populations, and Synthetics which are regularly reconstructed from selected source materials, as in seed propagated lines or, in the case of perennials, clones. Either or both approaches produce somewhat heterogeneous material which is widely based enough to avoid inbreeding depression, and all the resultant individuals are heterozygous. Population improvement can be further subdivided into two methods; Mass Selection which is quick but relatively inefficient if heritability is low, and Progeny Testing, which is slower but generally more efficient, and of which there are many variants. Maize is well adapted to sophisticated methods of Progeny Testing because of the ease with which it can be reliably both selfed and cross-pollinated. These methods are not appropriate or feasible in some other crops, such as the diploid Brassicas, in which self incompatibility forbids selfing, or in sunflower in which selfing is laborious. Both methods contribute to or are involved in Recurrent Selection which is a very widely used procedure for population improvement. There are four types of Recurrent Selection, i) Simple Recurrent Selection which is essentially Mass Selection with one or two years per cycle, ii) Recurrent Selection for General Combining Ability (GCA) with half sib progeny testing using a widely based variety as a tester, iii) Recurrent Selection for Specific Combining Ability (SCA), which is similar to that for GCA but uses an inbred line as a tester, and iv) Reciprocal Recurrent Selection with its objectives as the mutual adaptation of two populations. Choice of the method best suited to particular situations will depend on the availability of suitable material and accumulated knowledge of the genetic behaviour of the traits being considered.

The production of Synthetics involves the isolation of potential parents as lines or clones and testing for general and specific combining ability. The method is similar to that used in the production of hybrids when restricted to the use of two parents only in the absence of inbreeding depression.

3.4.3 Hybrids

The third breeding scheme, Hybrids, was first developed in maize and has been in use for the past 50 years or more. It depends on being able to isolate numerous inbred lines for use as parents with high general and specific combining abilities. The parents used are mostly homozygous so the resulting progeny are uniformly highly heterozygous. The hybrid technique has become highly sophisticated in maize but there is increasing interest in hybrid production in many other crops these days with the discovery that heterosis or hybrid vigor is considerably more widespread than expected. As already mentioned elsewhere in this chapter, the magnitude of any heterotic effects will depend to some extent on the genetic distances or diversity of the parents used. Attempts are being made to exploit hybrid vigor in many inbreeding species where heterosis has been observed, for example in safflower [52, 53, 54], in sesame [55, 56] and in poppy [57]. Indeed, linseed was one of the first crops of commercial value in which cytoplasmic male sterility (CMS) was identified with potential use in hybrid production [58]. However, all inbreeding species require some mechanism to prevent selfing to make hybrid production a viable commercial possibility in inbreeders, and self-pollination in outbreeders used for this purpose must be at a sufficiently low level to avoid selfed seed contaminating hybrid stocks. Recurrent selection and backcrossing techniques are sometimes used to generate populations of improved lines for use as parents in hybrid production.

Some crops, notably cotton and maize, retain a certain degree of heterozygosity even in rigorously inbred lines [59, 60]. This can cause problems with uniformity of both the lines themselves and any hybrids produced from them. It appears that the origin of this residual heterozygosity is so far unknown but may be due to the preservation of substantial linkage blocks that normally persist unchanged but recombine under enhanced frequencies promoted by homozygosis. This poses problems for the breeder in the maintenance of pure lines.

3.4.4 Clones

The Clones method of breeding involves selecting amongst vegetative descendants of variable F_1 families produced by crossing heterozygous parents. This is usually a very efficient method since it exploits all the genetic variability available amongst the material used. The method usually includes selection for high general and specific combining abilities and includes opportunities for somatic, clonal selection. The clone method and clonal selection are particularly important in the perennial oilseed crops, olive, oil palm and coconut, in which the establishment of clones plays an important role in the breeding of these crops. Hardon *et al* [16] list amongst the advantages of clonal selection in oil palm: more rapid selection progress, early selection for combinations of characters, early introduction of new genetic material for commercial planting, more precise testing of selected parent palms and improved prediction of field performance of planting material, and more uniform planting material which allows for precise definition of planting density and fertiliser regimes. They add a note of caution, however: since the variation within clones is genetically uniform, there may be a risk of epidemic pest or disease outbreak, although this can be reduced by planting mixtures of clones rather than monoblocks; in addition, material propagated through tissue culture will be comparatively expensive. There has also been concern expressed from time to time about clonal degeneration, a general term used to describe the loss of vigor and deterioration of material passed through a number of cycles of vegetative propagation. This appears to be due not to some mysterious biological or genetic decline of the cloned stocks as was always thought, but to the presence of somatic mutations, or diseases, especially viruses. Where the latter is the case, techniques of meristem tip culture can be used to clean up infected stocks. Somatic mutations, where they do occur, can sometimes be used to advantage in clonal section.

3.4.5 Backcrossing

An additional method to be included here, though already mentioned elsewhere in the chapter, is Backcrossing. This is widely used in all four schemes as a means of transferring one or a few genes from a donor parent by repeated backcrossing to the recipient or recurrent parent. It is particularly useful in the transfer of dominant genes from unadapted alien stocks, such as wild species, to adapted parents, such as those already included in a breeding program or commercial cultivars; an example is the transfer of the dominant cotton blackarm resistance from *Gossypium barbadense* to G. *hirsutum*. Recessive genes can also be transferred by inserting a cycle of inbreeding and selection between each backcross, or by crossing or making each backcross on a scale sufficient to ensure statistically that the gene is retained. However, in practice, all backcrossing is used for dominant genes or at least for characters with adequate heterozygote expression.

With the increasing availability of molecular markers, an extremely efficient form of backcrossing, referred to as "directed backcrossing", is being introduced. Where sufficient markers are readily available, backcross progeny can be screened for the presence of relevant portions of the donor and recipient (recurrent) chromosomes and the desired genotypes picked out on the basis of the markers rather than relying on assessment of plant phenotypes, as in conventional backcross methods. This method, along with other applications of RFLP mapping in plant breeding, are discussed in Chapter 6 and also by Tanksley *et al* [61].

3.4.6 Selection

Mention has already been made of the importance of selection in breeding programs, together with the number of characters a breeder has to consider in addition to those identified specifically in the breeding objectives, and the ease, or otherwise, with which they can be assessed and manipulated. There is also the possibility of selecting for two or more characters simultaneously, called 'index selection'. This can be employed either consciously through direct assessment, or indirectly using marker-aided selection as, for example, in selecting seedling traits in coconut which are highly correlated with important mature characters such as yield, thus saving time and space in the program. Co-selection of traits closely related through pleiotropy, when two or more characters are correlated for physiological or developmental reasons rather than genetic linkage, is sometimes successful. Indeed, developmental genetic correlations are very much more common than those due to genetic linkage. However, index-selection and co-selection must be pursued with care as the genetic and environmental influences on the different characters under consideration can be very different and lead to the wrong conclusions and the development of unsuitable material.

Vegetative propagation, whether natural or via tissue culture, does not usually alter the selection objectives in a program but it should accelerate progress [16]. The overall objectives remain the same whether selection is practised on a population of cloned individuals or a genetically segregating population.

The efficiency of the selection procedure, including the trialling associated with it, will greatly influence progress in the improvement of the crop whatever breeding scheme is followed. The importance of correct procedures being adopted during trialling, using properly replicated and randomised designs, cannot be overstated since considerable weight is usually placed on results from trials during selection programs. Consideration must be given to the heritability of the trait, to genetic and environmental effects and, above all, to genotype-environment interactions. The latter are particularly important when trials are conducted over a number of diverse sites and seasons.

Although the possibilities offered by marker aided selection, using molecular markers such as isozymes and RFLPs, are undoubtedly an exciting development in the application of molecular biology to oil crop breeding (see Chapter 6), some notes of caution have been sounded. Dudley *et al* [62] investigated the use of molecular markers in the grouping of parents in maize breeding programs to predict yield potential of hybrids, to assign inbreds to heterotic groups and to determine to which of the two inbred parents of a single cross a donor line is most closely related. They concluded that their study did not support the use of marker polymorphism to predict the yield of hybrids but, if data are available to measure the value of genotypes at a locus, these values can be used to predict the relative yields of hybrids. Also, caution should be used in assigning inbreds to heterotic groups based on the marker information. Although marker information agreed with yield information of the hybrid when the parents were diverse, poor agreement was obtained when parents were more closely related. As with most new techniques, the accumulation of knowledge about reliability and how widely they can be applied is crucial to their success, but until then their use must be employed, results interpreted and conclusions drawn, with caution; their potential, however, is undoubtedly great.

3.5 Future prospects

Something of the promise and practise for plant breeders of the new technology from biotechnology, molecular biology and genetic engineering is discussed by Gardner [63], who considers how breeders might make the best use of the advances in biotechnology. The initial step, he suggests, is to develop, or develop access to, a tissue culture system that allows regeneration of plants for the best cultivars or breeding lines. Some thought must also be given to what genes or traits might be useful to target in particular crops, without in any way limiting this choice to presently available germplasm. He also suggests it is important for breeders to keep in touch with the scientific community and to become aware of these new techniques and genes as they become available.

Austin *et al* [64] consider the potential for the application of molecular biology to crop improvement and discuss in considerable detail how they anticipate that molecular biology will enable the efficiency of breeding programs to be increased, and that it will also make it possible to achieve hitherto unattainable objectives. With specific reference to plant breeding, they suggest that conventional programs could be extended to include the transfer of genes to existing or new backgrounds, following genetic and physiological analysis leading to the identification of the specific genes involved, and the modification of genes *in vitro* and their subsequent insertion into existing or new backgrounds following their isolation by recombinant DNA technology. They list several components of the plant breeding process which Bartels *et al* [65] discuss in relation to the possibility of increasing efficiency. These include: a) the introduction of new genetic variability using plant transformation, somatic hybridisation and somaclonal variation, b) increasing the precision and speed of selection through the use of biochemical and molecular probes, selection *in vitro* and the use of haploid plants, c) modifications to the breeding or reproductive system through the conversion of self-incompatible into self-compatible plants, and the use of genes to induce male sterility, d) decreasing the generation time by the induction of *in vitro* flowering, and e) improved definition of breeding objectives through cloning and studies of genes at the base of specific genotypes. They go on to discuss and give examples of the application of genetic engineering to breeding strategies. Although none of these relates specifically to oil crops, with the possible exception of maize, they illustrate sufficient progress in other crops to give considerable encouragement that, given time, effort and application, such procedures will become available in most important crops, including many of the oil crops.

Tanksley *et al* [61] reviewed the applications to plant breeding of a particularly important tool to have emerged in recent years, that of genome mapping using restriction fragment length polymorphisms (RFLPs). Among the advantages of these molecular markers they list: a) facilitating the movement of desirable genes among varieties, b) enabling the transfer of novel genes from related wild species, c) the analysis of complex polygenic characters as ensembles of single Mendelian factors, and d) establishing genetic relationships between sexually incompatible crop plants. They also suggest that high density RFLP maps may also make it possible to clone genes whose products are unknown, such as those involved in disease resistance or stress tolerance. Such RFLP maps are well developed in a number of crops, including soybean [66] and rapeseed [67], but are still in their infancy in others, such as linseed [68]. However, the enormous potential offered by these biotechnological approaches is now well recognised, even for crops where their development is slow [69]. Initial studies using RFLP markers in peanut have indicated that there is little variability present amongst American cultivars [70]. Without sufficient variability for a large number of markers, it would not be feasible to construct a genetic linkage map based on conventional RFLP analysis. It is suggested that breeders could usefully increase the use of related wild species in breeding programs - RFLP analysis will facilitate this introgression process [71].

With particular reference to linseed but applicable more widely, Friedt *et al* [69] list the following range of biotechniques with likely applications in plant breeding programs: a) somatic tissue or cell culture for rapid propagation of plants from meristems or calluses, b) anther or microspore culture for regeneration of haploid plants, c) culture of hybrid-embryos out of wide crosses of cultivated species to their wild relatives, d) culture and regeneration of hybrid- protoplasts for the establishment of "asexual" interspecific or intergeneric hybrids, e) genetic engineering with isolated plant protoplasts ("direct gene transfer"), and f) application of genetic engineering via specific vector systems, eg. *Agrobacterium tumefaciens*.

Traditionally, there has been a tendency for breeders of crops requiring similar approaches to isolate themselves from those involved with different groups of crops, largely because the breeding techniques and procedures are often very different. Indeed, interaction between breeders involved in the different groups is often so limited to specific crops that we have professional associations concerned only with certain crops. This has been increasingly the case over the past 50 years of plant breeding during which time breeding techniques have steadily become more specialised with the evergrowing knowledge base *within* the particular crop groups. Special techniques have been evolved to cope with difficulties in handling certain crops, or even certain crops in certain situations, such as with Winter or Spring wheat and barley. This increased specialisation has led to barriers arising between the different groups of breeders, or even within groups where different objectives are being considered. This can lead to restrictions in approach and thinking; breeders who disregard these barriers can benefit from broadened horizons and bring fresh new ideas to their programs. This has never been more true than in these days of new and exciting techniques emerging from biotechnology, molecular biology and genetic engineering, and particularly in the oil crops where the diversity of plant types and approaches have a common goal - that of the provision of oil (and other storage products), its improvement in quality and quantity.

For most of the oil crops, the application of these techniques from biotechnology, molecular biology and genetic engineering is either in its infancy or is non-existent (Chapter 6). There is, thus, a vast and as yet largely untapped potential waiting in the wings to aid the improvement of oil crops, when and where required. Where these techniques have been applied, there have been notable successes, for example in the Brassicas, especially oilseed rape, in maize, soybean and sunflower. It appears that, given sufficient time and effort, funds, justification and incentive, there is little that cannot be achieved with the facilities provided by this new technology. The major obstacles, other than costs and funding, are likely to be more to do with restricted thinking and limited approaches than what can or cannot be achieved. The introduction of new breeding methods based on these techniques will depend as much on the decision by plant breeders to utilise the new methods as the availability of the new technology as it develops.

References

1 Beddigan, D. and Harlan, J.R., *Econ. Bot.* **1986**, *40*, 137- 154.
2 Simmonds, N.W., *Principles of Crop Improvement*, London and New York: Longmans, **1979**, pp. 123–204.
3 Fehr, W.R. and Hadley, H.H. (eds.), *Hybridisation of Crop Plants*: Am. Soc. Agronomy and Crop Science Soc. America, Publishers, Madison, WI, **1980**.
4 Wood, D.R., *Crop Breeding*: Am. Soc. Agronomy, Crop Sci. America, Madison, WI., **1983**
5 Downey, R.K., Robbelen, G, and Ashri, A. (eds.), *Oil Crops of the World*: New York: McGraw-Hill, **1989**.

6 Jolliff, G.D.,Tinsley, I.J., Calhoun,W. and Crane, J.M., *Meadowfoam (Limnanthes alba): its research and development as a potential new oilseed crop for the Willamette Valley of Oregon.* Oregon Agricultural Experiment Station Bulletin No. 648, Oregon State University, Corvallis, Oregon, **1981**.

7 Riley, K.W. and Belayneh, H., in: *Oil Crops of the World*: Downey, R.K., Robbelen, G, and Ashri, A. (eds.) New York: McGraw-Hill, **1989**; pp. 394–403.

8 Van der Plank, J.E., *Plant Diseases: Epidemics and Control*, London and New York: Academic Press, **1963**.

9 Van der Plank, J.E., *Disease Resistance in Plants*, London and New York: Academic Press, **1968**.

10 Innes, N.L., *Ann. Appl. Biol.* **1974**, *78*, 89–98.

11 Flor, H.H., *Annual Rev. Phytopathol.* **1971**, *9*, 275–296.

12 Day, P.R., *Genetics of host-parasite interaction*, San Francisco: Freeman, **1974**.

13 Klein-Lankorst, R., Rietveld,P., Machiels, B.,Verkerk, R.,Weide, R., Gebhardt, C., Koornneef, M. and Zabel, P., *Theor. and Appl. Genet.* **1991**, *81*, 661–667.

14 Anon., in: *Sesame: Status and Improvement*: Ashri, A. (ed.) Rome: FAO Plant Production and Protection, **1981**; Paper 29, pp 192–195.

15 Anon., in: *Sesame and Safflower: Status and Potentials*: Ashri, A. (ed.) Rome: FAO Plant Production and Protection, **1985**; Paper 66, pp 218–220.

16 Hardon, J.J., Corley, R.H.V. and Lee, C.H., in: *Improving Vegetatively Propagated Crops*: Abbott, A.J. and Atkin, R.K. (eds.) London: Academic Press, **1987**; Chap. 3, pp. 63–81.

17 Imrie, B.C. and Knowles, P.F., *Crop Sci.* **1970**, *10*, 349- 352.

18 Imrie, B.C. and Knowles, P.F., *Crop Sci.* **1971**, *11*, 6–9.

19 Broue, P., Douglass, J., Grace, J.P., Marshall, D.R., *Euphytica* **1982**, *31*, 715–724.

20 Grant, J.E., in: *Plant Breeding Symposium DSIR 1986*: Agronomy Society of New Zealand, **1986**, Special Publication No. 5, Paper 5, pp. 27–29.

21 Kondra Z.P. and Stefansson, B.R., *Can. J. Plant Sci.* **1970**, *50*, 643–647.

22 Thompson, K.F. and Hughes, W.G., in: *Oilseed Rape*: Scarisbrick, D.H. and Daniels, R.W. (eds.) London: Collins, **1986**; pp. 32–82.

23 Rucker, B. and Rudloff, E., *Proceedings of the 8th International Rapeseed Congress, Saskatoon, Canada, **1992**, pp 191–196.

24 McMichael, S.C., *Agron. J.* **1960**, *46*, 385–386.

25 Orf, J.H. and Hymowitz, T., *Crop Sci.* **1979**, *19*, 107–109.

26 Verbiscar, A.J. and Banigan,T.F., in: *Jojoba and its uses through 1982*. Proceedings of the 5th International Conference, Elias-Cosnik, A. (ed). Univ. Arizona,Tuscon, Arizona, **1983**, pp. 267–281.

27 Eyres, L. and Fenton, D., *Chemistry in New Zealand*, April **1984**.

28 Brush, M., *Treating PMS with Evening Primrose Oil* - Leaflet 4,Women's Health Concern Ltd., London, **1985**.

29 Jones, L.H., in: *Improving Vegetatively Propagated Crops*: Abbott, A.J. and Atkin, R.K. (eds.) London:Academic Press, **1987**, Chap. 23, pp. 385–405.

30 Corley, R.H.V., Wong, C.Y., Wooi, K.C. and Jones, L.H., in: *The Oil Palm in Agriculture in the Eighties*: Pushparajah, H. and Chew, P.S. (eds.) Kuala Lumpur: ISP, **1982**; Vol.1, pp. 173–196.

31 Burdon, R.D., in: *Plant Breeding Symposium DSIR 1986*, Agronomy Society of New Zealand, **1986**, Special Publication No. 5, Paper 25, pp.137–143.

32 Abbott, A.J. and Atkin, R.K., (eds.) *Improving Vegetatively Propagated Crops*, London: Academic Press, **1987**

33 Chen, B.Y., Heneen, W.K. and Jonsson, R., *Plant Breeding* **1988**, *101*, 52.

34 Urie, A.L., *Agron. Abstr.*, **1976**, American Society of Agronomy, Madison, WI., p.65.

35 Urie, A.L., *Crop Sci.* **1986**, *26*, 493–498.

36 Chen, B.Y. and Heneen, W.K., *Hereditas* **1989**, *111*, 255- 263.

37 McVetty, P.B.E., Austin, R.B. and Morgan, C.L., *Ann. Bot.* **1989**, *64*, 87–94.

38 Meunier, J. and Gascon, J-P., *Oleagineux* **1972**, *27*, 1–12.

39 Webb, D.M. and Knapp, S.J., *Crop Sci.* **1991**, *31*, 621- 624.

40 Lay, C.L. and Dybing, C.D., in: *Oil Crops of the World*: Downey, R.K., Robbelen, G, and Ashri, A. (eds.) New York:McGraw-Hill, **1989**; pp. 462–474.

41 Coffelt, T.A., in: *Oil Crops of the World*: Downey, R.K., Robbelen, G, and Ashri, A. (eds.) New York: McGraw-Hill, **1989**; pp. 319–338.

42 Hooks, J.A., Williams, J.H. and Gardner, C.O., *Crop Sci.* **1971**, *11*, 651–655.

43 Arunachalam, V., *Ind. J. Genet. Plant Breeding* **1982**, *41*, 226–236.

44 Brousse, G., in: *Oil Crops of the World*: Downey, R.K., Robbelen, G., and Ashri, A. (eds.) New York: McGraw- Hill, **1989**; pp. 462–474.

45 Coffelt, T.A. and Hammons, R.O., *Oleagineux* **1974**, *29*, 23–27.

46 Jain, S.K. and Abuelgaism, E.H., *Euphytica* **1981**, *30*, 437- 443.

47 Anon., *Seed Protein Improvement by Nuclear Techniques*. Proceedings of the 4th Research Co-ordination Meeting of the Seed Protein Improvement Programme and the 2nd Research Co-ordination Meeting on the Use of Aneuploids for Protein Improvement in Wheat. Vienna: International Atomic Energy Agency, **1978**.

48 Law, C.N., Snape, J.W. and Worland, A.J. Chromosome manipulation and its exploitation in the genetics and breeding of wheat. Stadler Genet. Symp., Columbia, MO, University of Missouri Agricultural Station *15*, 5–23, **1983**.

49 Estilai, A. and Knowles, P.F., *Crop Science* **1980**, *20*, 516- 518.

50 Sadanaga, K. and Grindeland, R.L., *Crop Sci.*, **1984**, *24*, 147–151.

51 Norden, A.J., in: *Peanuts - Culture and Uses*: Wilson, C.T. (ed.) American Peanut Research and Education Association Inc., Still Water, Okla., **1973**, pp.175–208.

52 Rubis, D.D., Safflower Utilisation Conf., USDA, Agric. Res. Service Publ. 74–93, **1967**, Albany, Calif., pp. 23–28.

53 Urie, A.L. and Zimmer, D.E., *Crop Sci.* **1970**, *12*, 545- 546.

54 Heaton, T.C. and Knowles, P.F., *Crop Sci.* **1982**, *22*, 520- 522.

55 Brar, G.S. and Ahuja, K.L., *Ann. Rev. Plant Science* **1979**, *1*, 245–313.

56 Osman, H.E., in: *Sesame and Safflower: Status and Potentials*: Ashri, A. (ed.) Rome: FAO Plant Production and Protection, **1985**; Paper 66, pp. 157–162.

57 Miczulska, I., *Nauk. Roln. Ser. A Prod. Rosl.* **1967**, *93A*, 197–204.

58 Bateson, W. and Gairdner, A.E., *J. Genetic.* **1928**, *11*, 269- 275.

59 Justus, N., *Agron*. J. **1960**, *52*, 555–559.
60 Thomson, N.J., *J. Agric. Sci. Camb*. **1973**, *80*, 135–146, 147–160, 161–171.
61 Tanksley, S.D.,Young, N.D., Paterson, A.H. and Bonierbale, M.W., *Bio/Technology* **1989,** *7*, 252–265.
62 Dudley, J.W., Saghai-Maroof, M.A., and Rufener, G.K., *Crop Science* **1991**, *31*, 718–723.
63 Gardner, R.C., in: *Plant Breeding Symposium DSIR 1986*: Agronomy Society of New Zealand, **1986**, Special Publication No. 5, Paper 60, pp. 333–339.
64 Austin, R.B., Flavell, R.B., Henson, I.E. and Lowe, H.J.B. (eds.) *Molecular Biology and Crop Improvement*. Cambridge: Cambridge University Press, **1986**.
65 Bartels, D., Gebhardt, C., Knapp, S., Rhode, W., Thompson, R., Uhrig, H. and Salamani, F., *Genome* **1989**, *31*, 1014–1026.
66 Keim, P., Diers, B.W., Olson, T.C. and Shoemaker, R.C., *Genetics* **1990**, *126*, 735–742.
67 Landry, B.S., Hubert, N., Etoh, T., Harada, J.J. and Lincoln, S.E., *Genome* **1991**, *34*, 543–552.
68 Marshall, G., *Agro-Food Industry Hi-Tech*, May/June **1992**.
69 Friedt, W., Nichterlein, K. and Nickel, M., in: *Flax: Breeding and Utilisation*: Marshall, G. (ed.) Kluwer Academic Press for the EC, **1989**.
70 Kochert, G.D., Halward, T.M., Branch, W.D. and Simpson, C.E., *Theor. and Appl. Genet*. **1991**, *81*, 661–667.
71 Halward, T.M., Stalker, H.T., LaRue, E.A. and Kochert, G., *Genome* **1991**, *34*, 1013–1020.

4 Biochemistry of Oil Synthesis

J. L. Harwood and R. A. Page

4.1 Use of photosynthate for oil synthesis

The ultimate source of carbon for fatty acid and lipid synthesis in plants is photosynthesis. Fixation of carbon by ribulose *bis*phosphate carboxylase/oxygenase allows the formation of hexose sugars. For most oil-accumulating tissues, transported sugars, such as sucrose or mannose, provide the carbon necessary for lipid formation in the seed. At early stages in seed maturation, the seed-containing structure (e.g. pod or silique) may have significant numbers of chloroplasts although the extent to which these can contribute to overall lipid synthesis in the seeds is not known. However, in several important oil crops, (avocado, olive, palm), the major triacylglycerol stores accumulate in a fruit which itself has significant photosynthetic capacity. The nature of photosynthesis in fruits is somewhat different from that of the leaves of either C_4 or C_3 plants [1]. Again, we do not know what proportion of total lipid synthesis comes from fruit-derived photosynthate although, since estimates of fatty acid and acyl lipid synthesis *in vitro* show that olive fruits are very active, the contribution by the fruit itself is likely to be very significant. Interestingly, because some of the characteristics of fatty acid (and complex lipid) [2] formation in the outer epicarp are different from those of the mesocarp (which is non-photosynthetic) then the balance of carbon used from fruit photosynthesis *versus* that which comes from leaf photosynthesis may well influence the quality of the final oil [3]. In the case of avocado, the fruit is also very active in photosynthesis and lipid synthesis. Indeed, avocado has been used extensively for the isolation and study of many lipid biosynthetic enzymes, as will be seen later.

Radiolabelled acetate is frequently used as a precursor for lipid synthesis. Its effectiveness relates both to the ready diffusion of free acetate into cells and organelles [4], and that it is metabolically inert in most situations but can be activated by acetyl-CoA synthetase, which has high activity in chloroplasts [5] where fatty acid synthesis is concentrated. In seeds, this enzyme is found in the cytosol as well as the proplastid [6]. Significant amounts of acetate have been detected in plant tissues (e.g. 1mM in mature spinach leaves [5]) and the role of acetate has been discussed by Givan [7]. This acetate is derived from the hydrolysis of acetyl-CoA by a hydrolase [8,9] and acetyl-CoA itself is considered generally to be produced by the pyruvate decarboxylase/dehydrogenase complex.

In developing seeds and other non-photosynthetic tissues, the proplastids normally contain a complete glycolytic pathway for yielding pyruvate from glucose as well as an active pyruvate dehydrogenase [6]. In some cases, however, the activities of certain glycolytic enzymes, such as enolase and phosphoglycerate mutase, have been rather low in purified plastid preparations from such tissues [10,11,12,13]. More recent reports indicate that developing embryos of rapeseed contain all of the glycolytic enzymes and that these have activities well in excess of those required to support known rates of acetyl-CoA generation for storage oil biosynthesis [14].

In the chloroplasts from some photosynthetic tissues, there may be problems with this mechanism of acetyl-CoA generation because either phosphoglyceromutase [15,16] or pyruvate dehydrogenase activity is low. In spinach leaves it has been proposed that chloroplast acetyl-CoA originates through a collaboration of organelles [17]. Firstly, in the mitochondrion, pyruvate is converted to acetyl-CoA which is then hydrolysed to free acetate. Diffusion of acetate from the cytosol into the chloroplast then takes place. Finally, acetate is converted to acetyl-CoA by chloroplastic acetyl-CoA synthetase [9]. Alternatively, it has been suggested that, at least in the cotyledons of germinated pea embryos, carnitine may act as a shuttle in the transfer of acetate groups from mitochondria to chloroplasts [18], although more recent investigations failed to find such a mechanism in the leaves of pea or spinach [19]. So far as the photosynthetic fruits (avocado, etc) are concerned, we do not know whether they follow the spinach leaf route or can convert hexoses into acetyl-CoA entirely within their chloroplasts. This problem deserves attention.

Another method of forming acetyl-CoA via an ATP and CoA-dependent citrate lyase which is used extensively in animals [20] does not appear to be important in most plants [21]. For a full discussion of the source of acetyl-CoA for plastid fatty acid synthesis see Liedvogel [22], Harwood [23] and Murphy *et al* [14].

4.2 *De novo* synthesis of fatty acids

De novo synthesis of fatty acids is carried out in the proplastids of developing seeds by two enzyme complexes, namely acetyl-CoA carboxylase and fatty acid synthetase. Over the last decade, our knowledge on the individual enzyme steps involved in *de novo* fatty acid synthesis has increased due to intensive research leading to the purification of the proteins involved. Fatty acid synthesis involves the sequential addition of 2 carbon units derived from

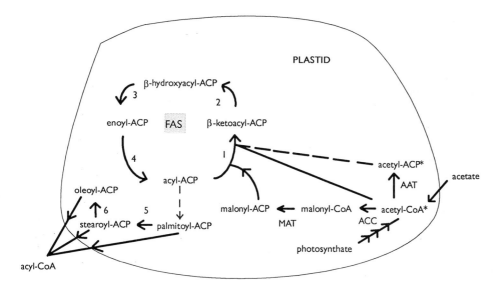

Figure 4.1. Schematic diagram of the reactions involved in *de novo* fatty acid synthesis.
AAT = acetyl-CoA:ACP transacylase; MAT = malonyl-CoA;ACP transacylase; ACC = acetyl-CoA carboxylase. The fate of acetate and photosynthate in fatty acid synthesis is described in section 4.1. *Acetyl-ACP and acetyl-CoA represent the 2 possible substrates for the initial condensation reaction (see section 4.2.3). Condensation of acetyl-CoA with malonyl-ACP is catalysed by the short chain condensing enzyme. Plant fatty acid synthetase (FAS) consists of the cyclic process of a condensation reaction (1) catalysed by ß-ketoacyl-ACP synthase, a reduction reaction (2) catalysed by ß-hydroxyacyl-ACP reductase, a dehydration reaction (3) catalysed by ß-hydroxy-ACP dehydratase and finally another reduction reaction (4) catalysed by enoyl-ACP reductase. The final product after 7 cycles is palmitoyl-ACP which is then further elongated to stearoyl-ACP by ß-ketoacyl-ACP synthase II (5) in conjunction with enzymes (2)-(4). Desaturation of stearoyl-ACP to oleoyl-ACP by a Δ_9 desaturase (6) also occurs in the plastid stroma. Further modification of palmitoyl-, stearoyl-, and oleoyl-ACPs is described in section 4.3.

malonyl-CoA (product of acetyl-CoA carboxylase reaction) onto a growing acyl chain, as depicted in Figure 4.1.

4.2.1 Acetyl-CoA carboxylase

The carboxylation of acetyl-CoA to form malonyl-CoA is an ATP- dependent reaction catalysed by acetyl-CoA carboxylase. This reaction is the first committed step of *de novo* fatty acid synthesis providing the essential substrate for fatty acid synthetase.

Plant acetyl-CoA carboxylase is a Type I biotin-containing enzyme [24] and, like animal and bacterial systems, catalyses a two step reaction:

$$ATP+CO_2+BCCP \xrightarrow{\text{Biotin carboxylase}} CO_2\text{-}BCCP+ADP+Pi$$

$$CO_2\text{-}BCCP+acetyl\text{-}CoA \xrightarrow{\text{Carboxyltransferase}} BCCP+Malonyl\text{-}CoA$$

In bacteria, the enzyme readily dissociates into its constituent proteins, biotin carboxylase, biotin carboxyl-carrier protein (BCCP) and carboxyltransferase, all of which have been purified to homogeneity [25]. In contrast in yeast and animals, acetyl-CoA carboxylase exists as a multifunctional protein (240kDa) which consists of three domains (biotin carboxylase, BCCP and carboxyltransferase) present on one single polypeptide chain [26,27]. The purification studies outlined below have shown that the major plant acetyl-CoA carboxylase is also a high molecular weight multifunctional protein.

Acetyl-CoA carboxylase has been purified from a number of plant sources: barley [28], wheat germ [29,30], parsley [30], castor bean [10], spinach [31], maize [32], oilseed rape [33,34], soybean seed [35] and pea [36]. However, the polypeptide sequences from these sources appeared to vary considerably. It is now thought that acetyl-CoA carboxylase is sensitive to proteolytic attack leading to isolation of the degradation products and that this accounts for the different products yielded in the above work [37]. Another possible reason for the different sized polypeptides observed in the purification of acetyl-CoA carboxylase is the presence of other biotin carboxylases [38,39]. Egin-Buhler and Ebel, [40] found that introduction of a proteinase inhibitor, PMSF (phenylmethylsulphonyl fluoride) and the use of monovalent avidin-sepharose to speed their purification procedure, led to the isolation of a high molecular mass form of acetyl-CoA carboxylase from parsley and wheat germ. Using similar methods, Slabas and Hellyer [33] and Charles and Cherry [35] managed to purify acetyl-CoA carboxylase from oilseed rape and soybean seed respectively. In both cases a major polypeptide of 220–240kDa (as seen on polyacrylamide gel electrophoresis (PAGE) under denaturing conditions) was isolated. The kinetic constants for oilseed rape [33] and soybean [35] were determined and were similar to those evaluated for the parsley [40], castor bean [41] and maize enzymes [32].

To date, little is known about the role plant acetyl-CoA carboxylase plays in control of fatty acid metabolism, although it is usually assumed that it is rate-limiting [23,37,42]. Changes in the acetyl-CoA carboxylase activity correlate well with lipid accumulation in maturing seeds of castor bean [10], rape [43] and *Cuphea* [44], indicating it may have an important regulatory role in seed lipid biosynthesis, but no experiments have been performed on determining the flux control co-efficients. Studies in avocado [45], castorbean [41] and soybean [35], show that acetyl-CoA carboxylase is not activated by citrate as is observed with animal carboxylases [46]. Also, there is no evidence to suggest that plant acetyl-CoA carboxylases are regulated by phosphorylation/dephosphorylation as are animal acetyl-CoA carboxylases [47]. However, it is interesting to note that malonyl-CoA, the product of acetyl-CoA carboxylase and palmitoyl-CoA, a major product of fatty acid synthetase, inhibits plant acetyl-CoA carboxylase [48,49]. This suggests that acetyl-CoA carboxylase may be an important factor in controlling the rate of fatty acid synthesis.

In leaves, acetyl-CoA carboxylase is localised in the chloroplasts [50,45,51,52] and its activity is stimulated by light. It has been suggested that light-dependent energy changes could modulate acetyl-CoA carboxylase activity [34]. It has been shown in wheat germ [53], maize [32], spinach [54,55] and soybean seeds [35] that acetyl-CoA carboxylase activity in light is high due to an increase in the pH, increased concentrations of ATP, Mg^{2+} and decreased levels of ADP induced by illumination. Recently, Jaworski *et al* [56] have provided the first *in vivo* evidence that acetyl-CoA carboxylase controls the rate of fatty acid synthesis during the normal light/dark cycle of the plant. Their studies involved the identification (using urea PAGE) and *in vivo* measurements of acetyl-ACP pool sizes [57]. It was shown that levels of acetyl-ACP were higher in the leaves from plants in the dark than from those in the light. Since there were no corresponding increases in the malonyl-ACP/CoA levels or any other acyl-ACP levels, this suggested that acetyl-CoA carboxylase provides light/dark control over the flux of *de novo* fatty acid synthesis. This work has been further supported by *in vitro* studies using isolated chloroplasts of spinach and pea [57].

Jarworski *et al* [58] then investigated the *in vivo* levels of acyl-ACPs in the developing seeds of spinach and castorbean. They observed a different pattern of the seed acyl-ACPs to that seen in leaf tissue. Firstly, the malonyl-ACP levels were similar to, or greater than acetyl-ACP, whereas in leaf, acetyl-ACP was much higher than malonyl-ACP. Secondly, the long chain acyl-ACPs and malonyl-ACP represented the major acyl-ACP pools. The high levels of malonyl-ACP suggested that regulation of *de novo* fatty acid synthesis in seeds could be different to that in leaf. The fact that, during seed maturation, levels of acetyl-CoA carboxylase decrease rapidly once maximal lipid accumulation occurs [43] while the levels of all other lipid metabolising enzymes remain high [59], also supports the idea that acetyl-CoA carboxylase regulation in seeds is different to that in leaves.

4.2.2 Acyl carrier protein

Acyl carrier protein (ACP) is a small multifunctional protein that plays a key role in fatty acid and lipid metabolism. Fatty acyl moieties are linked to a phosphopantetheine prosthetic group of ACP through a thioester linkage to form acyl-ACP which takes part in transfer, reduction, dehydration and elongation reactions.

The properties of ACP (a small molecule, highly acidic, and relatively stable enzymically) have aided the purification of this essential co-factor of fatty acid synthesis. Indeed, ACP was the first protein of the plant fatty acid synthetase complex to be purified to homogeneity. Avocado mesocarp and spinach leaf were the first plant tissues from which ACP was purified [60]. Acyl carrier protein has been isolated from the seed tissues of soybean [61], castor oil [62] and oilseed rape [63,59]. The latter group had difficulty in purifying rapeseed ACP from developing seeds compared to that from leaves. They found that the seed ACP was less stable and not soluble. In the end, by radiolabelling the ACP using an *E.coli* synthetase [64], they were able to purify rapeseed ACP to apparent homogeneity.

The presence of isoforms of ACP was first suggested in 1968 by Matsumara and Stumpf [65], based on their work in spinach. Further studies in castorbean, spinach and soybean [62], oilseed rape [63] and barley [64] have clarified the existence of different forms of ACP. The two castorbean ACP isoforms appear to have significant structural differences due to their different electrophoretic mobilities. It has also been noted that the isoforms of ACP are expressed differently in the leaves and seeds [62]. Ohlrogge and Kuo [62] observed that ACP I is the major isoform in spinach leaves whereas in developing spinach seeds, ACP II was the predominant species. Similar results were found with soybean leaves and developing cotyledons and castorbean leaves and endosperm. Whether these isoforms of ACP relate to different physiological functions is not yet known [23].

Comparison of the amino acid sequences of the ACPs from the various plant sources [23,65], show several regions of conservation. One major domain is that around the phosphopantetheinylated serine where formation of the thioester bond with the fatty acid moiety occurs. However, differences in the amino acid composition and N-terminal sequences of the ACP isoforms suggest that the different forms of ACP may be encoded by a multigene family. Indeed, de Silva *et al* [67] have shown that oilseed rape alone has approximately 35 seed-expressed ACP genes [68]. The results of Ohlrogge *et al* [69] on subcellular localisation of spinach ACP as well as those of Dorne *et al* [70] using a plastid ribosome-deficient barley mutant led to the suggestion that ACP was a nuclear-encoded protein which was synthesised as a precursor polypeptide containing a transit peptide. This has been confirmed in oilseed rape where the isolation and sequencing of several ACP cDNA clones has demonstrated a 51 amino acid N-terminal extension [71,67].

In *E.coli*, during biosynthesis of ACP, the prosthetic group, phosphopantetheine, is transferred from CoA to apo-ACP [72]. A holo-ACP synthase has been detected in the cytosol of spinach leaves and castorbean oilseeds [74]. The presence of this synthase in the cytosol suggests that the precursor protein is taken up by plastids and processed proteolytically to the mature holo-ACP [73]. However, work on over-expression of ACP in transgenic plants [68] has shown the presence of both holo-ACP and apo-ACP in the chloroplasts of transgenic plants. This suggests that the attachment of the prosthetic group is not necessary for the uptake of ACP into chloroplasts.

Finally, the rate of fatty acid synthesis *in vivo* correlates well with the levels of ACP in developing seeds of soybean [61] and oilseed rape [63]. Acyl carrier protein appears just before the onset of storage lipid biosynthesis. This suggests that ACP may be encoded for by genes subject to both temporal and tissue-specific regulation. It also suggests that the levels of the proteins involved in the fatty acid synthetase complex may be important in regulating the rate of lipid synthesis.

4.2.3 Fatty acid synthetase

The fatty acid synthetase (FAS) in plants is a Type II dissociable enzyme complex, i.e., it consists of individual proteins that can be isolated in an active enzyme form. The ACP polypeptide is readily dissociated from the synthetase. Plant fatty acid synthetase contains proteins catalysing the following reactions:

acetyl-CoA:ACP transacylase, malonyl-CoA:ACP transacylase, ß-ketoacyl-ACP synthetase, ß-ketoacyl-ACP reductase, ß-hydroxyacyl-ACP dehydratase, and enoyl-ACP reductase. The reactions catalysed by these enzymes are shown in Figure 4.1. It is only in the last decade that the isolation of the individual components of plant FAS has occurred. Recently, emphasis has shifted to isolation of the genes coding for the different enzymes in the hope of being able to manipulate the levels of fatty acid biosynthesis genetically and hence influence the oilseed metabolism [68,74,65].

Acetyl-ACP, which is the product of acetyl-CoA:ACP transacylase has long been considered to be the primer for fatty acid synthetase. The acetyl transacylase enzyme has been partially purified from barley chloroplasts [75], spinach [76], *Brassica* [77] and safflower seeds [78]. Originally, there was evidence that this enzyme might have been rate-limiting for the overall synthesis of fatty acids in plants [76]. However, Jackowski and Rock [79] have discovered the presence of a specific condensing enzyme in *E. coli* that is responsible for the formation of short chain acyl-ACPs. This enzyme, now denoted as short chain condensing enzyme or ß-ketoacyl-ACP synthetase III (KAS III), is unique with respect to the other condensing enzymes of *E. coli* fatty acid synthase in that it is cerulenin-insensitive. The enzyme catalyses two extra reactions: malonyl-ACP decarboxylation and acetyl-CoA:ACP transacylation and is the "slowest" step in fatty acid synthesis; hence refuting the idea that acetyl transacylase is important in control of fatty acid metabolism. The presence of KAS III has also been demonstrated in spinach leaves [80] and a number of other plant sources [81]. Another special property of the short-chain condensing enzyme is that it can use acetyl-CoA directly as a substrate [80]. Therefore, the importance of acetyl-ACP as a substrate and of the acetyl-CoA:ACP transacylase reaction in fatty acid synthesis is now in question. Further evidence to support a minor role for acetyl-ACP in fatty acid synthesis comes from recent studies of Jaworski *et al.* [58]. They showed that acetyl-ACP was the poorest primer for fatty acid synthesis when compared to acetyl-CoA, butyryl-ACP and hexanoyl-ACP. Pulse-chase experiments also showed that acetyl-ACP lost its label very slowly whereas the label was chased rapidly, and at the same rate, from acetyl-CoA and butyryl-ACP pools. All of the above evidence now suggests that KAS III is rate-limiting for the overall reaction of fatty acid synthetase.

Two other isoforms of ß-ketoacyl-ACP synthase (KAS) have been identified. The first isoform, KAS I, is responsible for the synthesis of fatty acids up to 16 carbons in length and is highly sensitive towards cerulenin but insensitive towards arsenite. The other isoform, KAS II catalyses the formation of stearoyl-ACP from palmitoyl-ACP, is moderately sensitive towards cerulenin and strongly inhibited by arsenite [23]. ß-Ketoacyl-ACP synthetase I has been purified to homogeneity in oilseed rape [82] and soybean seeds [83] and partially purified from barley seedlings [75], parsley cell suspensions [84] and spinach leaves [85]. Purification of KAS II from spinach [75] and oilseed rape [82,86] has been reported. Considerable work has been carried out on barley KAS I which is composed of α and β subunits [87]. There is 35% sequence homology between the primary amino acid sequence of the β subunit and *fabB*-encoded β-ketoacyl-ACP synthase from *E. coli* indicating that these proteins have a common evolutionary origin [88]. Oligonucleotide probes based on the N-terminus of soybean KAS I have been used to screen soybean cDNA libraries for the KAS I gene [83]. The same approach has been taken in oilseed rape for isolation of the full-length cDNA coding for KAS II [86].

The end product of fatty acid synthetase is normally palmitoyl-(C_{16})-ACP which can then be elongated further to stearoyl-(C_{18})-ACP by KAS II. The storage lipids in oil crops show a considerable diversity in the chain length of their fatty acid residues. For example, the relative proportions of C_{16}:C_{18} acids can vary from 45:42 (palm) through to 4:82 (rape) and 11:82 (soya), indicating that KAS II may have an important role in regulating the ratio of C_{16}:C_{18} products in *de novo* synthesis. Interest in the condensing enzymes has centered on two aspects. Firstly, KAS III is now thought upon as the enzyme that catalyses the priming reaction of plant fatty acid synthase hence it offers a potential site for metabolic control. Secondly, the activity of KAS II can control the proportion of C_{16}:C_{18} products of *de novo* synthesis and, therefore, may have a key role in regulating seed oil quality. Isolation of the genes encoding these two enzymes would potentially enable the manipulation of seed oil quality by genetic engineering as discussed in Chapter 6.

Malonyl-CoA:ACP transacylase has been purified from a number of plant tissues [89]. Interestingly, tissue-specific isoforms of the enzyme have been detected in soybean [90], although these isoforms have not as yet been ascribed any functional significance [65].

Little work has been done on the β-hydroxyacyl-ACP dehydratase enzyme. The enzyme has been purified to homogeneity from spinach leaves [91] and partly purified from developing safflower seeds [78]. For further details see reviews [23] and [89].

β-Ketoacyl-ACP reductase catalyses the first reductive step in fatty acid biosynthesis. This enzyme has been purified to homogeneity from spinach [91] avocado [92] and oilseed rape [93]. All of these enzymes are NADPH-specific. Amino acid sequence data have been obtained for the NADPH-specific reductase for avocado [94] and oilseed rape [95]. A cDNA clone has been isolated for the rapeseed enzyme and used to clone a full-length cDNA from *Arabidopsis* [65]. There is a high degree of nucleotide sequence homology between the rape and *Arabidopsis* cDNAs [65]. Topfer *et al* [96] have isolated cDNA clones from a *Cuphea lanceolata* cDNA library using a 340bp cloned PCR fragment that has an amino acid homology to the β-ketoacyl-ACP reductase of avocado. A fusion protein has been constructed and isolated from *E. coli*. This fusion protein shows high substrate specificity towards NADPH.

Enoyl-ACP reductase catalyses the second reductase step in fatty acid synthesis. Both NADH- and NADPH-specific enoyl-ACP reductases have been detected in safflower, castor bean and rape seeds [97]. An NADH-specific enzyme has been purified from avocado [98], spinach [91] and rape [99]. A partial amino acid sequence of the NADH-linked-enoyl-ACP reductase from rapeseed has been obtained [95]. The sequence of a cDNA from a rapeseed embryo library shows that a putative transit peptide of 73 amino acids is present [100]. It is interesting to note that in rapeseed, this enzyme is induced immediately prior to lipid deposition which is comparable to what happens to ACP. Hence, the NADH-specific enoyl-ACP reductase could very well be a seed-specific and temporally-regulated gene. Another interesting point is that the formation of stearate in rapeseed requires an NADPH-specific enoyl-ACP reductase since the NADH-specific enoyl-ACP reductase has no activity towards a C_{18} substrate [97].

Advances in purification of the individual components of fatty acid synthesis and isolation of some of the genes encoding for these individual proteins, especially in oilseed rape have now occurred. However, we are still no closer to stating how fatty acid synthesis *de novo* is regulated. Several groups are now concentrating on examining the regulation of gene expression of the FAS complex proteins, with the long-term goal being the use of genetic engineering to manipulate oilseed lipid composition. Moreover, our knowledge of the overall control of biochemical pathways in lipid metabolism is still minimal. With the ability to identify and measure the acyl-ACP pool sizes [101] a different approach to examining regulation of lipid metabolism could be taken. Determination of the elasticity coefficients and flux control coefficients [102,103] of the individual enzymes would tell us which enzymes are important in control of lipid biosynthesis in oilseeds and may suggest alternative methods for their manipulation.

4.3 Modification of the products of fatty acid synthetase

4.3.1 Short-chain fatty acids

As discussed above, for most plant species, stearate (about 75%) and palmitate (about 25%) are the main products of fatty acid synthase. However, a few actual or potentially important oilseed crops - such as palm kernel, coconut and *Cuphea* - produce medium chain fatty acids of 8–14 carbons in length. Recently, it has been demonstrated that this premature termination was due to the activity of a specific medium-chain thioesterase. The enzyme was first identified by Pollard *et al.* [104] as responsible for medium chain fatty acid production in seeds of the California Bay tree. The enzyme and its gene have now been purified and characterised [105]. The same mechanism for medium-chain acid production appears to operate in *Cuphea* [106] and may well do so in palm and coconut.

4.3.2 Fatty acid desaturases

The most abundant fatty acids in many of the major edible oilseed crop species are unsaturated C_{18} acids, such as oleate, linoleate and α-linolenate (see Chapter 1). These unsaturated fatty acids are produced by aerobic desaturation. Important desaturases (and hydroxylases) for oil production are shown in Figure 4.2. The first desaturase to act on C_{18} fatty acids is a Δ_9 enzyme which utilises stearoyl-(or palmitoyl-) acyl carrier protein (ACP). This soluble stromal enzyme was first purified from safflower seeds by McKeon and Stumpf [107]. It was shown to be highly specific for stearoyl-ACP (stearoyl-CoA substrate having only about 5% of the activity). Moreover, palmitoyl-ACP was also a poor substrate, which probably provides the explanation for the much higher amounts of palmitate rather than palmitoleate in plant tissues. It also ensures that palmitate and oleate are the main products of *de novo* fatty acid synthesis in most plants.

Stearoyl-ACP desaturase has now been isolated from several sources [108,83,109] and cDNAs corresponding to the genes from castor bean, cucumber, safflower, rapeseed and soybean have been cloned [37]. Active desaturase protein was produced upon expression of the safflower cDNA in *E.coli* [83] and of the castor bean cDNA in yeast [109]. There seems to be

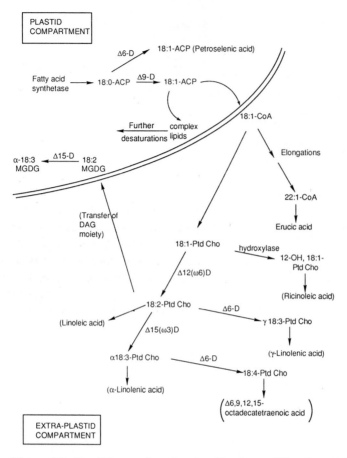

Figure 4.2. Possible reactions involved in the modification of the products of fatty acid synthetase.
Abbreviations: PtdCho = phosphatidylcholine; MGDG = monogalactosyldiacylglycerol; DAG = diacylglycerol; D = desaturase (with the position of hydrogen removal indicated).

considerable sequence conservation between stearoyl-ACP desaturases from different plants but not with the stearoyl-CoA desaturases of vertebrates or fungi, nor with the cyanobacterial Δ_{12} desaturase [37]. This is not surprising since the plant stearate desaturases are soluble and use an ACP substrate, while the animal and fungal enzymes are membrane-bound and use a CoA substrate.

The major substrate for the conversion of oleate to linoleate is the complex lipid, phosphatidylcholine [79,23]. However, as pointed out by Harwood [23], this is not an exclusive situation and other lipid substrates are certainly used particularly in connection with the generation of polyunsaturated species of chloroplast lipids. Nevertheless, in oilseeds, linoleate (and α-linolenate) accumulating in the storage tissue is almost exclusively derived from an extra-chloroplastic Δ_{12} desaturase using a phosphatidylcholine substrate.

Several processes are involved in the transfer of oleate from oleoyl-ACP in the plastid to the glycerol backbone of phosphatidylcholine (Figure 4.3). First, acyl-ACP thioesterase in the plastid liberates free (non-esterified) oleic acid which is re-esterified to coenzyme A on the plastid envelope [110]. The oleoyl-CoA then becomes part of the extra-plastidic acyl-CoA pool from which acyl groups can be drawn for esterification or acyl-exchange reactions. The exchange of fatty acids between acyl-CoA and phosphatidylcholine is reviewed by Stymne and Stobart [111]. There are two potential mechanisms. First, incorporation could proceed via acyl-CoA:lysophosphatidylcholine acyltransferase, the latter substrate being generated by phospholipase A_2 activity. However, seed phospholipase A_2 has poor activity in most tissues, particularly toward "normal" fatty acids. (The role of a specific castor bean phospholipase A_2 in the synthesis of triricinolein is discussed below). Moreover, radiolabelling experiments in developing soybean showed transfer of oleate at rates greater

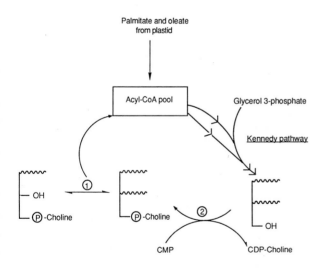

Figure 4.3. Entry of acyl products into diacylglycerol.
The *sn*-1 and *sn*-2 positions of diacylglycerol are esterified successively in the Kennedy pathway. Reaction 1 (lysophosphatidylcholine: acyl-CoA acyltransferase) allows the *sn*-2 position to be acylated in the forward direction or, alternatively, the desaturation product to be removed from phosphatidylcholine. The freely reversible reaction 2 (cholinephosphotransferase) allows equilibration of the glycerol backbone of phosphatidylcholine with the diacylglycerol pool.

than could be accounted for by the pre-existing pool of lysophosphatidylcholine [112]. The second mechanism is a direct acyl exchange and this was confirmed in safflower [113], sunflower [114] and linseed [115]. The mechanism of acyl exchange is discussed by Stymne and Stobart [111].

Evidence that phosphatidylcholine (or other complex lipids) are important substrates for fatty acid desaturation came originally from the results of radiolabelling experiments in leaves and the alga *Chlorella*, by Nichols, James and co-workers [116,117]. A similar mechanism was found by Pugh and Kates [118] to operate in the fat-accumulating yeast, *Candida lipolytica*. In plants, Stymne and Appelqvist [119] concluded that desaturation of oleate in maturing safflower used a phosphatidylcholine substrate and this was confirmed by Stymne and Glad [112]. More recent work has identified cytochrome b_5 and NADH:cytochrome b_5 reductase as required components [120]. The desaturase enzyme is cyanide-sensitive. The oleate desaturase from spinach chloroplast envelopes has been solubilised with Triton X-100 and purified more than 100-fold by three simple chromatographic steps. Purification was very much helped by the fact that the enzyme could be conveniently assayed with non-esterified fatty acid (oleic acid) substrates. Examination of the substrate specificity showed that the chloroplast enzyme acted as an ω_6-desaturase rather than a Δ_{12} desaturase. It was also tentatively identified as a 40kDa protein [121].

As mentioned above, phosphatidylcholine appears to be the major substrate for oleate desaturation to linoleate, although in safflower preparations, phosphatidylethanolamine may also function thus [122]. In photosynthetic tissues, isolated chloroplasts have significant capacity for oleate desaturation which may take place on phosphatidylglycerol or monogalactosyldiacylglycerol [79]. Certainly in "16:3-plants" (those accumulating hexadecatrienoate at the *sn*-2 position of their monogalactosyldiacylglycerol) phosphatidylcholine does not appear to be involved in chloroplastic linoleate formation [123,124]. However, the situation in most oil seeds or fruits remains uninvestigated and the best one can say is that, although phosphatidylcholine is definitely the major substrate, it would be premature to assume that it is the **only** substrate involved in linoleate production for triacylglycerol synthesis.

Some seeds accumulate high levels of α-linolenate (e.g. linseed) and there is indirect evidence that phosphatidylcholine can also serve as a substrate for the Δ_{15} desaturase that converts linoleate to α- linolenate [125]. However, because the Δ_{15} desaturase is so unstable, direct *in situ* desaturation has not been demonstrated [126]. This contrasts significantly with leaf tissues where monogalactosyldiacylglycerol has been shown clearly to be the most important substrate for this desaturase [23].

There has been some interest recently in Δ_6 desaturase activity. In the first place, Δ_6 desaturation is involved in the synthesis of γ-linolenate (18:3, $\Delta_{6,9,12}$)

which is important nutritionally and therapeutically [127]. The formation of γ-linolenate on phosphatidylcholine has been shown conclusively *in vitro* [128]. Furthermore, the Δ_6 desaturase is stereospecific for linoleate at the *sn*-1 position [129]. More highly unsaturated acids such as octadecatetraenoic acid (18:4, $\Delta_{6,9,12,15}$) are also found in some leaf tissues and seeds [130] and again are present in the phospholipid fraction. It has been found that octadecatetraenoic acid is synthesised, not from the γ-isomer, but the α-isomer of linolenic acid [131]. The Δ_6 desaturase is the final desaturase to act on the acyl chain and, hence, it is likely that this takes place at the *sn*-2 position of phosphatidyl-choline.

A second place where Δ_6 desaturase activity is important is in the generation of petroselinic acid (18:1, Δ_6). Petroselinic acid is of industrial interest (see Chapters 2 & 6) because cleavage of the acid gives rise to lauric and adipic acids which can be used for detergent and nylon manufacture, respectively. Recently, petroselinic acid synthesis has been examined in seed extracts from *Umbelliferae* species. The desaturase appears to use stearoyl-ACP substrate and, possibly, has a molecular mass of 36kDa [132,133]. The cloning of the Δ_6 desaturase genes and their expression in transgenic rapeseed plants is now underway in laboratories in the UK and USA.

4.3.3 Fatty acid hydroxylases

As an alternative to desaturation, the straight-chain fatty acids can be derivatised with various functional groups [134]. There has been some work on the generation of ricinoleic acid (12-hydroxy-9-octadecenoic acid) which makes up in excess of 90% of the total fatty acids of castor oil [135]. Triricinolein has a host of valuable uses in the chemical and pharmaceutical industries (see Chapter 2). *In vitro* studies with microsomes from castorbean seeds [136] showed that the reaction involved a mixed-function oxygenase system, which required O_2 and NADH. The substrate appeared to be oleoyl-phosphatidyl-choline [137,138,139] and recent work has shown that the ricinoleate product is released from the phospholipid by a phospholipase A_2 which is highly selective for oxygenated fatty acids [140]. The hydroxylase appears to use cytochrome b_5 as an electron donor [42].

4.3.4 Fatty acid elongases

The important fatty acid elongase enzymes in plants appear to be concentrated in microsomal membranes and to use NADPH and malonyl-CoA as substrates for two-carbon addition [23]. Three systems have been used for experiments on the elongation of saturated fatty acids, namely pea seeds, aged potato slices and leek epidermal cells. None of these is directly relevant to oil seeds although the general characteristics of the reactions, intermediates and subcellular location may be applicable. Summarizing, one can say that there seem to be chain length-specific elongases and that these can be located differentially in the endoplasmic reticulum or Golgi membranes [23,141].

So far as oilseeds are concerned, interest in fatty acid elongation has centred on tissues where monounsaturated acids accumulate. Older varieties of oilseed rape (*Brassica napus*), turnip rape (*Brassica campestris*) and related brassicas accumulated large amounts of erucate (22:1, Δ_{13}). Because of concern with the results of animal feeding experiments, where fatty lesions were observed in heart muscle of mice with high levels of dietary erucate, efforts were made to breed new rapeseed varieties with low erucate (see Chapters 1 and 3). This was largely successful and most modern rapeseed varieties are of the so-called zero erucate or "canola" type [142]. Ironically, erucic acid, which is a desirable feedstock for the chemical industry, is now in short supply and alternative cruciferous crops, such as *Crambe*, or new high erucic rapeseed varieties are being developed (see Chapters 6 and 9). Erucic acid is synthesised by elongation of oleoyl-CoA via eicosenoate [143] and this pathway is the same in other plants that accumulate erucate [144,145,146]. High-erucate varieties of rapeseed possess readily detectable oleoyl-CoA elongase whereas low-erucate cultivars do not [146]. One and two gene loci, respectively, are involved in controlling the level of erucate in turnip rape and oilseed rape [148,149,150].

Seeds of many other cruciferous plants are a rich source of monounsaturated fatty acids and the elongation systems have been studied in mustard (*Sinapsis alba*), honesty (*Lunaria annua*) and in the non-crucifer, nasturtium (*Tropaeolum majus*) [151,152,153,154]. Reaction intermediates corresponding to keto-, hydroxy- and 2-*trans*-unsaturated derivatives have been identified. Proteins catalysing the partial elongation reactions of ß-ketoacyl-CoA synthetase and ß-ketoacyl-CoA reductase were partially purified from solubilised microsomal preparations [154]. Activity of the condensing enzyme was strongly inhibited by iodoacetamide but not by cerulenin [155] in contrast to the condensing enzyme for saturated fatty acids in leek which was sensitive to cerulenin [156]. Interestingly, in developing seeds of *Limnanthes alba*, which accumulates two major very long chain unsaturated acids (Δ_5 eicosenoate and $\Delta_{5,13}$ docosadienoate), there seemed to be separate elongation systems for saturated and monounsaturated fatty acids, respectively [157].

4.4 Acyl lipid formation

A complete discussion of polar lipid synthesis in oil crops is beyond the scope of this chapter. A summary of the pathways involved in the metabolism of the major polar phosphoglyceride and glycosylglyceride lipids are shown in Figures 4.4 and 4.5, respectively. For further details the Fig.4.5 the reader is referred to chapters in Stumpf and Conn [158], to reviews by Harwood [159] and Harwood and Griffiths [126] and to the general text of Gurr and Harwood [20]. Since the ability of an oilseed to accumulate reserves depends on the biogenesis and maintenance of its membrane systems, the metabolism of the above membrane lipids is very important. In addition, as discussed above, the major extra-chloroplastic membrane lipid, phosphatidylcholine, is intimately involved in controlling the quality of the triacylglycerol stores by serving as a substrate for Δ_{12} and Δ_{15} desaturases, for the Δ_{12} hydroxylase and probably for the Δ_6 linoleate desaturase.

4.4.1 Timing and mechanism of oil accumulation

Oil deposits in seeds accumulate as spherical, electron-dense bodies, typically one to a few microns in diameter [160]. Oil deposition occurs at markedly different rates in different tissues. For example, in safflower it takes about 7 days whereas in olive the maturation period is around 7 months. The development of an oil-bearing seed typically follows three phases [161,162]. The first phase involves rapid cell division during which little triacylglycerol is synthesised. That which is made has a fatty acyl composition which tends to resemble that of membrane phospholipids and glycolipids rather than the final storage oil. Thus, the triacylglycerol in very young embryos will tend to be high in linoleate and α-linolenate irrespective of plant species [126]. During the second phase, when triacylglycerol synthesis is at its maximum, those fatty acids characteristic of the final oil are made. Thus, the formation of erucate in *Crambe* [160] and of stearate in cocoa [163] is seen. The third phase is a desiccation period where there is little further synthesis of seed oil [162].

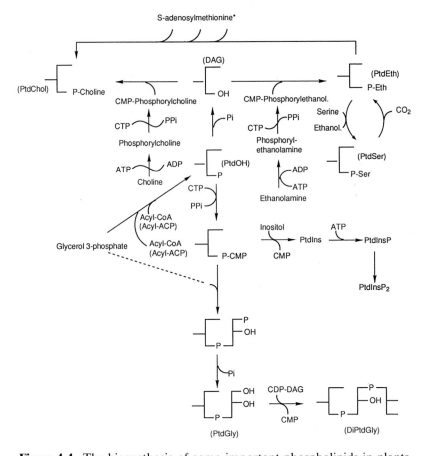

Figure 4.4. The biosynthesis of some important phospholipids in plants.
Abbreviations: PtdCho = phosphatidylcholine; PtdEth = phosphatidylethanolamine;
PtdSer = phosphatidylserine; PtdGly = phosphatidylglycerol; DiPtdGly = diphos-
phatidylglycerol [cardiolipin]; DAG = diacylglycerol; PtdIns = phosphatidylinositol;
PtdInsP = phosphatidylinositol 4-phosphate; PtdInsP$_2$ = phosphatidylinositol 4,5-
*bis*phosphate; PtdOH = phosphatidic acid. *In some plants S-adenosylmethionine is
used to methylate ethanolamine while it is part of water-soluble compounds such as
phosphorylethanolamine or CDP-ethanolamine and not to convert PtdEth directly to
PtdCho.

Many oilseeds synthesise triacylglycerols with unusual fatty acids that are
never found to be esterified to membrane lipids. In such oilseed species, there
must be a mechanism for channelling such unusual acyl moieties towards
storage oil accumulation, rather than towards membrane lipid formation. The
mechanism for this specific channelling of unusual fatty acids is at present
unknown. Triacylglycerol biosynthesis and phospholipid biosynthesis both

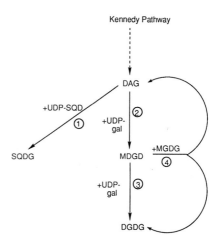

Figure 4.5. Biosynthesis of plant glycosylglycerides.
Abbreviations: DAG = diacylglycerol; MGDG = monogalactosyldiacylglycerol; DGDG = digalactosyldiacylglycerol; SQD = sulphoquinovose; SQDG = sulphoquinovosyldiacylglycerol. Reaction (1) = UDP-SQD:diacylglycerol sulphoquinovosyltransferase. Reaction (2) = UDP-galactose:diacylglycerol galactosyltransferase. Reaction (3) = UDP-galactose:MGDG galactosyltransferase. Reaction (4) = Galactolipid:galactolipid galactosyltransferase.

involve the same series of acyltransferase steps, with the exception of diacylglycerol acyltransferase, which is specific for triacylglycerol formation. One possibility is that there are seed-specific isoforms of each of the Kennedy pathway enzymes which have appropriate substrate specificities to ensure that unusual and potentially membrane-toxic fatty acids are channelled away from membrane formation and towards storage oil accumulation. It has been shown that in species such as *Cuphea*, which accumulate large amounts of short and medium chain fatty acids in their seed oil, there is a mechanism to prevent these fatty acids from entering phosphatidylcholine and hence membrane lipids [164]. It has also been found that even seeds which do not normally make "unusual" fatty acids, contain a similar mechanism for excluding unusual fatty acids from their membrane lipids. For example, the seed oil of safflower contains mostly oleic and linoleic acids. When homogenates from developing safflower seeds were fed lauric and erucic acids, these were incorporated into triacylglycerols, but were efficiently excluded from the membrane lipids [165].

The mechanism by which the oil bodies are formed is still a matter of controversy. Early theories such as the production of vesicles from the endoplasmic reticulum which, in oil storing tissues, differentiated into oil bodies have been discussed by Gurr [162]. Other ideas also invoked the inclusion of endoplasmic reticulum and, hence, endoplasmic reticulum proteins in the bounding layer of oil bodies [166]. In contrast, several workers have failed to observe boundary layers during oil droplet formation in castor bean [167] and mustard [168]. Indeed, the later accumulation of mature oil bodies in rapeseed and mustard [169,170,171,172] was found to coincide with the acquisition of a distinct boundary membrane. These observations agree with the proposal of Stobart *et al* [173] that protein-free oil droplets are released from the endoplasmic reticulum into the cytosol where they are, later, coated by appropriate proteins synthesised on free ribosomes. One class of protein which plays a vital role in the ontogeny and stabilisation of oil bodies is the oleosins. The structure and function of these proteins has been well reviewed recently [174,175].

4.4.2 Triacylglycerol biosynthesis

Triacylglycerols are made by the Kennedy pathway, named in honour of the pioneering lipid biochemist who elucidated the process in animals [176]. It is illustrated in Figure 4.6. Two acylation steps convert glycerol 3-phosphate into phosphatidate. Phosphatidate phosphohydrolase then generates diacylglycerol which is finally acylated to triacylglycerol. It is generally assumed that the biggest constraint on the overall flux through the pathway is the supply of acyl-CoAs but suggestions that phosphatidate phosphohydrolase and, especially, diacylglycerol acyltransferase may exert significant flux control have been made (see below). In addition, the acyltransferases of the Kennedy pathway play an important role in the control of oil quality.

The "backbone" of the acyl lipid molecule, glycerol 3-phosphate, is produced by the activity of dihydroxyacetone oxidoreductase [162,111]. The first enzyme involved in the Kennedy pathway is glycerol 3-phosphate acyltransferase (EC 2.3.1.15) which generates 1-monoacylglycerol 3-phosphate (lysophosphatidate). The equivalent enzyme from plastids, which uses acyl-ACPs instead of acyl-CoAs, has been purified and its gene cloned from some plants [177]. However, the glycerol 3-phosphate acyltransferase used in triacylglycerol formation is a membrane-bound enzyme and, so far, has eluded complete purification although some progress has been made recently by means of an affinity chromatography of the solubilised avocado enzyme [178]. On the other hand, the substrate specificity of the enzyme has been well studied in a number of microsomal preparations. In studies with safflower, glycerol 3-phosphate acyltransferase was selective for saturated fatty acyl-CoA esters [179,180]. Some discussion of the details of these experiments and general problems with substrate selectivity studies using plant microsomes is given by Stymne and Stobart [111]. Stereospecific analysis of triacylglycerol reveals that, although glycerol 3-phosphate acyltransferase does prefer saturated moieties, its specificity is much less than the second acyltransferase [181].

The second enzyme in the Kennedy pathway, lysophosphatidate acyltransferase (1-acylglycerol 3-phosphate acyltransferase : EC 2.3.1.51) exhibits a strong substrate specificity in most oilseed species. In safflower, this enzyme is

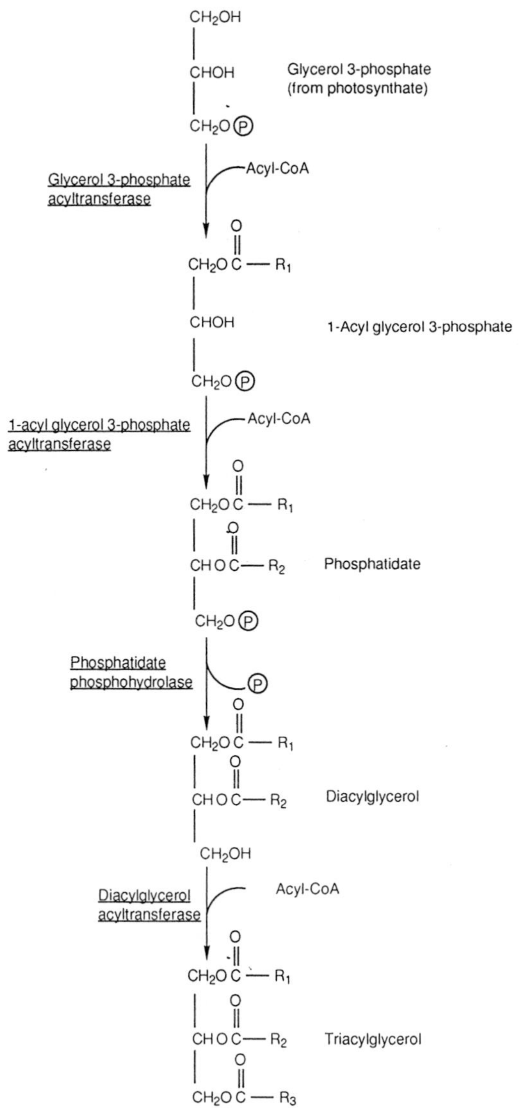

CH₂OH
CHOH — Glycerol 3-phosphate (from photosynthate)
CH₂O(P)

Glycerol 3-phosphate acyltransferase — Acyl-CoA

CH₂O C(=O) — R₁
CHOH — 1-Acyl glycerol 3-phosphate
CH₂O(P)

1-acyl glycerol 3-phosphate acyltransferase — Acyl-CoA

CH₂O C(=O) — R₁
CH O C(=O) — R₂ — Phosphatidate
CH₂O(P)

Phosphatidate phosphohydrolase — (P)

CH₂O C(=O) — R₁
CH O C(=O) — R₂ — Diacylglycerol
CH₂OH

Diacylglycerol acyltransferase — Acyl-CoA

CH₂O C(=O) — R₁
CH O C(=O) — R₂ — Triacylglycerol
CH₂O C(=O) — R₃

Figure 4.6. The Kennedy pathway for triacylglycerol synthesis.

much more specific towards linoleate than to oleate and discriminates almost totally against saturated fatty acids [180]. In rapeseed, erucoyl-CoA was an extremely poor substrate in comparison to oleoyl-CoA [182,183]. In addition, the same enzyme can discriminate against unusual fatty acids of industrial interest [165], whereas the other acyltransferases are less strict. Together, the first two acyltransferases of the Kennedy pathway are responsible for the non-random distribution of the acyl chains at the *sn*-1 and *sn*-2 positions on the glycerol backbone and, hence, of the phospholipids and triacylglycerols formed subsequently from them.

Phosphatidate phosphohydrolase (EC 3.1.3.4) hydrolyses phosphate from phosphatidate to yield diacylglycerol. Phosphatidate itself stands at a branch-point in metabolism (see Figure 4.4 and Gurr and Harwood [20]) and would be a good candidate for a metabolic control point in the pathway. Indeed, phosphatidate phosphohydrolase has been shown to be regulatory for fat synthesis in animals [184] and yeasts [185]. Its regulation seems to involve a translocatory mechanism i.e. the conversion of an inactive cytosolic form to an active membrane-bound enzyme [184] although the situation has been complicated recently by the discovery of two isoforms, one suggested to be involved in overall lipid synthesis and one in signal transduction [185]. Recently, Ichihara [187] found that the amount of membrane-bound phosphatidate phosphohydrolase could be varied by altering fatty acid levels in safflower and suggested that this could regulate triacylglycerol synthesis in plants. Other *in vitro* experiments using microsomes suggest that phosphatidate phosphohydrolase may be rate-limiting although phosphatidate never accumulates in tissues *in vivo* [129]. In safflower seed extracts, phosphatidate phosphohydrolase activity can be partly inhibited by EDTA and otherwise manipulated. Addition of Mg^{2+} to EDTA-inhibited microsomes stimulated the conversion of phosphatidate to triacylglycerol [180]. However, in cocoa, although Mg^{2+} was found to increase lipid formation *in vitro*, this resulted in the accumulation of phosphatidate rather than its being metabolised further to diacyl- or triacyl-glycerols [188]. Thus, the evidence that phosphatidate phosphohydrolase exerts significant flux control over triacylglycerol formation in plants is not yet conclusive. In animals, dietary fluctuations make it very important to control the rate at which triacylglycerols are made or catabolized and, under these conditions, phosphatidate phosphohydrolase would have an obvious role in the control of fat metabolism [20]. In contrast, the maturing oilseed is geared to provide maximum rates of fat synthesis and a putative role of phosphatidate phosphohydrolase in fine-tuning rates of formation may not be required.

The diacylglycerol formed by phosphatidate phosphohydrolase has two major fates. It can be used for triacylglycerol formation or it can be used by phosphotransferase enzymes for the final steps of zwitterionic phosphoglycer-ide synthesis. Studies with choline phosphotransferase have shown that it catalyses a freely-reversible reaction [189]. It was envisaged that oleate initially esterified to diacylglycerol (from the acylation of glycerol 3-phosphate) would be able to re-enter phosphatidylcholine and be desaturated [190]. Return of the linoleoyl-diacylglycerol moiety to the diacylglycerol pool would then make it available for triacylglycerol formation. Thus, for seed oils rich in polyunsatu-rated fatty acids, this might be one method for enriching the polyenoate content and, indeed, in linseed this appears to be a major pathway [189]. In other species, such as the stearate-rich cocoa, there is relatively little flow of diacylglycerol through phosphatidylcholine during active fat deposition, hence its low polyunsaturate content [188].

The final step in triacylglycerol synthesis is that catalysed by diacylglycerol acyltransferase (EC 2.3.1.20). This is the only step unique to triacylglycerol synthesis. Several reports have dealt with the substrate specificity of the enzyme using single or mixed acyl-CoA substrates and exogenously supplied emulsions of diacylglycerol [191,192]. The enzyme has some degree of selectivity although, in general, has a broad specificity. For example, only in plant species which were completely unable to utilise erucate did such acids fail to accumulate in triacylglycerol [192]. The rapeseed enzyme which **could** use erucate did, however, prefer shorter chain acids. Stereospecific analysis of endogenous triacylglycerols reveals that, in many plants, diacylglycerol acyltransferase has a similar acyl selectivity to glycerol 3-phosphate acyltransferase [126]. However, in borage, where γ-linolenic acid accumulates at the *sn*-3 but not the *sn*-1 position [130], and in *Anchusa officinalis* where octadecatetraenoate has a similar distribution [126], the two acyltransferases clearly have different selectivities. There are preliminary reports that diacylglycerol acyltransferase has been purified to apparent homogeneity from spinach leaves although these remain to be confirmed [193].

It has been suggested from both *in vivo* labelling studies [129], pool size measurements and *in vivo* [194] and *in vitro* results with microsomal fractions [195,194] that diacylglycerol acyltransferase may exert significant flux control over fat synthesis. In rapeseed this was suggested to be especially so at times of maximal triacylglycerol synthesis [196]. In animal tissue experiments using the inhibitor 2-bromooctanoate to modulate the flux, diacylglycerol acyltransferase has been shown to exert strong flux control [197] over the pathway. 2-Bromooctanoate has also been shown to inhibit diacylglycerol acyltransferase in rapeseed, a tissue from which the enzyme has been partly solubilised [198]. Thus, diacylglycerol acyltransferase appears to be an interesting enzyme from three standpoints. It is the only Kennedy pathway enzyme unique to triacylglycerol synthesis. Second, it exerts control over the quality of acyl groups attached at the *sn*-3 position, at least in some tissues. Third, it may exert significant flux control over the overall pathway. Further research on this enzyme, particularly its regulation, would be very welcome.

References

1 Blanke, M.M., Lenze, F., *Plant, Cell and Environment* **1989**, *12*, 31–46.

2 Sanchez, J., Del Cuvillo, M.T., Harwood, J.L., in : *Metabolism, Structure and Utilization of Plant Lipids* : Cherif, A. (ed.), CNP, Tunis **1992**, pp 39–42

3 Harwood, J.L., Rutter, A.J., del Cuvillo, M.T., De la Vega, M., Sanchez, J., *Proc. Brighton Crop Prot. Conf.* **1992**, in press.

4 Jacobson, B.S., Stumpf, P.K., *Arch. Biochem. Biophys.* **1972**, *153*, 656–663.

5 Kuhn, D.N., Knauf, M., Stumpf, P.K., *Arch. Biochem. Biophys.* **1981**, *209*, 441–450.

6 Dennis, D.T., Miernyk, J.A., *Annu. Rev. Plant Physiol.* **1982**, *33*, 27–50.

7 Givan, C.V., *Physiol. Plant.* **1983**, *57*, 311–316.

8 Murphy,D.J., Stumpf, P.K. *Methods in Enzymology.* **1981**, *72*, 768–773.

9 Liedvogel, B., Stumpf, P.K., *Plant Physiol.* **1982**, *69*, 897–903.

10 Simcox, P.D., Garland, W., Deluca, V., Canvin, D.T., Dennis, D.T., *Can. J. Bot.* **1979**, *57*, 1008–1014.

11 Journet, E.P., Douce, R. *Plant Physiol.* 1985, *79*, 458–467.

12 Denyer, K., Smith, A.M. *Planta* **1988**, *173*, 172–182.

13 Frehner, M., Pozueta-Romero, J., Akazawa, T. *Plant Physiol.* **1990**, *94*, 538–544.

14 Murphy, D.J., Rawsthorne, S., Hills, M.J. *Seed Sci. Res.* **1993**, *3*, 124–139.

15 Stitt, M., Ap Rees, T., *Phytochemistry* **1979**, *18*, 1905–1911.

16 Botha, F.C., Dennis, D.T., *Arch. Biochem. Biophys.* **1986**, *245*, 96–103.

17 Murphy, D.J., Walker,D.A. *Planta* **1982,** *156*, 84–88.

18 Masterson, C., Wood, C., Thomas, D.R. *Plant Cell & Environ.* **1990**, *13*, 767–771.

19 Roughan, G.J.H., Terning, L.I., Moloney, M.M., *Plant Mol. Biol.* **1992**, *18*, 1177–1179.

20 Gurr, M.I., Harwood, J.L., *Lipid Biochemistry*, Chapman and Hall, London, 4th edition, **1991**.

21 Stumpf, P.K., in : *The Biochemistry of Plants* : Stumpf, P.K., Conn, E.E. (eds.) New York, Academic Press, **1987**, Vol. 9, pp. 121–136.

22 Liedvogel, B. in: *The Structure, Metabolism and Function of Plant Lipids:* Stumpf, P.K., Mudd, J.B. and Nes, W.D., (eds.) Plenum, New York, **1987** pp 509–511.

23 Harwood, J.L., *Annu. Rev. Plant Physiol.* **1988**, *39*, 101–138.

24 Knowles, J.R., *Ann. Rev. Biochem.* **1989**, *38*, 195–221.

25 Wood, H.G., Barden, R.E., *Ann. Rev. Biochem.* **1977**, *46*, 385–413.

26 Hardie, P.G., Cohen, P., *FEBS Lett.* **1978**, *91*, 1–7.

27 Mishina, M., Roggenkamp, R., Schweitzer, E., *Eur. J. Biochem.* **1980**, *111*, 79–87.

28 Brock, K., Kannangara, C.G., *Carlsberg Res. Commun.* **1976**, *41*, 121–129.

29 Nielson, N.C., Adee, A., Stumpf, P.K., *Arch. Biochem. Biophys.* **1979**, *192*, 446–456.

30 Egin-Buhler, B., Loyal, R., Ebel, J., *Arch. Biochem. Biophys.* **1979**, *203*, 90–100.

31 Stumpf, P.K., in : *The Biochemistry of Plants* : Stumpf, P.K., Conn, E.E. (eds.) New York, Academic Press, **1980**, Vol. 4, pp. 177–203.

32 Nikolau, B.J., Hawke, J.C., *Arch. Biochem. Biophys.* **1984**, *228*, 86–96.

33 Slabas, A.R., Hellyer, A., *Plant Sci.* **1985**, *39*, 177–182.

34 Hellyer, A., Bambridge, H.E., Slabas, A.R., *Biochem. Soc. Trans.* **1986**, *14*, 565–568.

35 Charles, D.J., Cherry, J.H., *Phytochemistry* **1986**, *25*, 1067–1071.

36 Bettey, M., Ireland, R.J., Smith, A.M. *J. Plant Physiol.* **1992**, *140*, 513–520.

37 Browse, J., Somerville, C., *Annu. Rev. Plant Physiol.* **1991**, *42*, 467–506.

38 Focke, M., Hoffmann, S., Motel, A., Lichlenthaler, H., in : *Metabolism, Structure and Function of Plant Lipids* : Cherif, A. (ed.), CNP Tunis, **1992**, pp 103–112.

39 Wurtele, E.S., Nikolau, B.J., *Arch. Biochem. Biophys.* **1990**, *278*, 179–186.

40 Egin-Buhler, B., Ebel, J., *Eur. J. Biochem.* **1983**, *133*, 335–339.

41 Finlayson, S.A., Dennis, D.T., *Arch. Biochem. Biophys.* **1983**, *225*, 576–585.

42 Smith, R.F., Gauthier, D.A., Dennis, D.T., Turpin, D.H., *Plant Physiol.* **1992**, *98*, 1233–1238.

43 Turnham, E., Northcote, D.N., *Biochem. J.* **1983**, *212*, 223–229.

44 Deerberg, S., von Twickel, J., Forster, H-H, Cole, T., Fuhrmann, J., Heise, K-P, *Planta* **1990**, *180*, 440–444.

45 Mohan, S.B., Kekwick, R.G.O., *Biochem. J.* **1980**, *187*, 667–676.

46 Numa, S., Bortz, W., Lynen, F., *Adv. Enzyme Res.* **1965**, *3*, 407–423.

47 Carlson, C.A., Kim, K.H., *Arch. Biochem. Biophys.* **1974**, *164*, 478–489.

48 Livne, A., Sukenik, A., *Plant Cell Physiol.* **1990**, *31*, 851–858.

49 Roessler, P.G., *Plant Physiol.* **1990**, *92*, 73–78.

50 Kannangara, C.G., Jensen, C.J., *Eur. J. Biochem.* **1975**, *54*, 25–30.

51 Nikolau, B.J., Hawke, J.C., Slack, C.R., *Arch. Biochem. Biophys.* **1981**, *211*, 605–612.

52 Thomson, L.W., Zalik, S., *Plant Physiol.* **1981**, *67*, 655–661.

53 Eastwell, K.C., Stumpf, P.K., *Plant Physiol.* **1983**, *72*, 50–55.

54 Nakamura, Y., Yamada, M., *Plant Sci. Lett.* **1979**, *14*, 291–295.

55 Sauer, A., Heise, K., *Plant Physiol.* **1983**, *73*, 11–15.

56 Jaworski, J.G., Clough, R.C., Barnum, S.R., Post-Beittenmiller, D., Ohlrogge, J.B., in : *Plant Lipid Biochemistry, Structure and Utilization* : Quinn, P.J., Harwood, J.L. (eds.) London, Portland Press, **1990**, pp. 97–104.

57 Post-Beittenmiller, D., Roughan, P.G., Ohlrogge, J.B., : in *Metabolism, Structure and Utilization of Plant Lipids* : Cherif, A. (ed.) CNP, Tunis **1992**, pp 117–120.

58 Jaworski, J.G., Post-Beittenmiller, D., Ohlrogge, J.B., in : *Metabolism, Structure and Utilization of Plant Lipids*: Cherif, A (ed.), CNP, Tunis, **1992**, pp 113–116.

59 Slabas, A.R., Harding, J., Hellyer, A., Roberts, P., Bambridge, H.E., *Biochim. Biophys. Acta* **1987**, *921*, 50–59.

60 Simoni, R.D., Criddle, R.S., Stumpf, P.K., *J. Biol. Chem.* **1967**, *242*, 573–581.

61 Ohlrogge, J.B., Kuo, T.M., *Plant Physiol.* **1984**, *74*, 622–625.

62 Ohlrogge, J.B., Kuo, T.M., *J. Biol. Chem.* **1985**, *260*, 8032–8037.

63 Slabas, A.R., Hellyer, A., Sidebottom, C., Bambridge, H., Cottingham, I.R., Kessell, R., Smith, C.G., Sheldon, P., Kekwick, R.G.O., de Silva, J., Windust, J., James, C.M., Hughes, S.G., Safford, R., in : Plant Molecular Biology : von Wettstein, D., Chua, N.H. (eds.) *NATO ASI Series A*, **1987**, Vol. 140, pp. 265–277.

64 Hoj, P.B., Svendsen, I.B., *Carlsberg Res. Commun*, **1984**, *49*, 483–492.

65 Slabas, A.R., Fawcett, T., *Plant Molecular Biology* **1992**, *19*, 169–191.

66 Matsumara, S., Stumpf, P.K., *Arch. Biochem. Biophys.* **1968**, *125*, 932–941.

67 De Silva, J., Loader, N.M., Jarman, C., Windust, J.H.C., Hughes, S.G., Safford, R., *Plant Molecular Biology* **1990**, *14*, 537–548.

68 Ohlrogge, J.B., Browse, J., Somerville, C.R., *Biochim. Biophys. Acta* **1991**, *1082*, 1–26.

69 Ohlrogge, J.B., Kuhn, D.N., Stumpf, P.K., *P.N.A.S.* **1979**, *76*, 1194–1198.

70 Dorne, A.J., Corde, J.P., Joyard, J., Borner, T., Douce, R., *Plant Physiol.* **1982**, *69*, 1467–1470.

71 Safford, R., Windust, J.H.C., Lucas, C., De Silva, J., James, C.M., Hellyer, A., Smith, C.G., Slabas, A.R., Hughes, S.G., *Eur. J. Biochem.* **1988**, *174*, 287–295.

72 Alberts, A.W., Vagelos, P.R., *J. Biol. Chem.* **1988**, *241*, 5201–5204.

73 El-Hussein, S.A., Miernyk, J.A., Ohlrogge, J.B., *Biochem. J.* **198**8, *252*, 39–45.

74 Somerville, C., Browse, J., *Science* **1991**, *252*, 80–87.

75 Hoj, P.B., Mikkelson, J.D., *Carlsberg Res. Commun.* **1982**, *47*, 119–141.

76 Shimakata, T., Stumpf, P.K., *J. Biol. Chem.* **1983**, *258*, 3592–3598.

77 Wolf, A.M.A., Perchorowicz, J.T., in : *The Metabolism, Structure and Function of Plant Lipids* : Stumpf, P.K., Mudd, J.B., Nes, W.D. (eds.) New York, Plenum, **1987**, pp. 433–436.

78 Shimakata, T., Stumpf, P.K., *Arch. Biochem. Biophys.* **1982**, *217*, 144–154.

79 Jaworski, J.G., Rock, C.O., *J. Biol. Chem.* **1987**, *262*, 7927–7931.

80 Jaworski, J.G., Clough, R.C., Barnum, S.R., *Plant Physiol.* **1989**, *90*, 41–44.

81 Walsh, M.C., Klopfenstein, W.E., Harwood, J.L., *Phytochemistry*, **1990**, *29*, 3797–3729.

82 Mackintosh, R.O.W., Hardie, D.G., Slabas, A.R., *Biochim. Biophys. Acta* **1989**, *1002*, 114–124.

83 Kinney, A.J., Hitz, W.D., Yadav, N.S., in : *Plant Lipid Biochemistry, Structure and Utilization* : Quinn, P.J., Harwood, J.L. (eds.) London, Portland Press, **1990**, pp. 136–128.

84 Schuz, R., Ebel, J., Hahlbrock, K., *FEBS Lett.* **1982**, *140*, 207–209.

85 Shimakata, T., Stumpf, P.K., *Arch. Biochem. Biophys.* **1983**, *220*, 39–45.

86 Page, R.A., Harwood, J.L., in : *Metabolism, Structure and Utilization of Plant Lipids* : Cherif, A. (ed.), CNP, Tunis, **1992**, pp 201–204.

87 Siggaard-Andersen, M., Kauppinen, S., von Wettstein-Knowles, P., *Proc. Natl. Acad. Sci.*, U.S.A. **1991**, *88*, 4114–4118.

88 Kauppinen, S., in : *Plant Lipid Biochemistry, Structure and Utilization* : Quinn, P.J., Harwood, J.L. (eds.) London, Portland Press, **1990**, pp. 450–452.

89 Harwood, J.L., Walsh, M.C., Walker, K.A., in : *Methods in Plant Biochemistry* : Dey, P.M., Harborne, J.B. (eds.) London, Academic Press, **1990**, Vol. 3, pp. 193–216.

90 Guerra, D.J., Ohlrogge, J.B., *Arch. Biochem. Biophys.* **1986**, *246*, 274–285.

91 Shimakata, T., Stumpf, P.K., *Arch. Biochem. Biophys.* **1982**, *218*, 77–91.

92 Sheldon, P.S., Safford, R., Slabas, A.R., Kekwick, R.G.O., *Biochem. Soc. Trans.* **1988**, *16*, 392–393.

93 Sheldon, P.S., Ph.D. Dissertation, University of Birmingham, U.K., **1988**.

94 Sheldon, P., Kekwick, R., Sidebottom, C., Smith, C., Slabas, A., in : *Plant Lipid Biochemistry, Structure and Utilization* : Quinn, P.J., Harwood, J.L. (eds.) London, Portland Press, **1990**, pp.120–122.

95 Fawcett, T., Chase, D., Mackintosh, R., Nishida, I., Murata, N., Slabas, A., in : *Metabolism, Structure and Utilization of Plant Lipids* : Cherif, A. (ed.), CNP, Tunis, **1992**, pp 417–420.

96 Klein, B., Topfer, R., in : *Metabolism, Structure and Utilization of Plant Lipids* : Cherif, A. (ed.), CNP, Tunis **1992**, pp 156–159.

97 Slabas, A.R., Harding, J., Hellyer, A., Sidebottom, C., Gwynne, H., Kessell, R., Tombs, M.P., *Dev. Plant Biol.* **1984**, *9*, 3–10.

98 Caughey, I., Kekwick, R.G.O., *Eur. J. Biochem.* **1982**, *123*, 553–561.

99 Slabas, A.R., Sidebottom, C.M., Hellyer, A., Kessell, R.M.J., Tombs, M.P., *Biochim. Biophys. Acta* **1986**, *877*, 271–280.

100 Kater, M.M., Koningstein, G.M., Nijkamp, H.J.I., Stuitje, A.R., *Plant Mol. Biol.* **1991**, *17*, 895–909.

101 Post-Beittenmiller, D., Jaworski, J.G., Ohlrogge, J.B., *J. Biol. Chem.* **1991**, *266*, 1858–1865.

102 Kacser, H., Burns, J.A., *Symp. Soc. Exp. Biol.* **1973**, *32*, 65–104.

103 Heinrich, R., Rapoport, T.A., *Eur. J. Biochem.* **1974**, *42*, 107–120.

104 Pollard, M.R., Anderson, L., Fan, C., Hawkins, D.J., Davies, H.M., *Arch. Biochem. Biophys.* **1991**, *284*, 306–312.

105 Davies, H.M., Anderson, L., Fan, C., Hawkins, D.J., *Arch. Biochem. Biophys.* **1991**, *284*, 37–45.

106 Dormann, P., Schmid, P.C., Robers, M., Spener, F., *Biol. Chem. Hoppe-Seyler* **1991**, *372*, 528.

107 McKeon, T.A., Stumpf, P.K., *J. Biol. Chem.* **1982**, *257*, 12141–12147.

108 Cheesbrough, T.M., Cho, S.H., in : *Plant Lipid Biochemistry, Structure and Utilization* : Quinn, P.J., Harwood, J.L. (eds.) Portland Press, London, **1990**, pp. 129–130.

109 Shanklin, J., Somerville, C., *Proc. Nat. Acad. Sci.*, U.S.A. **1991**, *88*, 2510–2514.

110 Joyard, J., Stumpf, P.K., *Plant Physiol.* **1981**, *67*, 250–256.

111 Stymne, S., Stobart, A.K., in : *The Biochemistry of Plants* : Stumpf, P.K., Conn, E.E. (eds.) Vol. 9, New York, Academic Press, **1987**, pp. 175–214.

112 Stymne, S., Glad, G., *Lipids* **1981**, *16*, 298–305.

113 Stymne, S., Stobart, A.K., *Biochim. Biophys. Acta*, **1983**, *752*, 198–208.

114 Stymne, S., Stobart, A.K., *Biochem. J.* **1984**, *220*, 481–488.

115 Stymne, S., Stobart, A.K., *Planta* **1985**, *164*, 101–104.

116 Nichols, B.W., *Lipids* **1968**, *3*, 354–360.

117 Nichols, B.W., James, A.T., Breuer, J., *Biochem. J.* **1967**, *104*, 486–496.

118 Pugh, E.L., Kates, M., *Biochim. Biophys. Acta* **1973**, *316*, 305–316.

119 Stymne, S., Appelqvist, L-A., *Eur. J. Biochem.* **1978**, *90*, 223–229.

120 Smith, M.A., Cross, A.R., Jones, O.T.G., Griffiths, W.T., Stymne, S., Stobart, A.K., *Biochem. J.* **1990**, *272*, 23–29.

121 Schmidt, H., Heinz, E., in : *Metabolism, Structure and Utilization of Plant Lipids*: Cherif, A., (ed.), CNP, Tunis, **1992**, pp 140–143.

122 Sanchez, J., Stumpf, P.K., *Arch. Biochem. Biophys.* **1984**, *228*, 185–196.

123 Heinz, E., Roughan, P.G., *Plant Physiol.* **1983**, *72*, 273–279.

124 Roughan, P.G., Slack, C.R., *Trends Biochem. Sci.* **1984**, *9*, 383–386.

125 Stymne, S., Appelqvist, L-A., *Plant Sci. Lett.* **1980**, *17*, 287–294.

126 Harwood, J.L., Griffiths, G., in : *Advances in Plant Cell Biochemistry and Biotechnology* (Morrison, I.M., ed.) Vol. 1, JAI Press, London, **1993** in press.

127 Horrobin, D.F., *Prog. Lipid Res.* **1992**, *31*, 163–194.

128 Stymne, S., Stobart, A.K., *Biochem. J.* **1986**, *240*, 385–393.

129 Griffiths, G., Stymne, S., Stobart, A.K., *Planta* **1988**, *173*, 309–16.

130 Jamieson, G.R., Reid, E.H., *J. Sci. Food Agric.* **1968**, *19*, 628–631.

131 Griffiths, G., Brechany, E.Y., Christie, W.W., Stymne, S., Stobart, A.K., in : *The Biological Role of Plant Lipids* : Biacs, P.A., Gruiz, K., Kremmer, T. (eds.) Plenum Press, New York, **1989**, pp. 151–153.

132 Cahoon, E., Shanklin, J., Do, D., Ohlrogge, J., in : *Metabolism, Structure and Utilization of Plant Lipids* : Cherif, A. (ed.), CNP, Tunis, **1992**, pp 137–140.

133 Fairbairn, D., Ross, J.H.E., Bowra, S., Murphy, D.J., in: *Metabolism, Structure and Utilization of Plant Lipids* : Cherif, A. (ed.), CNP, Tunis, **1992**, pp 67–70.

134 Harwood, J.L., in : *The Biochemistry of Plants* : Stumpf, P.K., Conn, E.E. (eds.) Vol. 4, Academic Press, New York, **1980**, pp. 1–55.

135 Hilditch, T.P., Williams, P.N., *The Chemical Constitution of the Natural Fats*, 4th edition, Chapman and Hall, London, **1964**.

136 Galliard, T., Stumpf, P.K., *J. Biol. Chem.* **1966**, *241*, 5806–5812.

137 Moreau, R.A., Stumpf, P.K., *Plant Physiol.* **1981**, *67*, 672–676.

138 Bafor, M., Smith, M.A., Jonsson, L., Stobart, A.K., Stymne, S., *Biochem. J.* **1991**, *280*, 507–514.

139 Richards, D.E., Taylor, R.D., Murphy, D.J., *Plant Physiol. Biochem.* **1993**, in press.

140 Banas, A., Johanson, I., Stymne, S., in : *Metabolism, Structure and Utilization of Plant Lipids* : Cherif, A. (ed.), CNP, Tunis, **1992**, pp 209–212.

141 Cassagne, C., Moreau, P., Bessoule, J.J., Maneta-Peyret, L., Compere, P., Sturbois, B., Schneider, F., Morre, D.J., Lessire, R., in : *Metabolism, Structure and Utilization of Plant Lipids* : Cherif, A. (ed.), CNP, Tunis, **1992**, 459–462.

142 Gunstone, F.G., Harwood, J.L., Padley, F.B. (eds.) *The Lipid Handbook*, Chapman and Hall, London, **1986**.

143 Downey, R.K., Craig, B.M., *J. Am. Oil Chem. Soc.* **1964**, *41*, 475–478.

144 Pollard, M.R., McKeon, T.M., Gupta, L.M., Stumpf, P.K., *Lipids* **1979**, *14*, 651–662.

145 Pollard, M.R., Stumpf, P.K., *Plant Physiol.* **1980a**, *66*, 641–648.

146 Pollard, M.R., Stumpf, P.K., *Plant Physiol.* **1980b**, *66*, 649–655.

147 Stumpf, P.K., Pollard, M.R., in: *High and Low Erucic Acid Rapeseed Oils* (Kramer, J.K.G., Sauer, F.D., Pigden, W.J., eds.) Toronto, Academic Press, **1983**.

148 Stefansson, B.R., Houghen, F.W., *Can. J. Plant Sci.* **1964**, *44*, 359–364.

149 Krzymanski, J., Downey, R.K., *Can. J. Plant Sci.* **1969**, *49*, 313–319.

150 Jonsson, R., *Heriditas* **1977**, *86*, 159–170.

151 Murphy, D.J., Mukherjee, K.D. *FEBS Lett.* **1988**, *230*, 101- 104.

152 Fehling, E., Murphy, D.J., Mukherjee, K.D. *Plant Physiol.* **1990**, *94*, 492–498.

153 Whitfield, H.V., Murphy, D.J., Hills, M.J. *Phytochem.* **1992**, *32*, 255–258.

154 Fehling, E., Mukherjee, K.D., in : *Plant Lipid Biochemistry, Structure and Utilization* : Quinn, P.J., Harwood, J.L. (eds.) Portland Press, London, **1990**, pp. 151–153.

155 Schopker, H., Fehling, E., Mukherjee, K.D., in : *Metabolism, Structure and Utilization of Plant Lipids:* Cherif, A., (ed.), CNP, Tunis, **1992**, pp 197–200.

156 Lessire, R., Schneider, F., Bessoule, J.J., Cook, L., Cinti, D.L., Cassagne, C., in : *Metabolism, Structure and Utilization of Plant Lipids* : Cherif, A. (ed.), CNP, Tunis, **1992**, 144–147.

157 Lardans, A., Tremolieres, A., in : *Metabolism, Structure and Utilization of Plant Lipids* : Cherif, A. (ed.), CNP, Tunis, **1992**, pp 121–124.

158 Stumpf, P.K., Conn, E.E. (eds.) *The Biochemistry of Plants*, Vol. 9, New York, Academic Press, **1987**.

159 Harwood, J.L., *C.R.C. Crit. Revs. Plant Sci.* **1989**, *8*, 1–43.

160 Gurr, M.I., Blades, J., Appleby, R.S., *Eur. J. Biochem.* **1972**, *29*, 362–368.

161 Appelqvist, L-A, in : *Recent Advances in the Chemistry and Biochemistry of Plant Lipids* : Galliard, T., Mercer, E.I. (eds.) Academic Press, London, **1975**, pp. 247–286.

162 Gurr, M.I., in : *The Biochemistry of Plants* : Stumpf, P.K., Conn, E.E. (eds.) Academic Press, New York, Vol. 4, **1980**, pp. 205–248.

163 Lehrian, D.W., Keeney, P.G., *J. Am. Oil Chem. Soc.* **1980**, *57*, 61–65.

164 Bafor, M., Smith, M.A., Jonsson, L., Stobart, A.K., Stymne, S., *Biochem. J.* **1990**, *272*, 31–38.

165 Battey, J.F., Ohlrogge, J.B. *Plant Physiol.* **1989**, *90*, 835- 840.

166 Wanner, G., Formanek, H., Theimer, R.R., *Planta* **1981**, *151*, 109–123.

167 Harwood, J.L., Sodja, A., Stumpf, P.K., Spurr, A.R., *Lipids* **1971**, *6*, 851–854.

168 Bergfeld, R., Hong, Y-N, Kuhnl, T., Schopfer, P., *Planta* **1978**, *143*, 297–307.

169 Murphy, D.J., Cummins, I. *Plant Sci.* **1989**, *60*, 47–54.

170 Murphy, D.J., Cummins, I., *J. Plant Physiol.* **1989**, *135*, 63- 69.

171 Rest, J.A., Vaughan, J.G., *Planta* **1972**, *105*, 245–262.

172 Cummins, I., Murphy, D.J., in : *Plant Lipid Biochemistry, Structure and Utilization* : Quinn, P.J., Harwood, J.L. (eds.) Portland Press, London, **1990**, pp. 231–233.

173 Stobart, A.K., Stymne, S., Hoglund, S., *Planta* **1986**, *169*, 33–37.

174 Murphy, D.J., *Prog. Lipid Res.* **1990**, *29*, 299–324.

175 Huang, A.H.C., *Ann. Rev. Plant Physiol.* **1992**, *43*, 177–200.

176 Kennedy, E.P., *Fed. Proc. Fed. Am. Soc. Exp. Biol.* **1961**, *20*, 934–940.

177 Ishizaki, O., Nishida, I., Agata, K., Eguchi, G., Murata, N., *FEBS Lett.* **1988**, *238*, 424–430.

178 Eccleston,V.S., Harwood, J.L., in : *Metabolism, Structure and Utilization of Plant Lipids* : Cherif, A. (ed.), CNP, Tunis, **1992**, pp 79–82.

179 Ichihara, K., *Arch. Biochem. Biophys.*, **1984**, *232*, 685–698.

180 Griffiths, G., Stobart, A.K., Stymne, S., *Biochem. J.* **1985**, *230*, 379–388.

181 Ichihara, K., Noda, M., *Phytochemistry*, **1980**, *19*, 49–54.

182 Sun, C., Cao, Y-Z., Huang, A.H.C., *Plant Physiol.* **1988**, *88*, 56–60.

183 Berneth, R., Frentzen, M., *Plant Sci.* **1990**, *67*, 21–28.

184 Brindley, D.N., in : *Fats in Animal Nutrition* (Wiseman, J., ed.) London, Butterworths Scientific Ltd., **1989** pp. 85–103.

185 Hosaka, K., Yamashita, S., *Biochim. Biophys. Acta* **1984**, *796*, 102–111.

186 Day, C.P., Yeaman, S.J., *Biochim. Biophys Acta* **1992**, *1127*, 87–94.

187 Ichihara, K., Murata, N., Fujii, *Biochem. Biophys. Acta* **1990**, *1043*, 227–234.

188 Griffiths, G., Harwood, J.L., *Biochem. Soc. Trans.* **1989**, *17*, 688.

189 Slack, C.R., Campbell, L.C., Browse, J.A., Roughan, P.G., *Biochim. Biophys. Acta* **1983**, *754*, 10–20.

190 Stobart, A.K., Stymne, S., *Planta* **1985**, *163*, 119–125.

191 Ichihara, K., Noda, M., *Phytochemistry* **1982**, *21*, 1895–1901.

192 Cao, Y., Huang, A.H.C., *Plant Physiol.* **1987**, *84*, 762–765.

193 Kwanyuen, P., Wilson, R.F., *Biochim. Biophys. Acta* **1990**, *1039*, 67–72.

194 Perry, H.J., Harwood, J.L., in : *Metabolism, Structure and Utilization of Plant Lipids* : Cherif, A. (ed.), CNP, Tunis **1992**, pp 360–363.

195 Ichihara, K., Takahashi, T., Fujii, S., *Biochim. Biophys. Acta*, **1988**, *958*, 125–129.

196 Perry, H.J., Harwood, J.L., *Biochem. Soc. Trans.* **1991**, *19*, 243.

197 Mayorek, N., Grinstein, I., Bar-Tana, J., *Eur. J. Biochem* **1989**, *182*, 395–400.

198 Weselake, R.J., Taylor, D.C., Pomeroy, M.K., Lawson, S.L., Underhill, E.W., *Phytochemistry* **1991**, *30*, 3533–3538.

5 Transformation of Oil Crops

P. J. Dale and J. A. Irwin

5.1 Introduction

Transformation is genetic modification by the direct introduction of specific DNA sequences into plants. Genes controlling particular characters are introduced into plant cells by various methods, and modified plants regenerated from them. The genes inserted become integrated into the plant chromosomes and are inherited in a Mendelian manner. The dramatic developments in plant molecular biology, tissue culture and transformation methodology during the past decade now make it possible to modify plants in many novel ways. Current research programs will also widen the array of genes available for modifying our crop plants. The objective of this chapter is to contrast transformation with more traditional methods of genetic modification and to outline the state of the art of plant transformation, especially in the principal oil crop species. Inevitably the majority of the discussion will be on those species in which there has been greatest progress.

In traditional methods of crop improvement, new gene combinations are obtained by bringing together two whole nuclear genomes from selected parents by sexual hybridization. Genetic segregation in subsequent generations provides a range of genetic variation from which the plant breeder can select. The nature of the genetic variation in the progenies is directed to the extent that parents with desirable characters are chosen initially. Its success thereafter depends on the ability of the breeder to recognise and select desirable new gene combinations. Traditional plant breeding has been remarkably successful and has contributed substantially to the level of crop productivity enjoyed today. The disadvantages of traditional breeding are that very specific genetic changes are difficult to make while keeping other characters constant; and the genes available to the traditional plant breeder are limited to those from species that are sexually compatible with the crop to be improved.

As an alternative to conventional plant breeding which employs sexual hybridisation, parasexual hybridization can potentially be a means of transferring parts of a plant genome and therefore of achieving more limited gene transfer. This is accomplished using plant protoplasts. Protoplasts are cells that have had their cellulose cell walls removed by enzymatic digestion (Figure 5.1). Removal of the cell wall exposes the plasma membrane which bounds the remaining cell contents. Parasexual hybridization involves isolating protoplasts from two parental plant genotypes and partially destroying the genome of one parent before fusing the protoplasts and regenerating asymmetric hybrid plants in vitro [1,2]. Protoplast fusion can also be a means of obtaining new gene combinations within the cytoplasmic genome (cybrids), and, for example, has been a valuable way of combining the cytoplasmic male sterility character with triazine resistance in *B. napus* (controlled by genes in the mitochondria and chloroplasts, respectively; [3,4]). Somatic hybridization is of value for certain specific applications, but like sexual hybridization, the products of the genetic modification are poorly directed. There is no straightforward way to direct or influence the portion of the donor genome that contributes to the somatic hybrid. The regeneration of fertile and functional plants is also frequently difficult following protoplast fusion. A further discussion of the use of protoplast fusion in oilseeds can be found in Chapter 6.

In contrast to sexual and parasexual hybridization, transformation provides a means of introducing specific DNA sequences into plants, and therefore of modifying particular plant characters without attendant changes in other

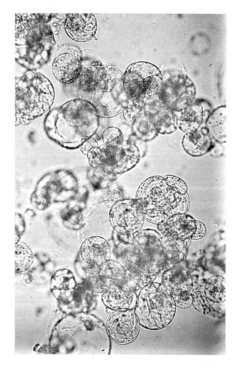

Figure 5.1. Dividing 5-day old protoplasts of *Brassica napus*. (Photograph courtesy of R.Mathias).

genes. In addition to providing a valuable tool to understand how genes work, by inserting them into plants and assaying phenotype (a method termed "reverse genetics"), it also provides a means of modifying plants for agricultural and industrial use. There are, of course, challenges in the area of plant transformation; it is necessary to isolate relevant genes and DNA control sequences which are able to regulate the plant character of interest, and it is important to have an efficient means of transformation that will cope with different species and genotypes. Progress towards isolating genes that control key steps in storage oil and protein biosynthesis is described in Chapter 6. Progress in transformation will be considered here.

5.2 Methods of transformation

The six major aims of any method of transformation are to obtain: a) the stable integration of specific DNA constructs into the plant genome without structural rearrangement, b) the desired level of transgene expression, c) insertion of the desired number of copies of the construct (usually one), d) the desired spatial and temporal expression of transgene action within the plant, e) structural stability of the construct, and f) stable expression of the transgenes over many vegetative and sexual generations. As will be discussed later, little is known so far about what determines the number of copies of the construct inserted or its position in the genome. In practice several independently transformed plants are created and individual plants are then selected which have the desired phenotypic modification.

Many methods have been explored to transform plants over the past decade. The major requirement is to overcome the physical barriers to entry of DNA into plant cells and to avoid degradation of the DNA before it becomes stably integrated into the chromosomes. Methods of transformation attempted have been reviewed comprehensively by Potrykus [5]. The three principal methods used for oilseed species, namely, a) via *Agrobacterium* infection, b) microprojectile bombardment, and c) direct DNA uptake into protoplasts, will be considered in more detail in sections 5.3 to 5.5. Two other methods that are receiving attention in rapeseed are microinjection of DNA, and the use of a microlaser beam to cut openings into plant cell walls to allow DNA to be introduced. Microlaser treatment of cells is in the very early stages of development [6], but the microinjection of 8-cell stage microspore-derived embryos with DNA has resulted in regenerated plantlets that were confirmed to contain integrated copies of the kanamycin resistance gene (NPTII) [7].

Although the efficiencies of transformation were quite high (27–51% of the regenerated plantlets assayed positive for the NPTII gene), the microinjection procedure is technically demanding for plant cells at present.

There have been a few reports of the use of viruses as vehicles for transferring DNA into plant cells. Most of these involve the Cauliflower mosaic virus (CaMV). It is known that within CaMV DNA, two of the open reading frames (ORFs) are dispensable in the lifecycle of the virus. Both can be completely deleted without changing the viability of CaMV under greenhouse conditions and, therefore, are obvious targets for substitution experiments [8]. Paskowski *et al* [9] replaced one ORF with a bacterial nptII gene. The resulting hybrid vector was found to be non-infective. However, in the presence of an unaltered CaMV genome acting as a helper plasmid, *Brassica rapa* protoplasts were transformed with the nptII hybrid vector and transformed cell cultures recovered under kanamycin selection although no plants were regenerated.

5.3 *Agrobacterium*

Requirements for successful transformation by *Agrobacterium* include the ability of the vector to deliver foreign DNA to the plant species of interest, and the possibility of regenerating plants from inoculated explants *in vitro*. Efficient regeneration of plant parts *in vitro* is important because whole plants must be regenerated from individual transformed cells [10]. For this purpose an

Figure 5.2: Transformation of *Brassica napus*. Regeneration of plants on kanamycin selective medium. One shoot is not transformed and is bleaching.

antibiotic resistance gene (commonly kanamycin resistance) is incorporated so that transgenic cells are resistant to that antibiotic. Inclusion of kanamycin in the plant regeneration medium will then confer a selective advantage to the transformed cells, and subsequently to the transgenic plants regenerated from them (Figures 5.2 and 5.3). If antibiotic selection does not operate efficiently, chimeric plants containing a mixture of transgenic and non-transgenic cells can develop [11–14]. In seed-propagated species, solid transgenic plants are then obtained by selection of transgenic progeny. In vegetatively-propagated species the production of solid transgenic plants might be obtained by further regeneration of plants from segments of the chimeric plant, but will depend on the nature of the chimerism.

One of the first priorities when developing a transformation system for a plant species of interest is to determine suitable explants that will undergo efficient shoot regeneration. The choice of explant used varies, but regeneration potential is generally better from seedling tissues than from adult plants although explants from mature plants can be responsive [15]. Various types of explant can be used, some with many cell types such as hypocotyls, seedling stems, flowering plant stems, meristems, immature embryos and petioles, and others with few cell types such as cotyledons and epidermal strips.

As a general rule, wild type *Agrobacterium* strains will infect dicotyledonous species but not monocotyledons [16]. There is evidence that *Agrobacterium* is able to deliver DNA to gramineous species [17] but so far no routine system of obtaining transgenic plants has been developed for species within this group. Two species of *Agrobacterium* are used for transforming plant, i.e. *A. tumefaciens* and *A. rhizogenes*.

Figure 5.3. Transformed *Brassica napus* plant transferred to soil in the glasshouse.

5.3.1 *Agrobacterium tumefaciens*

A. tumefaciens causes crown gall disease in many dicotyledonous species. The bacterium causes infection by transferring a defined piece of DNA (the transforming DNA or T-DNA) from its tumor-inducing (Ti) plasmid into the nuclear genome of cells at a wound site in the host plant. The movement of the T-DNA is mediated by virulence genes (*vir*) at another region of the Ti plasmid. The T-DNA is bounded by two 25-base pair direct repeat borders of DNA [18]. A wide variety of transformation vectors based on the Ti plasmid have been developed. These do not contain any of the oncogenic sequences responsible for crown formation, and hence, plant regeneration and normal plant growth can be obtained following the transfer of DNA into the nucleus of the recipient plant cell. Non-oncogenic vectors can be divided into two types, *cis* and *trans*, depending on whether the T-DNA region is carried on the same plasmid as the *vir* genes or on a separate plasmid.

Cis vectors

In *cis* vectors the *vir* genes are carried on the same plasmid as the T-DNA. The oncogenic T-DNA genes are replaced by a piece of DNA that has a region of homology to a small cloning vector that can replicate only in *E.coli*. This small vector is used to produce an intermediate vector incorporating a selectable marker gene that will function in the plant cell, an antibiotic resistance gene to select for transformed cell during the transformation process [19], and a multiple cloning site for the insertion of foreign DNA. Foreign DNA inserted into the intermediate vector is introduced into *Agrobacterium* by a

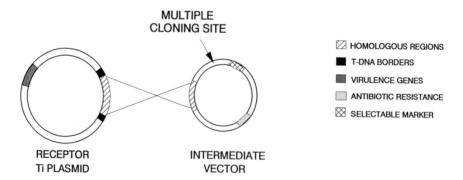

Figure 5.4. Cis vector system. Multiple cloning site, antibiotic resistance gene and plant selectable marker transferred from the intermediate vector to the Ti plasmid by homologous recombination. (The left or right T-DNA borders may be present in the intermediate vector).

two-step process of conjugation followed by homologous recombination between the disarmed Ti plasmid and the intermediate vector (Figure 5.4). The foreign DNA is transferred to the plant during transformation by the *vir* region acting in *cis* [20,21].

Trans vectors

In *trans* vectors the *vir* gene and the T-DNA are carried on different plasmids. These vectors are designed so that the T-DNA border sequences flank a multiple cloning site which allows the insertion of foreign DNA and marker genes that facilitate the direct selection of transformed plant cells. The plasmids can be manipulated in *E. coli* and transferred via conjugation to *Agrobacterium* strains which contain a disarmed Ti plasmid carrying the *vir* genes but in which the T-DNA and border sequences have been deleted (helper plasmid, Figure 5.5). Transfer of the foreign DNA inserted into the cloning vector is mediated by the *vir* region on the Ti plasmid acting in *trans* [22].

The oilseed species that can be transformed by *Agrobacterium tumefaciens* are shown in Table 5.1. The efficiency of transformation can vary considerably and it is often difficult to transfer protocols effectively between laboratories. Not all the factors affecting the efficiency of transformation have been identified and even experienced operators can have marked variation in success between experiments, but the main variables are: a) the regenerative potential and the pattern of regeneration of the plant species and genotype, b) the *Agrobacterium* strain, c) the plasmid vector and the efficiency of the antibiotic selection gene, d) the efficiency of the antibiotic to select cells within the particular plant species/genotype (some species are innately resistant to certain antibiotics) and e) the effect of the antibiotic on plant regeneration.

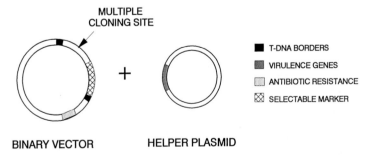

Figure 5.5. Trans vector system. Foreign gene transfer from binary vector to plant cell mediated by *vir* genes carried on the helper plasmid.

Table 5.1. The transformation status of oil crops

OIL CROP	METHOD	MARKER GENES	COMMENT (Source of cells treated & transformation efficiency)	REF.
Arachis hypogea. (peanut)	*Agrobacterium rhizogenes.*	*nptII* *gus*	Infection of hypocotyl segments. 3 transformed callus clones established. No transgenic plants regenerated.	[41]
B.carinata. (ethiopia mustard)	*Agrobacterium tumefaciens.*	*nptII* *hpt*	Stem explants. Transformation rate 1.5%[a]	[42]
B.juncea. (india mustard)	*Agrobacterium tumefaciens.*	*nptII* *gus*	Cotyledon and hypocotyl explants. Transformation rate 9–10%[a]	[43]
	Agrobacterium tumefaciens.	*nptII*	Cotyledons producing caulogenic callus infected with *Agrobacterium*. Transformation rate not accurately determined :- "1 event in 20 initial explants"	[44]
B.napus. (oilseed rape)	*Agrobacterium tumefaciens.*	*nptII*	Cut petioles of cotyledons, Transformation rate up to 55%[a]	[45]
	Agrobacterium tumefaciens.	*nptII* *pat*	Hypocotyl explants. Transformation rate up to 30%.[a] Note: 25% of transformed plants showed no NPTII activity.	[46]
	Agrobacterium tumefaciens.	*nptII*	Hypocotyl explants. Transformation rate 0.4–2.5%[a]	[47]
	Agrobacterium tumefaciens.	*nptII* *als*	Stem explants. Transformation rate 1%[b]	[48]
	Agrobacterium tumefaciens.	*nptII* *nos*	Stem segments. Transformation rate 7%[a]	[15]
	Agrobacterium tumefaciens.	*nptII* *nos*	Stem epidermal explants. Transformation rate c.2%[a].	[49]

Table 5.1. Continued

OIL CROP	METHOD	MARKER GENES.	COMMENT (Source of cells treated & transformation efficiency)	REF.
	Agrobacterium tumefaciens.	*nptII* *gus*	Inflorescence stalks cut into 0.5cm segments. Transformation rate 7%[d,g]	[50]
	Agrobacterium tumefaciens.	*dhfr*	Longitudinal stem sections. Transformation rate 10%[c].	[51]
	Agrobacterium tumefaciens.	*nptII* *nos*	Thin cell layer explants. Transformation rate c.6%[a]	[52]
	Agrobacterium tumefaciens.	*nptII*	Microspore derived embryos. Transformation rate 0.5%[e]	[53]
	Agrobacterium tumifaciens.	*nptII*	Microspore derived embryos. 3/600 produced transgenic plants. Transformation rate 0.5%[f].	[54]
	Agrobacterium tumefaciens.	*nptII* *hpt*	Microspores. Transformation rate up to 7.3%[d]	[55]
	Agrobacterium tumefaciens.	*nptII*	Hypocotyl protoplasts. Transformation rate 10%[c].	[56]
	Agrobacterium rhizogenes.	*nptII*	Hypocotyl segments. 5–15% shoot clones NPTII positive.	[57]
	Agrobacterium rhizogenes.	*nptII*	1% hairy root explants produced shoots, 90% of which were transgenic.	[50]
	Agrobacterium rhizogenes.		Infection of stem segments to produce hairy roots.	[23]
	Direct DNA uptake into protoplasts.	*nptII*	Electroporation. 21 colonies produced from 4×10^6 protoplasts. 2 independant plants regenerated.	[58]
B.nigra. (Black mustard).	*Agrobacterium tumefaciens.*	*hpt*	Protoplasts. 1–6.8% transformed calli produced depending on cultivar. No transgenic plants regenerated.	[59]

Table 5.1. Continued

OIL CROP	METHOD	MARKER GENES.	COMMENT (Source of cells treated & transformation efficiency)	REF.
	Direct DNA uptake into protoplasts.	*hpt*	Mesophyll protoplasts irradiated with X- rays. Number of resistant calli per protoplasts treated = 0.42% - 0.217%. No transgenic plants regenerated.	[60]
B.rapa. (turnip - syn. B.campestris)	*Agrobacterium tumefaciens.*	*nptII* *gus*	Hypocotyl explants. Transformation rate 1 - 9%[a] depending on cultivar.	[61]
	Agrobacterium tumefaciens.	*nptII* *gus*	Hypocotyl explants. Transformation rate 7 - 13%[a]	[62]
	Agrobacterium tumefaciens.	*nptII*	Hypocotyl and cotyledon protoplasts. Transformation rate:7–27%[h].	[63]
	Direct DNA uptake into protoplasts.	*nptII*	2×10^6 protoplasts per experiment. 2 transformed colonies produced. "very low efficiency". no plants regenerated.	[64]
Glycine max. (Soybean)	*Agrobacterium tumefaciens.*	*nptII*	3/9800 explants produced transgenic plants. All plants chimeric.	[11]
	Agrobacterium tumefaciens.	*nptII* *gus* *esps*	Cotyledon explants. 8/128 shoots regenerated were kanamycin resistant. Transformation rate 6%[d].	[65]
	Microprojectile bombardment.	*nptII*	Immature embryos. Transformation rate 10^{-5} [i]	[66]
	Microprojectile bombardment.	*gus*	Meristems. Transformation rate 15%[a] Germline transformation 0.5%	[67]
	Microprojectile bombardment.	*nptII* *gus*	Meristems. 2%[j] produced transformed chimeric plantlets.	[12]

Table 5.1. Continued

OIL CROP	METHOD	MARKER GENES.	COMMENT (Source of cells treated & transformation efficiency)	REF.
	Direct DNA uptake into protoplasts.	*gus* *hpt*	Electroporation. Transformation rate: $3 - 7 \times 10^{-4}$ of protoplasts[k]. 8% produced shoots.	[68]
	Direct DNA uptake into protoplasts.	*nptII* *gus* *pat* *cat*	Electroporation. Transformation rate : 10^{-4} to 10^{-5} [i]. Cotransformation rate of two genes introduced on independent plasmids : 18% - 27%[l]	[69]
	Direct DNA uptake into protoplasts.	*nptII*	Electroporation. "Number of kanamycin resistant calli increased from 4/47 after initial plating to 100 if plated 5 weeks after electroporation".	[70]
Gossypium hirsutum. (cotton)	*Agrobacterium tumefaciens.*	*nptII* *cat*	Hypocotyl explants. "Percentage transformation variable".	[71]
	Agrobacterium tumefaciens.	*nptII* *ocs*	Cotyledonary explants. 85–95% calli kanamycin resistant; 30% of calli regenerated.	[72]
	Microprojectile bombardment.	*gus* *hpt*	Embryogenic suspension culture. Transformation rate 0.7%[n]	[73]
Helianthus annus. (sunflower)	*Agrobacterium tumefaciens.*	*nptII* *gus*	Shoot meristems. Transformation rate 0.13%[a] Chimeric expression in plants.	[13]
	Agrobacterium tumefaciens.	*nptII*	Hypocotyl segments. 17 transformed plants grown to maturity.	[74]
	Microprojectile bombardment.	*nptII* *gus*	1/2000 apical explants produced stably transformed plant. Bombardment followed by cocultivation with *Agrobacterium* - 15% of plant produced were chimeric.	[14]

Table 5.1. Continued

OIL CROP	METHOD	MARKER GENES.	COMMENT (Source of cells treated & transformation efficiency)	REF.
	Direct DNA uptake into protoplasts.	*nptII*	48 transformed calli per 10^6 protoplasts. no plants regenerated.	[75]
Linum usitatissimum. (linseed)	*Agrobacterium tumefaciens.*	*nptII* *als*	Hypocotyl explants. More than 50 transgenic plants recovered	[76]
	Agrobacterium tumefaciens.	*nptII* *epsp*	Hypocotyl explants. Transformation rate 10%[a]	[77]
	Agrobacterium tumefaciens.	*nptII* *nos*	Hypocotyl explants. Kanamycin resistant shoots regenerated.	[78]
	Agrobacterium rhizogenes.		Cotyledonary explants. Transformed plants regenerated from hairy roots. Transformation rate : 0.5% - 4%[m] depending on cultivar.	[79]
Papaver somniferum. (Poppy).	*Agrobacterium rhizogenes.*		4 week old plants inoculated by wounding stem internodes. Gene transfer demonstrated, but no transgenic plants regenerated.	[80]
Zea mays. (maize)	*Agrobacterium tumefaciens.*	*nptII* *gus*	Shoot apex. Single injection of inoculum to meristematic region with hypodermic needle. 6/80 transformed plants produced.	[81]
	Microprojectile bombardment.	*hpt*	Embryogenic callus cultures. 3 transformed plants produced (1 per 10^8 cells).	[82]
	Microprojectile bombardment.	*gus* *luc* *als* *pat*	Embryogenic cells. 5 transgenic plants produced.	[83]

Table 5.1. Continued

OIL CROP	METHOD	MARKER GENES.	COMMENT (Source of cells treated & transformation efficiency)	REF.
	Microprojectile bombardment.	*gus* *pat*	Embryogenic cell suspension culture. 1 stable transformed plant produced per 1000 cellular foci[n].	[84]
	Microprojectile bombardment.	*gus* *pat*	Suspension cultures of Black Mexican sweet maize. 1 transformed callus produced per 10 - 20 cellular foci[o].	[85]
	Direct DNA uptake into protoplasts	*nptII* *hpt*	Electroporation. Transformation rate 0.08%[p].	[86]
	Direct DNA uptake into protoplasts.	*nptII*	Electroporation. Transformation rate up to 5%, typically 0.5 - 1%[i]. Shoots regenerated from 18% of transformed calli.	[34]
	Direct DNA uptake into protoplasts.	*nptII* *hpt*	Polyethylene glycol mediated DNA uptake. Transformation rate 0.3%[o]. No plants regenerated.	[30]
	Direct DNA uptake into protoplasts	*nptII* *gus*	Polyethylene glycol mediated DNA uptake. 1 transformed cell line per 10^4 protoplasts.	[31]
	Direct DNA uptake into protoplasts.	*nptII* *cat*	DNA uptake mediated by cationic liposomes. Recovered up to 8%[i] stable transformed microcalli.	[87]
	Direct DNA uptake.	*nptII*	Electroporation of immature embryos. 4%-28% embryogenic calli resistant to kanamycin. Transgenic plants regenerated.	[88]

Table 5.1. Continued

Oil crop species not yet transformed:
   Carthamus tinctorius (Safflower)
   Cocos nucifera (Coconut)
   Elaeis guineensis (Oil Palm)
   Guizotia abyssinica (Niger)
   Olea europaea (Olive)
   Ricinus communis (Castor)
   Sesamum indicum (Sesame)
   Simmondsia chinensis (Jojoba)

a.   Proportion of explants giving at least one transformed shoot.
b.   Transgenic plants selected by spraying with herbicide.
c.   Proportion of embryos surviving cocultivation that exhibited NPTII activity.
d.   Proportion of stem segment or leaf disc explants able to form callus on selective medium.
e.   Shoots allowed to regenerate for two weeks before selection applied.
f.   Proportion of treated embryos regenerating transgenic plants.
g.   Proportion of regenerated shoots found to be transgenic.
h.   The percentage of surviving cells that were transformed.
i.   Proportion of calli that were kanamycin resistant.
j.   Proportion of the estimated number of bombarded cells that gave stable transformed clones.
k.   Expressed as the number of microcolonies recovered after electroporation.
l.   Proportion of transformants in which both genes were expressed.
m.   Proportion of the buds from *A.rhizogenes* induced roots that were opine positive.
n.   Cellular foci : clusters of transiently expressing cells.
o.   Number of resistant calli per million protoplasts.
p.   Proportion of surviving cells that divided and were transformed.

nptII – neomycin phosphotransferase gene.
gus   – ß-glucuronidase gene.
hpt   – hygromycin phosphotransferase gene.
pat   – phosphinothrycin transferase gene.
cat   – chloramphenicol acetyl transferase gene.
nos   – nopaline synthase gene.
ocs   – octapine synthase gene.
als   – acetolacetate synthase gene.
esps – 5-endopyruvylshikimate synthase gene.
luc   – firefly luciferase gene.
dhfr – dihydrofolate reductase gene.

5.3.2 *Agrobacterium rhizogenes*

A. rhizogenes transforms plant cells in a similar way to *A. tumefaciens* except that in its wild type form it induces hairy roots (roots with a high density of root hairs) rather than galls [23,24]. These transformed roots can be excised from the host plant and grown independently in vitro where they continue to proliferate. The main use of *A. rhizogenes* has been to induce hairy roots to produce and to study the production of secondary plant metabolites. They have also been used as a means of supporting obligate parasites normally found in the rhizosphere [25]. Whole plants can regenerate from hairy roots either spontaneously or by manipulation of growth regulators *in vitro*. The regenerated transgenic plants contain the T-DNA from the Ri plasmid (root inducing as opposed to the tumour inducing plasmid of *A. tumefaciens*) which frequently confers a characteristic Ri plant phenotype of crinkled leaves, shortened internodes and altered flower morphology. Two approaches have been used with *A. rhizogenes* to introduce novel genes into plants: the novel gene is either introduced into the T-DNA of the Ri plasmid or it is introduced on to a second vector plasmid and inserted into the bacterium along with the Ri plasmid. The second approach relies on cotransformation of the same cell with both vector and wild type T-DNA. The Ri phenotype (crinkled leaves etc) is used to identify regenerants carrying the Ri plasmid transformation, and expression of the novel transgene is used to identify plants that are cotransformed. Progeny are obtained from the cotransformed regenerant, and following independent assortment of the two T-DNA, segregants are selected that have the novel gene but not the Ri phenotype. Although transgenic plants have been produced by this method, the regeneration of fertile plants from hairy roots and the removal of the Ri phenotype by segregation is not an insignificant complication. Also *A. rhizogenes* has undergone relatively little development as a plant gene vector system compared with *A. tumefaciens*.

5.4 Microprojectile bombardment

A range of devices has been used to accelerate DNA-coated, small metal particles, into plant cells [26]. One of the earliest, and probably the most widely used is the Biolistic™ particle accelerator (Figures 5.6 and 5.7) which initially employed a standard 0.22 calibre gun cartridge, but has now been modified to use compressed helium for propulsion [27]. The principle is that small tungsten or gold particles (1μm) are coated with the DNA to be delivered, and loaded onto the front of a cylindrical macroprojectile (cartridge device). The plant cells are placed beneath the end of the barrel in a chamber that is partially evacuated to reduce air resistance against the propelled particles. When the cartridge is fired, the macroprojectile travels down the barrel and hits the back of a stopper plate; the coated metal particles pass through a central hole in the stopper plate and enter into the plant tissues.

An alternative propulsion system involves discharging an electric current into a small droplet of water [28]. The steam generated from this is sufficient to accelerate particles to high velocity which forces them to enter plant tissues. A

Figure 5.6. Biolistic™ Particle Accelerator System PDS 1000 (Dupont).

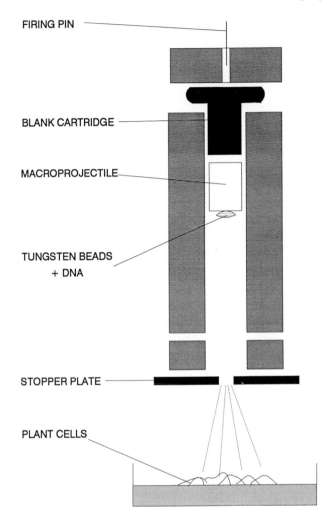

FIRING PIN

BLANK CARTRIDGE

MACROPROJECTILE

TUNGSTEN BEADS
+ DNA

STOPPER PLATE

PLANT CELLS

Figure 5.7. Diagrammatic representation of Particle Accelerator shown in Figure 5–6.

declared advantage of the electric discharge device is that velocities are easy to adjust and can be adapted to the needs of the plant tissue. The helium modification for the Biolistic™ accelerator has also enabled particle velocities to be varied more easily and reliably.

Although the principle of shooting small DNA coated fragments of metal into plant cells is crude, in practice it is giving a surprising level of success. Transgenic soybean and maize plants have been obtained and there are indications that other hitherto recalcitrant species may be on the verge of being transformed by this method. Because of its poor relative efficiency compared with *Agrobacterium*, particle bombardment is the method of choice only if transformation by *Agrobacterium* infection proves unsuccessful. It is likely, however, that the efficiency of particle bombardment will improve in the future, as the parameters for successful transformation become better understood.

5.5 Direct DNA uptake into protoplasts

Isolated DNA has been shown to be taken up by plant protoplasts under three sets of conditions : (a) by the use of a chemical helper (eg. Polyethylene glycol or PEG) that acts on both DNA structure and the plasma membrane that surrounds the plant protoplast, (b) by a procedure using a short electric pulse called electroporation, and (c) by the use of positively charged liposomes which interact with the negatively charged DNA. The first two methods are most commonly used to stimulate DNA uptake.

5.5.1 PEG-mediated DNA uptake

Paskowski *et al* [29] used PEG to demonstrate for the first time that a bacterial plasmid vector carrying a marker gene that is selectable in plant cells, could become integrated within the nuclear genome of cells following treatment of protoplasts with DNA in the absence of homology to either T-DNA border sequences or to host DNA. The PEG/$CaCl_2$ stimulated DNA uptake technique has subsequently been used to transfer genes into maize [30,31].

5.5.2 Electroporation

Treatment of plant protoplasts with short pulses of electric current produces transient holes in the plasma membrane. The electric pulse is generated by discharging a capacitor across an electroporation chamber which consists of two flat metal electrodes within a sterile plastic cuvette. Plant protoplasts are suspended between the electrodes in an ionic solution [32]. Successes have been reported in the transformation of rice and maize [33,34].

5.5.3 Liposomes

Antonelli and Stadler [35] showed that maize protoplasts could be transformed using cationic liposomes. This method works by promoting gene transfer either by the interaction of polycations with the anionic plasma membrane to give regions of positive charge which than react with negatively charged DNA; or, by mixing a suspension of positively charged liposomes with DNA before treatment of protoplasts. The latter treatment creates cationic complexes (surrounding the negatively charged DNA) which react with the anionic plasma membrane.

The most notable advances in transformation using this method has been in the Gramineae (maize and rice) which were considered to be recalcitrant to transformation for many years. Despite these successes, DNA uptake into protoplasts is unlikely to be a method that becomes firmly established for the routine transformation of cereals for the following reasons: a) cereals cannot be regenerated from protoplasts isolated directly from plants, so it is necessary to establish specialised suspension cultures which are frequently difficult to maintain, b) the ability to form suspension cultures and the regeneration of plants from protoplasts isolated from them, is highly dependent on plant species and even genotype, c) prolonged periods in suspension culture before protoplast isolation, and on solid culture afterwards, frequently leads to the induction of mutations (somaclonal variation) and unwanted genetic variability among regenerated plants [5].

5.6 Features of transformation

Although transformation procedures are sufficiently advanced to be able to create transgenic plants routinely in several important crop species, there are phenomena associated with the insertion and expression of DNA that are poorly understood. There is currently no control over the number of copies of a gene construct inserted into a plant or over their position of insertion in the plant genome. There is wide variation in the expression of transgenes in different independently transformed plants, which may be related to gene position effects [36]. Expression levels can also be down-regulated or switched off by a process associated with gene methylation. Structural rearrangements of constructs or their incomplete transfer to recipient plants is also possible [37].

There is emerging evidence that genes insert at a higher frequency in areas of DNA sequence homology between the construct and DNA in the chromosome [38]. This phenomenon may eventually give the opportunity of targeting gene insertion into particular regions of the chromosome that give more stable levels of transgene expression.

In practice, it is usual to create a range of transgenic plants and to identify by molecular and gene segregation analysis those carrying a single copy of the construct. Plants are then selected which have the desired phenotypic modification that is stable over several sexual generations in seed propagated crops or in asexual generation in vegetatively propagated species.

The long-term value of transformation for modifying plants for agricultural or industrial use depends on the isolation of genes important in biochemical pathways. But, just as important, is the isolation of gene promoters to regulate their action. Promoters are already available which give constitutive (or close to) expression or expression in embryos, stigmas, pollen, leaves and other parts of plants; it is likely that the modification of many processes can only be achieved by very fine and specific temporal and spatial expression of genes within plants, as discussed in Chapter 6.

Tissue culture-induced somaclonal variation associated with the transformation process is another feature that needs to be assessed in transgenic plants. This source of variation is independent of the construct inserted and can have a significant influence on plant performance, and frequently reduces it [39]. There is the potential for somaclonal variation among regenerated plants whenever plant cells are taken through a disorganised tissue culture phase. In general the longer the cells remain in a callus or suspension culture the greater is the extent of this variation. The nature of the variation can be epigenetic and non-heritable, but more significantly it is frequently genetic and caused by point mutations, alterations in chromosome structure, and changes in chromosome number including aneuploidy and polyploidy [40].

5.7 Future developments

Over the past ten years there has been a steady increase in the range of species that can be transformed and in the efficiency of that transformation; and this development and refinement process will certainly continue. It is important that there is progress in the important oil crop species that currently cannot be transformed. It is also necessary to be able to extend transformation to a wide range of plant genotypes within a crop species so there is the opportunity to insert transgenes into already proven cultivars or advanced breeding lines. It is unlikely that the problem of somaclonal variation among transgenic plants will ever be eliminated, because a tissue culture phase will probably remain an important part of the transformation process. It will continue to be necessary, therefore, to regenerate several independent transgenic plants and to select those which have no or minimal amounts of this source of variation. The creation of a range of transgenic plants followed by selection is also likely to remain an important feature, because of transgene position effects and the need to select transgenic plants with desired phenotypic modifications.

The choice of method adopted for an oilseed species will depend on the species to be transformed and Table 5.1 gives the present state of the art of transformation in different oilseed crop species. As a rule, the *Agrobacterium* method is the one of choice for dicotyledonous species but several important oilseed species have still to be transformed by this method. For monocotyledonous species some success has been achieved by DNA uptake into protoplasts, but there are significant drawbacks to this method for routine use. Particle acceleration is now showing promise and if delivery rates and particle penetration can be improved it is likely to continue to be an attractive method for transforming recalcitrant species, particularly the major perennial oil crops, coconut, olive and oil palm.

References

1. Dudits, D., Maroy, E., Praznovszky, T., Olah, Z., Gyorgyey, J. Cella, R., *Proc.Natl.Acad.Sci.USA*. **1987**, *84*, 8434–8438.
2. Bauer-Weston, B., Gleddie, S.,Webb, J. Keller,W., *Rapeseed in a Changing World*: McGregor, D.I., (ed.)Saskatoon: Organizing Committee of the Eighth International Rapeseed Congress, GCIRC, **1991**; pp.328.
3. Abe, T. Futsuhara, Y., *Theor.Appl.Genet*. **1986**, *72*, 3–10.
4. Barsby, T.L., Chuong, P.V., Yarrow, S.A., Wu, S-C., Coumans, M., Kemble, R.J., Powell, A.D., Beversdorf, W.D. Pauls, K.P., *Theor.Appl.Genet*. **1987**, *73*, 809–814.
5. Potrykus, I., *Physiol.Planta*. **1990**, *79*, 125–134.
6. Weber, G., Monajembashi, S.,Wolfrum, J. Greulich, K.O., *Physiol.Planta*. **1990**, *79*, 190–193.
7. Neuhaus, G., Spangenberg, G., Mittelsten Scheid, O. Schweiger, H-G., *Theor.Appl.Genet*. **1987**, *75*, 30–36.
8. Futterer, J., Bonneville, J.M. Hohn, T., *Physiol. Planta*. **1990**, *79*, 154–157.

9. Paszkowski, J., Pisan, B., Shillito, R.D., Hohn, T., Hohn, B. Potrykus, I., *Plant Mol.Biol.* **1986**, *6*, 303–312.

10. Dale, P.J. Ball, L.F., *Rapeseed in a Changing World*: McGregor, D.I., (ed.)Saskatoon: Proceedings of the Eighth International Rapeseed Congress, **1991**; pp.1122–1127.

11. Parrott, W.A., Hoffman, L.M., Hildebrand, D.F., Williams, E.G. Collins, G.B., *Plant Cell Rep.* **1989**, *7*, 615–617.

12. McCabe, D.E., Swain, W.F., Martinell, B.J. Christou, P., *Biotechnol.* **1988**, *6*, 923–926.

13. Schrammeijer, B., Sijmons, P.C., van den Elzen, P.J.M. Hoekema, A., *Plant Cell Rep.* **1990**, *9*, 55–60.

14. Bidney, D., Scelonge, C., Martich, J., Burrus, M., Sims, L. Huffman, G., *Plant Mol.Biol.* **1992**, *18*, 301–313.

15. Fry, J., Barnason, A. Horsch, R.B., *Plant Cell Rep.* **1987**, *6*, 321–325.

16. De Cleene, M. De Ley, J., *Bot.Rev.* **1976**, *42*, 389–466.

17. Dale, P.J., Marks, M.S., Brown, M.M., Woolston, C.J., Gunn, H.V., Mullineaux, P.M., Lewis, D.M., Kemp, J.M., Chen, D.F., Gilmour, D.M. Flavell, R.B., *Plant Sci.* **1989**, *63*, 237–245.

18. Barker, R.F., Idler, K.B., Thompson, D.V. Kemp, J.D., *Plant Mol.Biol.* **1983**, *2*, 335–350.

19. Bevan, M.W. Flavell, R.B., *Nature* **1983**, *304*, 184–187.

20. Zambryski, P., Joos, H., Genetello, C., Leemans, J., Van Montagu, M. Schell, J., *EMBO J.* **1983**, *2*, 2143–2150.

21. Fraley, R.T., Rogers, S.G., Horsch, R.B., Eichholz, D.A., Flick, J.S., Fink, C.L., Hoffman, N.L. Sanders, P.R., *Biotechnol.* **1985**, *3*, 629–635.

22. Bevan, M., *Nuc.Acids Res.* **1984**, *12*, 8711–8721.

23. Guerche, P., Jouanin, L., Tepfer, D. Pelletier, G., *Mol.Gen.Genet.* **1987**, *206*, 382–386.

24. Tepfer, D., *Physiol. Planta.* **1990**, *79*, 140–146.

25. Graveland, R., Dale, P.J. Mithen, R., *Mycol.Res.* **1992**, *96*, 225–228.

26. Batty, N.P. Evans, J.M., *Trans. Res.* **1992**, *1*, 107–113.

27. Sanford, J.C., *Physiol. Planta.* **1990**, *79*, 206–209.

28. Christou, P., *Physiol. Planta.* **1990**, *79*, 210–212.

29. Paszkowski, J., Shillito, R.D., Saul, M., Mandak, V., Hohn, T., Hohn, B. Potrykus, I., *EMBO J.* **1984**, *3*, 2717–2722.

30. Armstrong, C.L., Petersen, W.L., Buchholz, W.G., Bowen, B.A. Sulc, S.L., *Plant Cell Rep.* **1990**, *9*, 335–339.

31. Lyznik, L.A., Kamo, K.K., Grimes, H.D., Ryan, R., Chang, K.L. Hodges, T.K., *Plant Cell Rep.* **1989**, *8*, 292–295.

32. Draper, J., Scott, R., Kumar, A. Dury, G., *Plant Genetic Transformation and Gene Expression*: Draper, J., Scott, R., Armitage, P. Walden, R., (eds.) London: Blackwell, **1988**; pp. 161–198.

33. Toriyama, K., Arimoto, Y., Uchimiya, H. Hinata, K., *Biotechnol.* **1988**, *6*, 1072–1074.

34. Rhodes, C.A., Pierce, D.A., Mettler, I.J., Mascarenhas, D. Detmer, J.J., *Science* **1988**, *240*, 204–207.

35. Lucas, W.J., Lansing, A., de Wet, J.R. Walbot, V., *Physiol. Planta.* **1990**, *79*, 184–189.

36. Hobbs, S.L.A., Kpodar, P. DeLong, C.M.O., *Plant Mol.Biol.* **1990**, *15*, 851–864.

37. Zambryski, P.C., *Ann. Rev. Genet.* **1988**, *22*, 1–30.

38. Offringa, R., van den Elzen, P.J.M. Hooykaas, P.J.J., *Trans Res.* **1992**, *1*, 114–123.

39. Dale, P.J. McPartlan, H.C., *Theor. Appl.Genet.* **1992**, (In Press) .

40. Karp, A., *Oxford Surveys of Plant Molecular and Cell Biology*: Miflin, B.J., (ed.)Oxford: Oxford University Press, **1991**; Ed. 7th ,pp.1–58.

41. Lacorte, C., Mansur, E.,Timmerman, B. Cordeiro, A.R., *Plant Cell Rep.* **1991**, *10*, 354–357.

42. Narasimhulu, S.B., Kirti, P.B., Mohapatra,T., Prakash, S. Chopra,V.L., *Plant Cell Rep.* **1992**, *11*, 359–362.

43. Barfield, D.G. Pua, E-C., *Plant Cell Rep.* **1991**, *10*, 308–314.

44. Mathews, H., Bharathan, B., Litz, R.E., Narayanan, K.R., Rao, P.S. Bhatia, C.R., *Plant Sci.* **1990**, *72*, 245–252.

45. Moloney, M.M., Walker, J.M. Sharma, K.K., *Plant Cell Rep.* **1989**, *8*, 238–242.

46. De Block, M., De Brouwer, D. Tenning, P., *Plant Physiol.* **1989**, *91*, 694–701.

47. Radke, S.E., Andrews, B.M., Moloney, M.M., Crouch, M.L., Kridl, J.C. Knauf, V.C., *Theor.Appl.Genet.* **1988**, *75*, 685–694.

48. Miki, B.L., Labbe, H., Hattori, J., Ouellet,T., Gabard, J., Sunohara, G., Charest, P.J. Iyer,V.N., *Theor.Appl.Genet.* **1990**, *80*, 449–458.

49. Misra, S., *J. Exp. Bot.* **1990**, *41*, 269–275.

50. Boulter, M.E., Croy, E., Simpson, P., Shields, R., Croy, R.R.D. Shirsat, A.H., *Plant Sci.* **1990**, *70*, 91–99.

51. Pua, E.C., Mehra-Palta, A., Nagy, F. Chua, N.H., *Biotechnol.* **1987**, *5*, 815–817.

52. Charest, P.J., Holbrook, L.A., Gabard, J., Iyer,V.N. Miki, B.L., *Theor.Appl.Genet.* **1988**, *75*, 438–445.

53. Huang, B., *In vitro cell dev. biol.* **1992**, *28*, 53–58.

54. Swanson, E.B. Erickson, L.R., *Theor.Appl.Genet.* **1989**, *78*, 831–835.

55. Pechan, P.M., *Plant Cell Rep.* **1989**, *8*, 387–390.

56. Thomzik, J.E. Hain, R., *Plant Cell Rep.* **1990**, *9*, 233–236.

57. Damgaard, O. Rasmussen, O., *Plant Mol.Biol.* **1991**, *17*, 1–8.

58. Guerche, P., Charbonnier, M., Joianin, L.,Tourneur, C., Paszkowski, J. Pelletier, G., *Plant Sci.* **1987**, *52*, 111–116.

59. Sacristan, M.D., Gerdemann-Knorck, M. Schieder, O., *Theor.Appl.Genet.* **1989**, *78*, 194–200.

60. Kohler, F., Benediktsson, I., Cardon, G., Andreo, C.S. Schieder, O., *Theor.Appl.Genet.* **1990**, *79*, 679–685.

61. Radke, S.E., Turner, J.C. Facciotti, D., *Plant Cell Rep.* **1992**, *11*, 499–505.

62. Mukhopadhyay, A., Arumugam, N., Nandakumar, P.B.A., Pradhan, A.K., Gupta, V. Pental, D., *Plant Cell Rep.* **1992**, *11*, 506–513.
63. Ohlsson, M. Eriksson, T., *Hereditas* **1988**, *108*, 173–177.
64. Paszkowski, J., Pisan, B., Shillito, R.D., Hohn, T., Hohn, B. Potrykus, I., *Plant Mol. Biol.* **1986**, *6*, 303–312.
65. Hinchee, M.A.W., Connor-Ward, D.V., Newell, C.A., McDonnell, R.E., Sato, S.J., Gasser, C.S., Fischoff, D.A., Re, D.B., Fraley, R.T. Horsch, R.B., *Biotechnol.* **1988**, *6*, 915.
66. Christou, P., McCabe, D.E. Swain, W.F., *Plant Physiol.* **1988**, *87*, 671–674.
67. Christou, P., McCabe, D.E., Martinell, B.J. Swain, W.F., *Trends in Biotechnol.* **1990**, *8*, 145–151.
68. Dhir, S.K., Dhir, S., Sturtevant, A.P. Widholm, J.M., *Plant Cell Rep.* **1991**, *10*, 97–101.
69. Christou, P. Swain, W.F., *Theor. Appl. Genet.* **1990**, *79*, 337–341.
70. Christou, P., Murphy, J.E. Swain, W.F., *Proc. Natl. Acad. Sci. USA* **1987**, *84*, 3962–3966.
71. Umbeck, P., Johnson, G., Barton, K. Swain, W., *Biotechnol.* **1987**, *5*, 263–266.
72. Firoozabady, E., DeBoer, D.L., Merlo, D.J., Halk, E.L., Amerson, L.N., Rashka, K.E. Murray, E.E., *Plant Mol. Biol.* **1987**, *10*, 105–116.
73. Finer, J.J. McMullen, D., *Plant Cell Rep.* **1990**, *8*, 586–589.
74. Everett, N.P., Robinson, K.E.P. Mascarenhas, D., *Biotechnol.* **1987**, *5*, 1201.
75. Moyne, A.L., Tagu, D., Thor, V., Bergounioux, C., Freyssinet, G. Gadal, P., *Plant Cell Rep.* **1989**, *8*, 97–100.
76. McHughen, A., *Plant Cell Rep.* **1989**, *8*, 445–449.
77. Jordan, M.C. McHughen, A., *Plant Cell Rep.* **1988**, *7*, 281–284.
78. Basiran, N., Armitage, P., Scott, R.J. Draper, J., *Plant Cell Rep.* **1987**, *6*, 396–399.
79. Zhan, X., Jones, D.A. Kerr, A., *Plant Mol. Biol.* **1988**, *11*, 551–559.
80. Yoshimatsu, K. Shimomura, K., *Plant Cell Rep.* **1992**, *11*, 132–136.
81. Gould, J., Devey, M., Hasegawa, O., Ulian, E.C., Peterson, G. Smith, R.H., *Plant Physiol.* **1991**, *95*, 426–434.
82. Walters, D.A., Vetsch, C.S., Potts, D.E. Lunquist, R.C., *Plant Mol. Biol.* **1992**, *18*, 189–200.
83. Fromm, M.E., Morrish, F., Armstrong, C., Williams, R., Thomas J., Klein, T.M., *Biotechnol.* **1990**, *8*, 833–839.
84. Gordon-Kamm, W.J., Spencer, T.M., Mangano, M.L., Adams, T.R., Daines, R.J., Start, W.G., O'Brien, J.V., Chambers, S.A., Adams, W.R.Jr, Willets, N.G., Rice, T.B., Mackey, C.J., Krueger, R.W., Kausch, A.P. Lemaux, P.G., *The Plant Cell* **1990**, *2*, 603–618.
85. Spencer, T.M., Gordon-Kamm, W.J., Daines, R.J., Start, W.G. Lemaux, P.G., *Theor. Appl. Genet.* **1990**, *79*, 625–631.
86. Huang, Y.W. Dennis, E.S., *Plant Cell Tiss. and Organ Cult.* **1989**, *18*, 281–296.
87. Antonelli, N.M. Stadler, J., *Theor. Appl. Genet.* **1990**, *80*, 395–401.
88. D'Halliun, K., Bonne, E., Bossut, M., De Beuckleer, M., Lemans, J. *Plant Cell*, **1992**, 4, 1495–1505.

6 Biotechnology of Oil Crops

D. J. Murphy

6.1 Introduction

The term "Biotechnology" has taken on a number of new meanings in recent years. In the past, biotechnology was generally understood to be the employment of biological organisms to carry out useful technological transformations. Classical examples of biotechnological processes include the use of yeasts in the fermentation of sugars to alcohol for the manufacture of wines, beers and other ethanol-based beverages, or the use of different strains of yeasts to provide aeration in bread-making. Both of these examples are ancient uses of microbial organisms to carry out technological processes which would otherwise be difficult or impossible. Classical biotechnology was an empirical process, with little understanding of the molecular events underlying the transformations involved.

Modern biotechnology is something quite different. It involves the use of a number of new technological innovations, many of which are based on molecular biology, to alter the characteristics of biological organisms in order that they may be better utilised for a variety of applications. Such applications cover an enormous range, such as: the improvement of existing domestic animals and crop plants for human and animal nutrition; the adaption of animals and plants to serve as sources of useful non-edible products; the alteration of the characteristics of microbes such as yeasts, bacteria and viruses in order to produce useful products, either in fermentation chambers or by inoculating them into a host organism; and even the production of entirely new, often hybrid organisms, for example by cell-fusion, which may have useful exploitable properties. It should be stressed, therefore, that modern biotechnology encompasses a wide variety of scientific techniques and potential products. The techniques range from the complex and often expensive methods of recombinant DNA technology, through intermediate methodologies such as cell and tissue culture, to relatively simple and routine procedures such as chemical mutagenesis and screening. It is now apparent that biotechnology is being applied in many different areas of production, ranging from petrochemicals, pharmaceuticals, renewable energy, to general arable and pastoral agriculture, food processing, forestry, mining and environmental control. With each passing year, there is a gradually increasing number of products emerging from the new biotechnology. To a great extent, progress in this area has been limited by a relative lack of funding, but as successful products of modern biotechnology increasingly emerge into the market place, the pace of research and therefore of the discovery and commercial manufacture of yet further new products will inevitably accelerate.

6.2 General perspective of agricultural biotechnology

In a recent survey, it was estimated that there were more than 1000 companies involved in modern biotechnology research in the USA and about 500 in Japan [1]. A similar number of companies probably exists in Europe. As long ago as 1980, the US Supreme Court (Diamond vs Chakrabarty) made a landmark ruling to the effect that microorganisms could be patented. In 1985 the US Patent Office extended patent protection to genetically engineered plants and in 1988 to genetically engineered animals [2]. This provided a relatively secure legal framework for the protection of many early products of biotechnology and has been followed by similar rulings in Japan and Europe. The result was to encourage the development of biotechnology companies throughout the world, although much of the investment continues to the concentrated in the USA (50%), Europe (25%) and Japan (15%) [3].

The legal position of some biotechnology patents, however, remains uncertain. There is some dispute about the patenting of cDNA clones derived from genome sequencing projects like HUGO (human genome) or yeast. Nevertheless, with a number of test cases now being processed through the courts, it is likely that the patenting situation will be clarified over the next few years. Another uncertainty that may impede the commercialisation of biotechnological products is the regulatory framework governing aspects of biotechnology research (such as safety and containment), the release of genetically modified organisms into the environment and the transfer of products derived from the latter into the human (and to a lesser extent, domestic animal) food chain. Concerns about the lack of communication between the scientific research community and national or transnational agencies responsible for enacting and enforcing such regulations have been enunciated recently, particularly with regard to Western Europe:

"regulations are developing in part independently of the rest of the field. They tend to respond more to publicly expressed concerns about the possible environmental, public health and negative socio-economic impact than to the steady stream of data generated in the course of risk-assessment research. This may reflect a problem of perception and of defective communication: the data provided by the scientific community are perceived to be either ambiguous, or incomplete, or in the eyes of a small but very vocal group of opponents to the technology as irrelevant at least.

The situation has three directly observable effects on further development of biotechnology in Europe. First, a steady stream of regulations is being produced at both the EEC and national levels that takes little account of the facts and data and therefore is a serious brake on further research and development efforts. Secondly, it creates a perception of inherent riskiness of biotechnology, thereby leading to reinforcement of the demands of the public for further regulatory constraints. Thirdly, it increases the cost, and delays the development, of agricultural biotechnology." [4].

This important topic is discussed in greater detail by Levin & Strauss [5] and in Chapter 8 of this volume. Once again, the USA has provided a clear lead by the relaxation in guidelines governing the release of genetically engineered foods for human consumption announced by the US Vice-President in May 1992 [6]. The same policy has been pursued with equal or even greater vigor by the new US Administration during 1993. Interestingly, one of the first products to be affected by the 1992 ruling was a genetically engineered tomato variety from Calgene, termed "Flavr Savr"©[1] which contains an antisense gene to inhibit structural deterioration during fruit ripening [6]. The same company is now developing the first genetically engineered oil crops, including two new rapeseed varieties accumulating lauric-rich and stearic-rich oils respectively [7,8] for future commercialisation. Such developments will put pressure on European and other regulatory agencies to clarify their rules or risk the further erosion of the competitive position of their biotechnology companies *vis a vis* those from USA.

Despite the widespread perception that biotechnology has underperformed relative to investment, the world sales of biotechnology-derived end-products (excluding fermented foods and drinks) were about $US 10 billion in 1985, which was at least three times the volume of investment in the field made between 1980 and 1985. Industry estimates of demand for the year 2000 vary between $US 35 and 55 billion, representing a better than threefold increase

---

[1] "Flavr Savr" is copyrighted by Calgene Inc, USA

even according to the most cautious figures [9]. In the USA alone, the value of the biotechnology industry is projected to increase from US$ 8.67 billion in 1990 to US$37.87 billion in the year 2000, with an annual growth rate of about 16% [10].

The potential market for plant biotechnology is even greater than this. The global market for traded seeds and agrochemicals in 1987 was estimated at $US 113 billion in farm-level sales [11]. Of this, traded seeds represented 29%, fertilisers were 51% and pesticides were 20%. Estimates of future markets for the products of agricultural biotechnology range from $US 10 to $US 100 billion by the year 2000. The most cautious industry estimates of $US 10 billion total sales include about $US 7 billion for seeds, $US 2 billion for veterinary products and $US 1 billion for microbiological products. It is predicted that biotechnology products will tend to capture an increasing share of the existing markets rather than open up substantial additional markets for agricultural products. The dominance by the seed sector of potential plant biotechnology markets may explain the rush by many transnational chemical companies in the 1980s to acquire seed companies. It is estimated that chemical transnationals spent $US 10 billion over this period in purchasing seed companies which will enable them to market and distribute their future plant biotechnology products [12].

6.3 Oil crop production

Within the global market for agricultural products, oil crops represent a significant and increasing proportion both with regard to value and land use. At present, some 80% of total oil and fat production is utilised in the food sector, but oilseeds have the added attraction, compared with cereals and beans, of being actual or potential medium-high value industrial feedstocks. The estimated global production of oils and fats for 1991 was 85.3 million tonnes (MT) of which 59.1MT were derived from the eight principal oil crops, i.e. soybean, oil palm, rapeseed, sunflower, groundnut, cottonseed, coconut and olive [13]. About 20MT of the global oils and fats were derived from animal and marine fish sources. Over the past 20 years, the use of animal and fish oils has declined relative to vegetable oil and it is likely that this process will continue into the future as meat consumption falls and marine fish stocks are depleted. This represents an opportunity for the further expansion of edible vegetable oil production.

The worldwide consumption of vegetable oils from 1982–88 rose by an average of 4.3% per annum, far outstripping population increases over this period and leading to an average increase in *per capita* food oil consumption of 1.8% per annum (see Figure 6.1). Only four oil crops, i.e. soybean, oil palm, rapeseed and sunflower, account for three quarters of all traded edible vegetable oils. In terms of applying modern biotechnological methods such as the use of molecular markers in breeding programs and genetic engineering, the most promising major oil crop is undoubtably rapeseed. From being a relatively insignificant break crop in the 1970s, rapeseed is now the third most important worldwide oil crop, after soybean and oil palm, with an average annual growth rate of 9.4% from 1982–88. In addition to its quantitative importance, rapeseed has the advantage of being the only routinely transformable oil crop. Rapeseed is also very amenable to tissue culture *in vitro* and, unlike soybean and oil palm, it grows in a wide variety of climates from the Arctic Circle to the Equator. Of the other major oil crops, routine transformation methods will soon be available for soybean and sunflower. Oil palm transformation is, however, a more remote prospect.

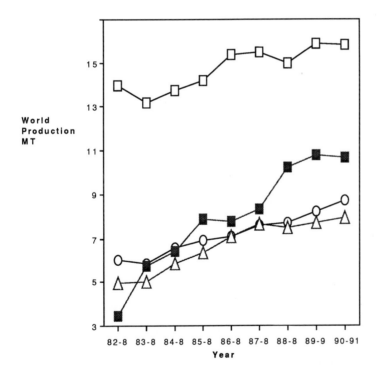

Figure 6.1. World production of major oil crops from 1982–1991 [13].
□——□, soybean; ■——■, oil palm; ○——○, rapeseed; △——△ sunflower.

6.4 Potential for oil crop biotechnology

Vegetable oils are currently produced at a rate of about 62MT per annum [13] and the demand by the year 2000 is estimated to rise to about 90MT. Of the current 62MT production, only about 13MT of vegetable oil is used for industrial purposes. With the prevalence of agricultural food surpluses - not only in developed countries like the USA and EEC, but also potentially in some developing countries [14,15] - there is considerable impetus for the expansion of industrial crop production. An industrial crop can be defined as any crop for which the major utilisation is in the non-edible sector, whether this be fuels, pharmaceuticals, cosmetics, bulk chemicals or specialised chemicals. Hence, even given the buoyant state of the general oil crop sector at present, with a 4.3% annual increase in production, the potential for industrial oil crops may be even more lucrative. Industrial oil crops can supply partially refined complex chemicals and other products that are impossible or prohibitively expensive to produce from other crops like cereals and pulses. In some cases, for example, ricinoleic acid from castor bean, industrial oil crops can supply products that are unavailable from any other sources, either biological or mineral. In many cases, however, industrial oil crop products must compete with those derived from fossil hydrocarbons. In the medium-long term, fossil hydrocarbons, which are a non- renewable resource, will lose their price advantage over vegetable oils and eventually the latter will be the sole source of the entire global oleochemical industry (see Chapter 9).

The prospects for oil crop production are therefore good in both the short, medium and long terms. In the short term, the continued rise in demand for edible oils and an incipient rise in demand for industrial oils will be matched by the availability of set-aside and other surplus land from cereal and animal production. In the medium and long term, there will be an accelerating demand for renewable oleochemicals and other products derived from vegetable oils. The application of modern biotechnology to oilseed crops will allow these opportunities to be grasped through an appropriate collaboration between the public and private sectors.

6.5 Goals for oil crop biotechnology

The major value component of an oil crop is the oil. Two obvious goals for biotechnologists are therefore to maximise the oil yield of the crop and to manipulate the oil quality to suit the various downstream applications. In addition to enhancing the yield and value of the oil, biotechnology can also be employed to improve the quality of byproducts, such as seed proteins. It can also help to reduce or eliminate undesirable seed components such as gossypol in cottonseed or glucosinolates in rapeseed. Biotechnology can be used to improve the disease resistance of the crop (hence decreasing the requirement for chemical inputs), and finally it can accelerate the development of improvements via hybrid seed. Looking further into the future, biotechnology may allow for the rational design of oil crop plant architecture in order to optimise seed yield, growing time, flowering time, desiccation rates, harvesting potential and other useful agronomic characters. An important additional goal for oil crop biotechnology is to translate its achievements to the agricultural systems of developing countries, many of which are currently importers of vegetable oils but which have the potential to become self-sufficient or even to export such oils in the future.

6.5.1 Designer oil crops

There are two major strategies available in order to achieve many of these goals. In the first strategy, "designer oil crops", a successful pre-existing, high yielding oil crop is adapted to suit various market applications using all the resources of

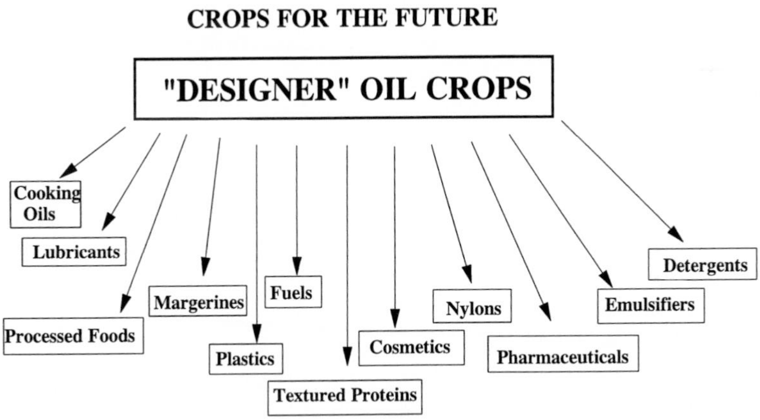

CROPS FOR THE FUTURE

Figure 6.2. Designer oil crops.
A single major oil crop species, such as rapeseed, soybean or oil palm, can potentially be modified by genetic engineering and breeding to produce a host of useful edible, industrial and pharmaceutical end-products.

modern biotechnology. For example, rapeseed can be engineered genetically to make available a series of new varieties with different seed oil compositions targeted to appropriate chemical, pharmaceutical and edible markets. This may mean, for example, that five adjacent farms could each grow a different rapeseed cultivar designed respectively as a source of oil for the manufacture of products such as detergents, plastics, margarines, cosmetics and lubricants (see Figure 6.2). To some extent, this is simply an extension of an existing situation in countries, such as the UK and Germany, where high oleic rapeseed is grown as an edible oil and high erucic rapeseed is grown as an industrial oil, often on adjacent farms. The problem of cross-pollination in rapeseed is minimised by the self-compatibility of the crop, coupled with the short distances travelled by its pollen - eg in recent field trials in the UK it was found that in rapeseed plants at an isolation distance of 47m, the incidence of bee-assisted cross-pollination was less than 3 plants per million (Dale, P.J., unpublished results).

The attraction of producing a "designer" oil crop that can be manipulated to suit many different markets is that all of the agronomic practices, such as input regimes, harvesting technology and seed processing, will be the same for each variety. Economies of scale in producing equipment for the management of the crop will be achieved, eg seed drills, sprayers, trailers with appropriate wheel spacing and ground clearance, harvesters, threshers and seed processors. Since they are simply growing a new variety of a familiar existing crop, it is not necessary for farmers to learn new husbandry skills. Equally, new agronomically-useful traits, such as disease resistance, that are developed for one variety can be introgressed relatively quickly into the other varieties.

Before companies invest in designer oil crops, it will be important to ensure that a commercially viable long-term market exists for each of the seed oils. To some extent this can lead to a vicious circle whereby a secure market cannot exist without a plentiful supply of the novel oil crop variety but investment to enable this oil crop variety to be produced is not forthcoming unless a secure market can be identified [16]. This cycle can only be broken by bold and decisive action by companies and governments. Some companies like Calgene in the USA, are already taking steps to produce and market designer rapeseed crops, [7,8], but in many cases the translation of fundamental discoveries towards near-market applications is deemed to be too long-term, particularly in the adverse economic climate of the early 1990s. The recognition of this problem had led to a number of government-sponsored initiatives in Europe and the USA whereby matching public funds are available for joint private/public sector research in such strategic areas as industrial crops [16, see also Chapter 7].

Providing that adequate funding for research and development is forthcoming, designer oil crops will be a reality within the next few years. Although they have many attractions, as outlined above, they do suffer from the drawback that they maintain the reliance of developed agricultural nations on a relatively few crop-species. Extensive monocultures may involve huge economies of scale but are inherently less ecologically stable than more diverse farming practices. The dangers to monocultures of pest infection or adverse climatic conditions are obviously higher than they are to multicultures. Concern has also been expressed about the narrow genetic resource base available for many of the older domesticated crops, such as wheat and for some newer crops, like rapeseed. The problem in rapeseed is that it is a relatively new species arising from a fusion of the two diploid genomes of *Brassica rapa* and *Brassica oleracea* to produce the amphidiploid *Brassica napus*, as shown in Figure 6.3 (see also Chapter 1). This genetic limitation can sometimes be overcome in rapeseed by producing new resynthesised lines of *Brassica napus*, using wild relatives of its *B.rapa* and *B.oleracea* progenitors. Hence wild *Brassica* species such as *Brassica macrocarpa* and *Brassica insularis* can still be found around the Mediterranean basin and useful traits from these species, such as disease resistance, have been crossed into *Brassica napus* [17].

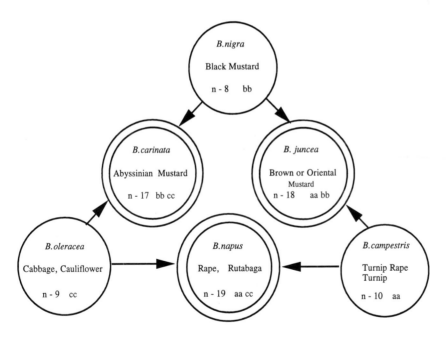

Figure 6.3. Genome relationships between the six major *Brassica* species. This shows the traignle of U [76] which consists of all the three "basic" diploid *Brassica* species at each corner and their "derived" amphidiploid or allotetraploid species on the sides of the triangle.

Whatever the scientific case, the topic of agricultural diversification has now become a political issue. The EEC has made it a matter of policy to encourage the adoption of new crops into European agriculture in order to diversify the types of products obtainable and the range of crop species available to farmers. This policy was recently outlined as follows:

> At present the majority of Europe's agriculture is based on less than 10 crop species; cereals sugar crops, potatoes, oil seed crops and grasses. The increasing trend towards "monoculture based production", primarily due to subsidies and guaranteed markets increases the dominance of these species and intensive production. With the reform of the CAP and inevitable reduction in subsidies, increased diversification and a decrease in intensification are stated Commission objectives. A more diversified production will give a broader spectrum of raw-materials and bring advantages to farmers, industry and the consumer. [18].

As discussed above, diversification of agricultural products is now possible from a single "designer" crop via genetic engineering. However, this will not increase *crop* diversification. The latter can only be achieved by the introduction, or in some cases the re-introduction, of novel crop species. Such a strategy can usefully complement and augment the policy of designer oil crops by allowing for the development of *novel* oil crops.

6.5.2 Novel oil crops

There is an enormous number of wild plants which are potential oil crops. In a survey carried out by the US Department of Agriculture from 1957 until the late 1970's, more than 7000 plant species were screened and over 75 new fatty acids discovered [19,20]. Numerous species were found that produced seeds containing industrially useful fatty acids and several of these are now under development as novel oil crops [20]. Some examples include *Crambe* and meadowfoam to provide very long-chain fatty acids and *Cuphea* spp for short to medium-chain fatty acids. The problems with introducing new crops include the very long timescales involved in their domestication and the requirements for re-tooling of agricultural machinery and downstream processing equipment. Many existing major crops have been domesticated over tens of thousands of years and, in order for new crops to compete with current crops with regard to yield, harvestability, disease resistance, etc, a huge investment in breeding is required. Modern biotechnology can assist in this process, particularly through techniques such as RFLP analysis and tissue culture. This may allow for the cultivation on an economic scale of new oil crops such as *Cuphea* and meadowfoam in the not-to-distant future (see Chapter 9). Ideally, the development of novel oil crops should proceed in parallel with that of designer oil crops. It is likely that each will eventually find specific market niches as well as allowing the agricultural sector to respond to political pressures which may distort pure economic or scientific considerations. The development of and future scope for new oil crops has been reviewed [21] and is discussed in more detail in Chapter 9 of this volume.

6.6 Enabling technologies

Biotechnology is assisting in the improvement of oil crops in two ways. Firstly, the potential and efficiency of classical breeding programs is being enhanced by increasing the genetic diversity within breeding lines and by using marker-assisted selection programs to transfer useful genes into an elite agronomic background in a relatively short period of time. Secondly, genetic engineering is being used to isolate genes directly from unrelated species and to transfer these into advanced breeding lines of oil crops [22]. These two approaches should be seen as complimentary: it is unlikely that genetic engineering will replace conventional breeding but it will undoubtably have a major role in the improvement of oil crops.

6.6.1 Introducing variation into oil crops from related species

The potential of classical breeding programs can be enhanced by several biotechnological methods which enable genetic variation to be introduced into breeding lines from related species and genera. Cross pollination of "sexually-incompatible" species often results in the development of hybrid embryos which abort maybe a few days after fertilisation has occurred. If these embryos are removed prior to abortion and grown *in vitro*, it is possible to develop hybrids which may be partially fertile and can be used in backcrossing programs to introgress variation into breeding lines from the otherwise sexually incompatible species. The classic example of the use of such embryo rescue techniques is in the development of synthetic *B.napus* lines for use in oilseed rape breeding. *Brassica napus*, one of the most important oil crops, is a spontaneous amphidiploid species derived from the interspecific hybridisation of *B.rapa* and *B.oleracea*, followed by chromosome doubling (see Figure 6.3). Partly due to its origins, there is little variation for important agronomic traits in this species. By using embryo rescue techniques, however, synthetic *B.napus* lines can be routinely developed which are sexually compatible with oilseed rape cultivars. This has enabled the extensive genetic variation of the cultivated and wild forms of the two diploid parental species to be used in breeding programs [23,17], although the full potential of synthetic *B.napus* lines in rape breeding programs is probably still to be realised. Embryo rescue techniques are also being used to attempt to increase the genetic variation of other oil crops. For example, hybrids have been developed between soybean and wild perennial *Glycine* species [24]. It is possible that these techniques will enable variation to be introduced into other oil crops which have closely related but sexually incompatible wild relatives, such as sesame.

Interspecific and intergeneric hybrids can also be generated by somatic fusion of protoplasts, prepared from the two parental lines by the enzymic digestions of cell walls. Fusion between protoplasts is induced by using polyethyleneglycol. The hybrid protoplasts develop into callus which can then be induced to differentiate by the appropriate use of plant growth regulators. An obvious problem with this technique is that the hybrid will contain the total genome of both parental species. In addition, several protoplasts may fuse together and it is frequently difficult to distinguish hybrid genotypes from the parental genotypes until the plants have been regenerated.

A more promising technique is asymmetrical protoplast fusion in which the protoplasts from one of the parental species (the donor) are irradiated with X-rays, UV or γ radiation in order to disrupt their nuclei, whereas the protoplasts of the other species (the recipient) are treated with iodoacetic acid, which disrupts cytoplasmic organelles such as mitochondria. Only hybrid cells will be able to develop, which will contain the nucleus of the recipient plus the cytoplasmic organelles and a variable amount of nuclear DNA from the donor, which it is hoped will integrate into the recipient nuclear DNA. This technique enables just part of the donor genome to be transferred into the recipient line.

True hybrid plants can be screened by visual phenotype, isozyme analysis, Restriction Fragment Length Polymorphisms (RFLPs) and other molecular markers. Since irradiation disrupts the donor nuclei to different degrees, a series of asymmetric hybrids is produced, ranging from plants indistinguishable from one or other parent to true intermediates between the two species. The advantage of this method is that, unlike transformation, the genes do not need to be identified and isolated before transfer. All that is necessary is that a plant is selected with a desirable attribute, eg a high value oil, disease resistance, high water-use efficiency etc, and that protoplasts can be prepared from such plants. Hybrids can then be selected that contain the desired trait (perhaps encoded by numerous genes and hence very difficult to transfer by normal transformation methods) in the desired genetic background. Backcrossing to a suitable elite cultivar can then give a commercial oil crop variety expressing the new trait.

While these techniques are being used extensively to increase the diversity available to the plant breeder, only genes within relatively closely related species and genera can be transferred at present. Unfortunately, this often leads to genetically unstable hybrids in which the introduced genes from the related species cannot integrate fully into the genome of the crop species and may be lost within a backcrossing program. Nevertheless, protoplast fusion already has considerable potential for oil crop improvement and this will increase further as the techniques continue to be refined in the future.

6.6.2 Induced variation

Attempts have frequently been made to induce genetic variation in breeding lines through the use of chemical mutagens such as ethylmethanesulphonate (EMS), or by radiation. The basis of the method is to produce a large number of mutations in a population of seeds and to screen for the desired phenotype in the M_1 generation. Since each resulting individual plant will contain many other mutations in addition to the desired one, an extensive backcrossing program using wild-type elite cultivars is then required in order to obtain plants containing single-gene mutations.

The potential of this method has recently been elegantly demonstrated with the development of low linolenic acid varieties of linseed, now termed "linola". Linseed is a good industrial oil crop producing seeds with high levels of α-linolenic acid-rich oil, suitable for a variety of applications, most notably as a drying agent in paints, varnishes, putty, inks etc. Linseed oil can contain up to 70% α-linolenic acid [24]. A decline in the demand for linseed oil during the 1970s and 1980s led to a search for lower linolenic acid varieties of linseed which could then be used as a source of edible oils, for which demand was still buoyant. A survey of naturally occurring variants from germplasm collections showed that the lowest level of α- linolenic acid was still relatively high at 45% total oil. The lack of natural low α-linolenic acid variants led to a mutagenesis program whereby seeds of the Australian cultivar Glenelg were exposed either to γ radiation or EMS. At maturity, over 7,000 plants were harvested and their seeds screened for a low α-linolenic acid phenotype by a rapid oxidation-based color test. Two recessive mutant lines, with low α-linolenic acid contents of 30% were obtained. Upon crossing these two mutant lines a very low α- linolenic acid double-recessive line was produced which contained only 2% of this fatty acid in its seed oil but normal levels in non seed lipids [26,27]. A similar approach was employed to reduce the α-linolenic acid content of the Canadian linseed cultivar, McGregor, from its normal range of 40–60% to about 2% of seed oil [28]. The results of these two mutagenesis programs are new "linola" linseed lines containing high linoleic (up to 70%) but very low α-linolenic acid (2%) seed oils. These new linseed varieties can serve as sources of premium grade high polyunsaturated edible oils on a par with the highest quality sunflower or safflower oils.

Induced mutation is a "sledgehammer" technique requiring the screening of as many as 100,000 seeds. It relies upon a rapid, facile screening method and an extensive backcrossing program to remove the large numbers of undesirable mutations. Also, being a mutagenesis method, it can only delete gene functions and cannot be used to add new genes to crops. Nevertheless it can produce dramatic results within a few years and its relatively low-technology methods make it a useful approach to be adopted where genetic engineering is impossible, e.g. due to the lack of a suitable transformation system, or difficult due to lack of resources, e.g. in developing countries.

In addition to inducing mutation through chemicals or radiation, somaclonal variation, i.e. variation induced by tissue culture, has been characterised in several oil crops, for example, soybean [29] and *Brassica* [30]. While this variation has been shown to be heritable, its causes are not fully understood, but could be due to the activation of latent transposable elements or chromosomal rearrangements [31,32]. The potential for the development of somaclonal variation for the production of new oil crop breeding lines has yet to be explored.

6.6.3 Molecular markers and their use in breeding programs

The "quiet revolution" in crop improvement may well be the use of molecular markers to assist conventional breeding programs. Once a useful agronomic character has been identified in a breeding line, it may take many years to transfer this character into an elite agronomic background. This is particularly true if the desired character is from an exotic source such as a wild relative or different crop type, and may be linked genetically to undesirable characters or if the character can only be assessed at maturity, as would be the case for oil content. Morphological markers have been used to assist in selection programs but these are of limited value due to their rather low abundance in most crops and their virtual absence from crops such as rapeseed. However, three classes of molecular markers, namely isozymes, RFLPs and, more recently, Random Amplified Polymorphic DNAs (RAPDs) are beginning to have a significant impact on the efficiency of plant breeding programs. Of these, RFLPs are likely to have the greatest effect and they are currently being used in breeding programs of numerous crops including several important oil crops such as rapeseed [33], soybean [34,35] and maize [36]. In the UK *Brassica* program, RFLPs have been used to develop genetic linkage maps of over 500 loci in *B.napus*, 200 loci in *B.oleracea* and 150 loci in *B.nigra* [37]. These RFLP-based genetic maps are not only invaluable tools for plant breeders, but are also of great interest for solving fundamental scientific problems relating to evolution, genetics, physiology, ecology, etc.

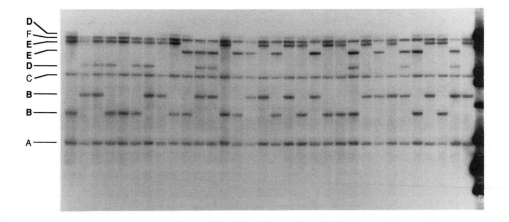

Figure 6.4. Restriction Fragment Length Polymorphism (RFLP) Film.
Each of the 32 lanes contains a series of restriction fragments generated by cleaving total DNA from 32 different rapeseed breeding lines and blotting the gel onto a nylon filter. The filter is then probed with radioactively-labelled DNA (in this case from a gene involved in oil synthesis). In some cases, i.e. A, C and F, the probe hybridises to the same band in each genotype demonstrating that these genes are monomorphic. In other cases however, i.e. B, D and E, the probe hybridises to different bands in different genotypes, showing that these genes are polymorphic. This allows for the rapid identification and mapping of genes of agronomic and scientific interest, e.g. for oil yield, disease resistance, chilling tolerance, etc., from only small samples of plant tissue in relatively large breeding populations. Such techniques can halve the time required for a traditional crop breeding program. Film produced courtesy of A Sharpe, C Batchelder and D J Lydiate, 1993.

To develop RFLPs, restriction fragments of various lengths are generated by cleaving DNA from breeding lines with endonucleases and separating the fragments by gel electrophoresis. Cloned radioactively-labelled DNA sequences are then used as probes. These hybridize to homologous sequences on one or more of the restriction fragments and can be visualised through exposure to photographic film as shown in Figure 6.4. The length of the fragment(s) to which the probe hybridises (and hence the position on the photographic film) will depend upon the position and number of restriction sites within the DNA donor. With the use of a large number of probes, and/or by varying the restriction site by using different endonucleases, many polymorphisms between different genotypes can be detected and loci identified. Within segregating populations, linkage relationships between these RFLPs and with loci regulating agronomically important traits can be deduced.

RFLPs have several useful characteristics as molecular markers;

1. The technology can be applied to all crops and extensive polymorphisms can usually be detected.

2. Many RFLPs can usually be detected in a single segregating population.

3. Markers are co-dominant. This enables heterozygotes to be detected in backcrossing programs so that recessive traits can be selected for without the need for test crosses.

4. There are no environmental effects on the occurrence of RFLPs and tissue can be sampled from any part of the plants. This enables loci regulating, for example, oil quality to be selected by examining the genotype of young seedlings well before flowering or seed set.

The most important application of RFLPs may be in the rapid introgression of traits from exotic germplasm into elite breeding lines and the selection of quantitative trait loci (QTLs). Frequently, important agronomic characters such as oil content, disease resistance and cold tolerance are inherited quantitatively and are thought to be controlled by numerous loci, each having a relatively small effect. Using conventional breeding methods, selection for these characters has often been very difficult. However, it is now becoming possible to identify linkages between QTLs and RFLPs [36,38] and hence select for these traits within breeding programs. For example, several QTLs which determine seed protein and oil content have been identified in soybean with the use of RFLPs [35].

More recently, RAPD markers have been proposed as a new type of genetic marker which may overcome some of the technological limitations of RFLPs [39,40]. DNA from the donor genotype is incubated with short oligonucleotides (usually 10-mers), which act as primers for DNA amplification via the polymerase chain reaction. The amplified DNA is separated by electrophoresis and visualised by staining with ethidium bromide. While the technological investment required for using these markers is considerably less than for RFLPs, their usefulness may be limited by other factors such as reproducibility and by problems with their application in complex genomes such as *Brassica* [41], although some preliminary RAPD linkage maps have now been developed for *Brassica* spp.

6.6.4 Gene isolation

Genetic engineering involving the transfer of a specific gene(s) from one organism to another is reliant on the success of three procedures. Firstly, the required gene(s) needs to be identified and isolated; secondly, it needs to be transferred into the crop species ("transformation") and thirdly, there must be appropriate temporal and spatial expression of the introduced gene. With regard to oil crop improvement, it may be desirable to alter the fatty acid composition of the oil by transferring a gene encoding one of the enzymes involved in oil biosynthesis (see Chapter 4). This gene may come from another plant, a microbe or an animal. If the enzyme which the gene encodes can be purified, antibodies can be raised against it to probe cDNA libraries in expression vectors such as the λ-bacteriophage. Alternatively, amino acid sequence data from the purified enzyme can be used to construct oligonucleotides which can then be used to probe cDNA libraries. In either case, a cDNA encoding part or all of the enzyme is then isolated. Such methods have been used to isolate a soluble Δ_9 stearate desaturase gene from *Brassica*. Transformation of an 'antisense' copy of this gene into a rapeseed relative (see 6.6.7) resulted in a seed oil with an elevated stearic acid content, which may be valuable in the manufacture of cocoa butter substitutes [7].

Unfortunately, however, it is often impossible to purify the target enzyme. This may be due to its low abundance, its instability or its association with membranes. A good example is the C_{18} Δ_{12} and Δ_{15} fatty acid desaturases of oil crops which have resisted purification for the last 35 years. Very recently, molecular genetic rather than biochemical approaches have proved successful in isolating the genes that encode these enzymes. In this case, the cruciferous species *Arabidopsis thaliana* has been used due to its ease of genetic manipulation and the considerable amount of genetic information that is available (see 6.6.8). Additionally, its close phytogenetic relationship to rapeseed (they are both members of the *Brassicaceae*) makes it an ideal candidate for gene isolation. Desaturase mutants possessing a low content of Δ_{12} and Δ_{15} fatty acids were isolated by screening a large number of chemically mutagenised seeds [42]. The locations of the mutations were mapped with respect to RFLP markers and, as the mutations mapped sufficiently close to several of these RFLP markers, it was possible to isolate the Δ_{15} desaturase gene by the technique of "chromosome walking" [43]. This method relies on the availability of an overlapping library of *Arabidopsis* genomic DNA in yeast artificial chromosome (YAC) vectors which allow very large pieces of DNA (50–70 kbp) to be cloned. By developing an ordered sequence of YACs along a chromosome it is theoretically possible to "walk" along the chromosome from an RFLP marker on one side of the relevant locus to another marker on the other side of the locus. The gene of interest will therefore be contained within one or more of the vectors. The identity of the putative gene (in this case a desaturase) can then be verified either by homology with other similar genes and/or by adding this gene back to the original mutant plant line in which it was defective. The result should be the restoration of the wild-type phenotype, i.e. the "rescuing" of the mutant.

If YAC vectors are not available, it is still possible to clone genes for which gene products are unknown or unavailable. Cloning can be achieved by various tagging approaches, including T-DNA tagging and transposon tagging. The former method relies upon the random insertion of a foreign piece of DNA, such as T-DNA from *Agrobacterium*, into the genome. Often only a single gene will be disrupted and, by screening for the required mutant phenotype (such as an altered fatty acid profile), it is possible to isolate the gene encoding for the desired character with the use of probes based upon the T-DNA sequence. This method has been used successfully to isolate a Δ_{15} desaturase gene from *Arabidopsis* [Browse, J. personal communication]. In a similar manner, transposon tagging relies upon the identification of a mutant phenotype caused by the insertion of a transposable element into the gene of interest with the consequent specific disruption of this gene. This method has been used successfully to isolate genes from maize and *Antirrhinum* which have several endogenous transposable elements. The maize *Ac* transposable element has also been transferred into other species such as tomato and *Arabidopsis thaliana* to explore the feasibility of transposon tagging approaches to gene isolation in species which do not have endogenous transposable elements.

6.6.5 Gene transfer

Transformation by the insertion of foreign genes is now routine for several experimental plant species, such as tobacco, petunia, tomato and *Arabidopsis*. Amongst the major oil crops, only rapeseed is able to be transformed in a routine manner. One reason for this is that rapeseed cell cultures and tissue slices respond well to infection by the bacterium *Agrobacterium tumefaciens*. Providing the correct vector is used, the transformed rapeseed cells can then be regenerated at reasonably high frequencies in the presence of antibiotic selection systems [44]. It is essential that a plant species to be transformed should be amenable to cell and tissue culture, to regeneration in selection media, and be capable of receiving foreign DNA either via a viral or bacterial vector, or directly by uptake of naked DNA or by particle bombardment. Sunflower [45], soybean [46], cottonseed [47] and maize [48,49] have also been transformed by various methods but these have yet to be refined as routine procedures and progress lags several years behind rapeseed. This topic is discussed in more detail in Chapter 5.

The relative ease of rapeseed transformation has resulted in the transfer of several genes of potential agronomic interest within the past few years. Several companies, including Monsanto and Dupont, have transferred genes into rapeseed conferring attributes such as herbicide tolerance (e.g. to the BASTA herbicides) and pest resistance (e.g. *Bacillus thuringensis* toxin or cowpea trypsin inhibitor, both of which may be toxic to pests of rapeseed). More recently, Calgene have announced the transfer of a medium chain acyl hydrolase gene from the California Bay plant into rapeseed, which may result in the production of a lauric acid-rich oil suitable for industrial use [8]. These and other recent breakthroughs in oil crop biotechnology will doubtless be followed by similar achievements within the next few years as the transformation technologies now in use for rapeseed begin to be adapted for the other major annual oil crops as discussed in Chapters 5 & 8.

6.6.6 Expression of transgenes

Once a suitable gene has been isolated from an organism, it is necessary to ensure that, when the new gene is inserted into the recipient crop plant, it is expressed in the appropriate quantity, in the correct tissue and at the right developmental stage. These three levels of gene regulation, i.e. quantitative, spatial and temporal, are effected by complex sets of interactions between non-coding regions of DNA at the 5' and 3' non-translated flanking regions around the open reading frame and, possibly, by elements within introns. Regulation of gene expression may occur at the transcriptional and/or translational levels and the protein product may itself be regulated, e.g. with regard to its enzymatic activity, by further changes such as covalent modification (e.g. phosphorylation, reduction/oxidation, acylation), or modulation by effectors (e.g. cofactors, inhibitors, activators) which may depend upon the developmental and physiological stage of the organism.

Relatively little research has been carried out to elucidate the detailed regulatory mechanisms that govern developmentally-related gene expression in plants. It is known that the 5' non-coding regions (promotors) of most genes are the most important determinants of regulated gene expression but they are not the only ones. There is some evidence that heterologous promotors, particularly between monocots and dicots, are less efficient at enabling the expression of genes in transgenic plants than are endogenous promotors. An example from rapeseed concerns the insertion of an oleosin gene from maize into transgenic rapeseed plants. Oleosins are seed-specific proteins involved in the stabilisation of storage oil bodies [50]. When the maize oleosin gene was driven by its own promotor in transgenic rapeseed plants, little expression of the transgene was observed. In contrast, the maize oleosin gene driven by a rapeseed promotor (isolated from the gene encoding the seed storage protein, napin) was efficiently expressed in a tissue-specific manner and the protein was correctly targeted to its normal cellular location, the storage oil bodies [51]. Therefore a foreign gene can be correctly regulated at least with regard to its spatial expression, if it is driven by a promotor from the same species into which the gene is inserted, even if that promotor normally regulates a totally different gene, albeit one expressed in the same tissue as the foreign gene.

Simply inserting a novel gene into a plant and obtaining developmentally-regulated expression is by no means the end of the story. As research in this field progresses, more and more examples are being found of post-transcriptional regulation of gene expression. An instructive recent example has been reported in developing oat seeds, which contain two classes of storage proteins, i.e. globulins and avenins. The globulin and avenin genes are expressed at similar levels and produce similar amounts of mRNA transcripts. Both classes of mRNA transcript are fully loaded with ribosomes and yet the seeds accumulate a ten-fold excess of globulin proteins over avenins [52]. Clearly there is a difference in the rate of translation elongation or termination reactions which favours the accumulation of globulins rather than avenins, even though both gene sets are expressed at similar rates. It is important that such post-transcriptional regulatory mechanisms are more closely understood if we hope to be able to predict the outcome of gene-insertion experiments in crops. Therefore, although foreign genes can be expressed at modest levels in a developmentally regulated manner in transgenic oil crops such as rapeseed [53], this regulation is still at a relatively crude level. To a great extent this is because genetic engineering, and particularly the regulation of gene expression is still relatively poorly understood, resulting in an essentially empirical approach to the problem. A more sophisticated appreciation of genetic control mechanisms will allow for a more considered approach to the regulation of alien gene expression in crop plants in the future.

6.6.7 Gene amplification or inhibition

In addition to inserting novel genes into oil crops, it is often desirable to upregulate or downregulate endogenous genes. For example, the upregulation of a gene encoding a rate-limiting step in lipid biosynthesis in seeds may lead to the accumulation of more oil, hence enhancing the oil-yield of the crop. Equally the downregulation of Δ_{15} desaturase genes in rapeseed embryos may lead to lower levels of α-linolenic acid, reducing the requirement for hydrogenation and adding to the value of the seed oil. Both up and downregulation of gene expression can sometimes be achieved by conventional breeding but this is often a long and expensive process which in many cases (like that of reducing α-linolenic acid in rapeseed) has yielded rather limited results to date.

The addition of extra gene copies into a plant may often lead to enhanced gene expression and the accumulation of more of the desired product. Rather confusingly, however, extra gene copies sometimes interact with their endogenous homologs to cause a reduction or even an abolition of gene expression, rather than its enhancement - a phenomenon termed "co-suppression" [54,55,56]. Once again, genetic engineers must rely upon an empirical approach to gene amplification until this phenomenon is understood in more detail.

The insertion of extra gene copies in the reverse, or antisense, orientation into a plant often leads to a considerable reduction in expression of the endogenous gene. The mechanism for antisense-mediated downregulation of gene expression is poorly understood at present but the important thing is that this technique seems to work fairly reliably. It allows for the selective attenuation (rather than the complete abolition) of the function of one specific gene in a specific tissue of interest and is therefore much more direct than "sledgehammer" techniques, such as chemical or radiation-induced mutagenesis. This method has already been used to inhibit the expression of the Δ_9 stearate desaturase gene in *Brassica* resulting in the accumulation of a seed oil enriched in stearic acid [7]. It was very important that the Δ_9 desaturase gene function was unaffected in non-seed tissues since the accumulation of relatively small (7–8%) proportions of stearic acid in membrane lipids of leaves has been shown to cause severe abnormalities at growth temperatures of 12°C and below (Browse et al, unpublished results). The problem was solved by using a seed-specific gene promotor, i.e. napin, to drive the expression of a Δ_9 stearate desaturase gene in the antisense orientation. The napin promotor ensured that the antisense gene was only expressed in the developing seeds with the result that, while the seed oil contained a 20-fold higher proportion of stearic acid, the levels of this fatty acid in membrane lipids of other tissues remained unaltered [7]. This result is encouraging for the genetic engineering of novel fatty acids into oil crops because it implies that changes in seed oil composition can be achieved without affecting the composition of the vital membrane lipids, as discussed in more detail in section 6.7.

6.6.8 Use of *Arabidopsis*

The advantage of *Arabidopsis* for the isolation of recalcitrant genes, such as desaturases, from oilseed crops has already been mentioned above. *Arabidopsis* has further advantages since it is the model system of choice for a great deal of modern plant molecular genetic research, particularly into aspects of many developmental processes. Additionally, *Arabidopsis* will be the first plant to have most or all of its genome sequenced thanks to an ongoing internationally coordinated program, mainly involving laboratories in the EEC and USA [57]. The program will result, over the next ten years, in the availability of huge amounts of DNA sequence data which will assist in the understanding of many agriculturally relevant processes in this model plant. Examples include flowering time, vernalization, disease resistance, canopy architecture, embryo development, drought and salt stress responses, etc. The translation of this knowledge to the major crop species will be the next stage and here rapeseed is at a considerable advantage over other oil-rich and non-oil-rich crops, such as soybean, sunflower, maize, wheat and rice. This is because *Arabidopsis* and *Brassica* are closely related genera and the protein-coding regions of their homologous genes are about 86% conserved (King, G.J., unpublished results). It is likely that much of the long-range genetic linkage and short-range genome architecture is maintained between *Arabidopsis* and *Brassica* (Lydiate,D.J., personal communication). This colinearity of the two genomes will have been disrupted by a) the contraction of the *Arabidopsis* genome, which will affect particularly colinearity in repeated sequences and intergenic regions and b) large-scale chromosomal rearrangements, such as translocations, deletions and inversions, which have even disrupted the collinearity of closely related *Brassica* species. These factors notwithstanding, it is likely that the maintenance of gene order over short genetic distances will allow for the ready transfer of information and technology from *Arabidopsis* to the important *Brassica* oil crops. As an example, the conservation of gene order over distances as short as 10-100 adjacent genes will allow for the use of ordered *Arabidopsis* YACs to facilitate chromosome walks to genes of agronomic interest in *Brassica* species.

6.7 Problems and limitations

6.7.1 Channelling of novel fatty acids to seed oils

A major goal of oil crop biotechnology is to change the fatty acid composition of the seed oil in an established crop such that it can be used as an industrial feedstock. Often this means that unusual fatty acids with very long or very short chain lengths, internal hydroxyl groups, epoxy groups or conjugated double bonds must be produced. It is essential that these unusual fatty acids are channelled towards storage triacylglycerols rather than membrane lipids since their incorporation into the latter may be toxic to the plant. For example, the medium chain acyl hydrolase gene of the California Bay was cloned into transgenic rapeseed resulting in elevated levels of lauric acid being produced in developing seed tissues [8]. Previous *in vitro* data had suggested that lauric acid may be rather poorly incorporated into triacylglycerols in rapeseed [58]. This meant that there was a danger that the lauric acid may instead accumulate on membrane lipids, where its short chain-length would cause it to disrupt membrane packing and act like a detergent, either destabilising or even fragmenting cell membranes. Luckily, the *in vitro* data did not predict events *in vivo* in the transgenic plants. Although lauric acid is never normally accumulated in rapeseed, it was effectively channelled towards triacylglycerols in the transgenic plants [8]. This finding gives encouragement to the view that developing seeds have a mechanism to divert unusual and potentially toxic fatty acids away from their cell membranes, even if these fatty acids are totally novel to them.

6.7.2 Flexibility of seed oil composition

It is often assumed that virtually any fatty acid can be incorporated into storage oils with little or no effect on the performance of the plant as a whole. Some support for this comes from the observation that oil-storing plants contain about a thousand different fatty acids in their storage oils [59]. These fatty acid types range from C_8 to C_{24} chain lengths, mono, di and trienoic unsaturates, conjugated multienoics, epoxy and hydroxy derivatives, and even wax esters. In contrast, membrane lipids contain a very limited range of fatty acids of C_{16} and C_{18} chain lengths. Also, seed oil composition can be changed dramatically with virtually no pleiotropic effects, e.g. on plant growth, seed yield, germination rate etc. An example of this was found when the original high erucic acid (45–55%) rapeseed cultivars were bred to produce a very low erucic (<1%) but high oleic acid (60%) variety [60]. The fact that rapeseed oil normally contained high erucic acid until the 1960s but has now been manipulated to contain high oleic [60], lauric [8], or stearic [7] acids without affecting germination rates or plant growth and habit means that at least this crop species is extremely malleable with respect to its seed oil composition. Although its is difficult to extrapolate such findings to all species, similar radical modifications have been made in the oil content of sunflower and safflower (high linoleic → high oleic) and linseed (high α-linolenic → high linoleic) with little or no effect on their overall agronomic performance. This gives grounds for considerable optimism that oil quality can be changed virtually at will in most or all of the major oil crops.

6.7.3 Homogeneous seed oils

Most seed oils contain a mixture of fatty acids, although one type may predominate, e.g. oleic acid constitutes 80% of olive oil, petroselinic acid is 80% of coriander oil, linoleic acid is 70% of sunflower oil and α-linolenic acid is up to 70% of linseed oil. In some cases it may be desirable to have a mixture of fatty acids in the seed oil. For example, the approximate 1:1:1 ratio of palmitate: oleate: stearate in cocoa butter confers the desirable properties for chocolate manufacture which give this oil its high value. The judicious mixture of oleic and linoleic acids in soybean, rapeseed and sunflower oils allow them to be used to manufacture solid but "spreadable" margarines containing sufficient polyun-saturates to address a particular "health conscious" consumer demand.

In the majority of cases, however, and certainly for most industrial applications, the end-use of a seed oil is targeted towards only one of its constituent fatty acids. Examples include ricinoleic acid from castorbean, erucic acid from crucifers, and lauric acid from coconut or palm kernel oil. It is obviously desirable, therefore, to maximise the amount of the desired fatty acid in a given seed oil. Not only will this increase the yield of the required product, but it can result in considerable savings in downstream processing costs. This can have dramatic effects on the price and hence the demand for a particular seed oil. For example erucic acid is a valuable industrial feedstock with a host of applications including polymer synthesis, coating agents, antifoaming agents in detergents and lubricants [61,62]. Currently available high-erucic rapeseed cultivars contain only about 50% erucic acid in their seed oil. It has been estimated by one of the principal European companies involved, Unichema, that the development of a rapeseed oil with >90% erucic acid would result in a ten-fold increase in the market for this oil and this view is mirrored by industry analysts in the USA [62].

It is likely that the demand for other industrial oils will be similarly price-elastic and this may be a serious impediment to their commercial development unless more homogeneous seed oils can be produced. The problem here is that some of the most important oil crops, including rapeseed, appear to be unable to form homogeneous seed oils for fundamental biochemical reasons - see Chapter 4 for a more detailed discussion of oil crop biochemistry. Rapeseed is only able to accumulate triacylglycerol oils with erucic acid on the *sn*-1 and *sn*-3 positions, resulting from the low erucic acid specificity of the acyltransferase responsible for inserting fatty acids on the *sn*-2 position. In rapeseed, this acyltransferase will only utilise erucic acid at 1% of the efficiency of oleic acid. No amount of conventional plant breeding is likely to resolve this problem. A possible solution is to transfer the gene encoding the homologous enzyme from another species where there is no discrimination against erucic acid on the *sn*-2 position. Possible candidates include nasturtium and meadowfoam (*Limnanthes*) and efforts are now in progress in several Research Institutes in the UK, Germany and USA to clone acyltransferases from these species for insertion into rapeseed with the goal of producing a rapeseed oil containing 90–95% erucic acid.

The *sn*-2 acyltransferase of rapeseed also discriminates against other fatty acids of industrial interest such as lauric and petroselinic acids [63,64]. Therefore, in order to produce very high lauric or petroselinic acid rapeseed oil, it may be necessary to isolate the *sn*-2 acyltransferase from non-discriminating species such as *Cuphea* for a lauric acid-using enzyme, and coriander for petroselinic acid-using enzyme, for transfer to rapeseed. At the same time, it may be necessary to downregulate the endogenous rapeseed *sn*-2 acyltransferase - but only in the seed. Therefore, in the case of lauric acid, the cloning and insertion into rapeseed of at least three genes would be required, i.e. the California Bay or *Cuphea* medium chain hydrolase (to produce lauric acid), the *Cuphea sn*-2 acyltransferase (to ensure lauric acid is esterified to the *sn*-2 position of the oil), and an antisense-orientated rapeseed *sn*-2 acyltransferase (to downregulate the endogenous *sn*-2 acyltransferase gene). All three genes would need to be under the control of seed-specific gene promotors with moderate to high levels of expression. Examples might include the napin [7], acyl-carrier-protein [8] or Δ_9 stearate desaturase [65] promoters. This illustrates the complexity of the problems facing oil crop genetic engineers but, as shown above, several solutions have been proposed and research is now underway to determine the best strategy for producing more homogeneous seed oils. Now that oil crop genetic engineering has been proved to be feasible, it is important that fatty acid levels in seed and fruit oils are increased towards 100% if these products are to increase their market share or even invade new markets in competition with fossil-derived oleochemicals.

6.8 Applications to developing countries

In general, the new biotechnologies can be regarded as being particularly appropriate for use in developing countries, as discussed by Meeusen [66] and Bollinger [67] and more recently by Altman [68]. More particularly, the development of oil crop biotechnology will provide numerous opportunities for developing countries although the type of opportunity will vary enormously depending upon the nature of the country and its agriculture. Many developing countries have little or no seed crushing capacity of their own and must rely on the import of expensive refined oils that are normally paid for in hard currencies which they can ill-afford. Collectively, developing countries import about 8MT vegetable oil a year at a cost of some $US 3 billion [69]. It is ironic that countries like Bangladesh, India and Pakistan with good local oil crops, especially rapeseed, are forced to import expensive foreign oils, such as US or Brazilian soybean oil. India and China alone are forecast to have annual net seed oil imports of 5.2 MT by the year 2000 at a cost of about $US 2 billion [70]. Probably the best way to rectify this imbalance between oil demand and production is to step up the development of existing oil crops and facilities for their processing in such countries.

One relatively easy way to achieve an increase in seed oil production is to increase oil crop yields from the low values obtained at present. In India and Pakistan, rapeseed and mustard yields are only 0.5–0.7 T.ha^{-1} in contrast to 1.1-1.4 T.ha^{-1} for Canadian spring rapeseed. With the use of irrigation, yields of *Brassica juncea* of 2.0–2.5 T.ha^{-1} have been achieved in India and even increasing these yields only as far as the relatively low Canadian values would reduce India's oil deficit by 70% [67]. This is achievable without relying on the substantial fertiliser and pesticide inputs that raise the yield of European winter rapeseed to in excess of 3.5 T.ha^{-1}. Apart from using irrigation, yields can be also raised by increasing the agronomic performance of the crop with respect to characters such as disease resistance. There are also opportunities here for the utilisation of new biotechnological approaches, such as RFLP markers, from more developed countries to accelerate the movement of such characters into crops from developing countries. This can be a two-way process whereby germplasm resources are pooled to the benefit of all countries concerned and the new techniques and expertise are made available to scientists from developing countries in order to solve their particular local problems. Scientists from the Brassica & Oilseeds Research Department of the John Innes Centre, Norwich, UK are now beginning to apply this mutually beneficial reciprocal transfer of germplasm and expertise in rapeseed biotechnology with colleagues in China and Pakistan.

While some developing countries are major net importers of seed oils, others are amongst the largest oil exporters. This applies particularly to countries like Malaysia and Indonesia which export vast amounts of palm and coconut oils. One of the major markets for edible palm oil in the USA has been threatened by the perceived disadvantages of high saturated oils in the diet. This has been used with great effect to promote the consumption of home-grown alternatives, such as soybean oil. The market for lauric acid-rich palm kernel and coconut oils may also be threatened in the future by the development of high-lauric acid rapeseed varieties, as announced recently in the USA [8]. This has focused the minds of palm and coconut producers towards the employment of biotechnology to manipulate oil compositions. At present this seems a daunting, but not impossible, task. It is likely to be many years before perennial tree crops such as palm, coconut and olive can be manipulated by genetic engineering. Not only is the technology completely undeveloped in these species, the regeneration time before new fruits and seeds are produced from seeds or grafts by the trees can be in excess of 12 years - contrast this with the 4–6 week timescale for the comparable process in *Arabidopsis*!

Although genetic engineering is not on the immediate horizon for the perennial oil crops, there are numerous other actual or potential applications of biotechnology which are available. Tissue culture propagation of oil palm and, to a lesser extent, coconut is being developed. Clonal propagation has been used for oil palm but serious problems involving abnormal flowering in clones subjected to large-scale production have severely set back the exploitation of the method [71]. If this method could be made more reliable, the availability of variants with a wide range of different oil compositions would allow for large-scale propagation of clones with novel (for palm) fatty acids within 5-10 years. Other potential methods, including the use of haploids, protoplasts and embryo rescue, are under development, in particular for palm, and RFLP technology has been under development since 1987 [72, for review see ref. 73]. In contrast to oil palm, the extension of modern biotechnological methods to the major Mediterranean perennial oil crop, olive, has scarcely even begun. One reason for this is that olive culture tends to be fragmented on small family farms rather than on the huge plantations typically used for oil palm cultivation. Hence, we have an ironical situation that a "developing" country like Malaysia is able to sustain a high quality coordinated biotechnology and conventional research program into oil palm using its own well-reputed institutes (such as PORIM - Palm Oil Research Institute of Malaysia) in collaboration with plantation owners and oil users like Unilever. In contrast, "developed" countries such as Italy and Spain have been unable to date to mount comparable programs to improve their indigenous olive crops. Coconut cultivation, like that of olive, is relatively fragmented. It is estimated that some 5 million smallholdings, 98% of which are of less than 2 hectares, make up the 10.7 million hectares of worldwide coconut plantings [74]. This is likely to retard attempts to use either conventional methods or modern biotechnology to improve coconut crops. Indeed, the 1.56% increase in coconut oil production during the past decade in the major Asian and Pacific countries is almost exactly matched by the 1.5% increase in planted area. There has evidently been very little yield improvement due to the development of better coconut varieties [74]. Clearly there is enormous scope for both qualitative and quantitative improvements in all of the tree oil crops but this applies particularly to olive and coconut.

It is important that producers of perennial oil crops take as much advantage as possible of the opportunities presented by modern biotechnology. This may allow them to maintain a competitive edge over the increasing threat from genetically-engineered annual crops, such as high-lauric rapeseed, or potential new crops such as *Cuphea*. In the longer term, it is quite possible that the ever-increasing rate of progress in molecular biology will enable perennial crops to be transformed and regenerated, possibly by the early decades of the next century. Whatever transpires in the short to medium term, the tropical perennial oil crops with their unmatched high oil yields eventually be an invaluable resource for the oleochemical industry as fossil oils become depleted and ultimately run out altogether. It will be important to foster and, if possible, develop and extend the uses of these crops, over the coming decades before they truly come to the fore as the major global resource for renewable oleochemicals for the 21st century and beyond.

6.9 Future challenges

Modern biotechnology is now at an exciting transition point, particularly with regard to oil crops. Over the past decade, enormous technical advances have been made in areas such as plant transformation, the use of molecular markers such as RFLPs and RAPDs, in tissue culture, and in obtaining appropriately regulated gene expression in genetically engineered crop plants. Many of the earlier misgivings concerning the consequences of manipulating factors such as fatty acid composition of seed oils have been laid to rest. It now appears that seed storage products such as oils and proteins may be particularly amenable to manipulation by genetic engineering because they are relatively inert end products of metabolic pathways sequestered in specific organelles within the cell. Unlike many biologically active compounds such as pesticides or hormones, seed storage compounds are not metabolic intermediates. Hence the fatty acid composition of seed oils can be altered quite drastically without in any way affecting the acyl composition of the cell membranes. Similarly, the amino acid composition of seed storage proteins can be altered in a specific manner without affecting the composition of any other cellular proteins. This means that radical alterations to the oil or protein content of a seed can be carried out without affecting any other aspects of the architecture, physiology or agricultural performance of the crop.

Biotechnology is now emerging from the research environment and into the world of commerce. Investment in biotechnology, by national governments, transnational agencies, and by private companies continues to increase. The first products of plant biotechnology are already nearing the market-place and many more are waiting in the wings. There is much research still to be done before genetic engineering moves from its present essentially empirical basis towards being a more exact and predictable scientific discipline. The excitement generated by a few initial successes, particularly in oil crop biotechnology, should not distract attention or funding from fundamental research aimed at understanding the biochemical and genetic processes that underlie those aspects of plant development that it is wished to modify by genetic engineering.

Another important consideration concerns the mechanism by which the scientific discoveries of biotechnology are transferred into the market-place and ultimately to the consumer. A continuous dialog between all interested partners, such as government departments, public funding agencies, fundamental research scientists in academic institutions, company scientists, working in research or development, engineers, and sales and marketing staff, is important for the efficient harnessing of intellectual resources and their translation into useful products for society.

Many of the products of modern biotechnology will be genetically manipulated organisms. It is imperative that government and transnational agencies face up to the challenge involved in developing regulations to manage these new technologies. A fundamental problem at the present time is the difference in perception of genetically engineered organisms between the two major global players in plant biotechnology, Europe and the USA. In the USA, the Food and Drug Administration regards genetically engineered organisms as essentially no different from other organisms. In contrast, the European community considers them to be a special category of organism [75]. Therefore, the regulations defining the release of such organisms into the environment tend to be more stringent in Europe than elsewhere [75]. This may lead to the transfer of biotechnology research and development expertise away from Europe unless a degree of convergence between the different international regulatory regimes can be arrived at. Such an exodus of biotechnology companies from Germany to the USA has already started, due to the more favourable regulatory regime for such research and development in the USA. Another complication is that, in the USA, numerous agencies at the federal state or local level may be involved in approving the release of genetically manipulated organisms. Not only can this prolong the approval process, it can also result in different standards being adopted in different areas of the country. This lack of consistency at all levels from local authorities to global trading blocs can act as a major deterrent to companies wishing to invest in and eventually to implement biotechnological research and development.

Clearly there is a need to reassure the public about the safety of genetically manipulated organisms. This is particularly crucial if these organisms enter part of the domestic animal or human food chains. It is also important to have a rational, coherent and not overly complex set of regulations so that the development and marketing of biotechnological products is not made prohibitively time consuming or expensive. Providing that we can all live up to these challenges, there is enormous scope for the utilisation of modern technology to improve oil crops in order to supply a vast range of both edible and industrial feedstocks.

References

1 Anon. *International Business Week,* New York, **1990**, *26 February*, 31–36.
2 Wyke, A. *The genetic alternative: a summary of biotechnology. The Economist* London, **1988**, 30th April.
3 Persley, G.J., *Biotechnology in the Service of World Agriculture.* CAB International, UK, **1990**.
4 De Greef, W. *Agro-Ind Hi-Tech*, **1991**, *4*, 3–7.
5 Levin, M.A., Strauss, H.S., *Risk assessment in genetic engineering.* McGraw-Hill, New York, **1991.**
6 Gershon, D., *Nature*, **1992**, *357*, 352.
7 Knutzon, D.S., Thompson, G.A. Radke, S.E., Johnson, W.B., Knauf, V.C., Kridl, J.C., *Proc. Natl. Acad. Sci.*, **1992**, *89*, 2624–2628.
8 Voelker, T.A., Worrel, A.C., Anderson, L., Fan, C., Hawkins, D.J., Radke, S.E., Davies, H.M., *Science* **1992**, *257*, 72–73.
9 Anon. *Agro-Ind Hi-Tech.* **1991**, *4*, 39–56.
10 Anon. *Agro-Ind Hi-Tech.* **1992**, *5*, 47.
11 OECD. *Biotechnology - Economic & Wider Impacts, Organisation for Economic Cooperation & Development*, Paris, **1989**.
12 James, C., Persley, G.J., in: *Agricultural Biotechnology: Opportunities for International Development,* CAB International, Wallingford, UK., **1990**, 367–377.
13 Anon. *Oils & Fats International.* **1992**, *8*, 41.
14 Buttel, F.H. in: *Agricultural Biotechnology: Opportunities for International development.* Persley, G.J. (ed.) CAB International, Wallingford, UK. **1990**, pp 311–321.
15 Van den Doel, K., Junne, G., *Trends in Biotechnol*, **1986**, *4*, 88–90.
16 Murphy, D.J., *Trends in Biotechnol*, **1992**, *10*, 84–87.
17 Mithen, R.F., Magrath, R., *Plant Breeding*, **1992**, *108*, 60- 68.
18 Rexen, F., in: *Towards sustainable crop production systems.* Mongraph series, Royal Agric. Soc. England.**1992**, *11*, 79- 80.
19 Princen, L.H., *Econ. Bot.* **1983**, *37*, 478–492.
20 Princen, L.H., in: *Fats for the Future*, Cambrie, R.S. (ed.) Horwood, Chichester, UK, **1989.**
21 Anon. *Inform* **1990** 2: 678–699.
22 Murphy, D.J. *Industrial Crops and Products,* **1993**, *2*, 251- 259.
23 Chen, B.-Y., Heneen, W.K., *Hereditas*, **1989**, *111*, 255- 263.
24 Fehr, W.R., in: *Oil Crops of the World*, Robbelen, A., Downey, A.K., Ashri, A. (eds.), McGraw-Hill, **1989**, pp 283–300.
25 Oulaghan, S.A., Wills, R.B.H., *N.Z. Journal of Exp. Agric.* **1974**, *2*, 381–383.
26 Green, A.G., Marshall, D.R., *Euphytica*, **1984**, *33*, 321- 328.
27 Green, A.G., *Theor. Appl. Genet.*, **1986**, *72*, 654–661.
28 Rowland, G.G., *Can. J. Plant Sci.*, **1991**, *71*, 393–396.
29 Amberger, L.A., Shoemaker, R.C., Palmer, R.G., *Theor. Appl. Genet.* **1992**, *84*, 600–607.

30 Sacristan, M.D., *Theor. Appl. Genet.* **1982**, *62*, 193–200.

31 Larkin, P.J., Scowcroft, W.R., *Theor. Appl. Genet.* **1981**, *60*, 197–214.

32 Lee, M., Philips, R.L., *Annual Rev. Plant Physiol.* **1988**, *39*, 413–437.

33 Landry, B.S., Hubert, L., Etoh, T., Harada, J.J., Lincoln, S.E., *Genome*, **1991**, *34*, 543–552.

34 Keim, P., Diers, B.W., Olson, T., Shoemaker, R.C., *Genetics*, **1990**, *126*, 735–742.

35 Diers, B.W., Keim, P., Fehr, W.R., Shoemaker, R.C., *Theor. Appl. Genet.* **1992**, *83*, 608–612.

36 Edwards, M.D., Stuber, C.W., Wendel, J.F., *Genetics*, **1987**, *116*, 113–125.

37 Lydiate, D.J. *John Innes Centre Annual Report*, **1993**, Norwich, UK, in press.

38 Lander, E.S., Botstein, D., *Genetics*, **1989**, *121*, 185–199.

39 Rafalski, J.A., Tingery, S.V., Williams, J.G.K., *Ag. Biotech. News Inform.*, **1991**, *3*, 645–648.

40 Waugh, R., Powell, W., *Trends in Biotechnol*, **1992**, *10*, 186–191.

41 Devos, K., Gale, M.D., *Theor. Appl. Genet.*, **1992**, *84*, 567–572.

42 Hugly, S., Kunst, L., Browse, J., Somerville, C., *Plant Physiol.*, **1989**, *90*, 1134–1142.

43 Arondel, V., Lemieux, B., Hang, I., Gibson, S., Goodman, H.M. *Science* **1992**, *258*, 1353–1358.

44 Moloney, M.M., Walker, J.M., Sharma, K.K., *Plant Cell Rep.* **1989**, *8*, 238–242.

45 Everett, N.P., Robinson, K.E.P., Mascarehas, D., *Biotechnology,* **1987**, *5*, 1201–1204.

46 McCabe, D.E., Swain, W.F., Martinell, B.J., Christon, P., *Biotechnology* **1988**, *6*, 923–926.

47 Umbeck, P., Johnson, G., Barton, K., Swain, W., *Biotechnology,* **1987**, *5*, 263–266.

48 Batty, N.P., Evans, J.M., *Transgenic Res.*, **1992**, *1*, 107- 113.

49 Anon. *Agric. Biotech. News Inform.* **1992**, *3*, 412–413.

50 Murphy, D.J., *Prog. in Lipid Res.* **1990**, *29(4)*, 299–324.

51 Lee, W.S., Tzen, J.T.C., Kridl, J.C., Radke, S.E., Huang, A.H.C., *Proc. Natl. Acad. Sci.* **1991**, *88*, 6181–6185.

52 Boyer, S.K., Shotwell, M.A., Larkins, B.A., *J. Biol. Chem.*, **1992**, *267*, 17449–17450.

53 Stayton, M., Harpster, M., Brosio, P., Dunsmuir, P., *Aust. J. Plant Physiol.*, **1991**, *18*, 507–517.

54 Van der Krol, A.R., Mur, L.A., Beld, M., Moll, J.N.M., Stuitje, A.R., *Plant Cell*, **1990**, *2*, 291–299.

55 Napoli, C., Lemieux, C., Jorgensen, R., *Plant Cell*, **1990**, *2*, 279–289.

56 Smith, C.J.S., Watson, C.F., Bird, C.R., Ray, J., Schuch, W., Grierson, W., *Mol. Gen. Genet.* **1990**, *224*, 477–481.

57 Various. in: *The Multinational Coordinated Arabidopsis thaliana* Genome Research Project. National Science Foundation publication, NSF92–112, USA, Progress Report: Year Two **1992**.

58 Oo, K.-C., Huang, A.H.C., *Plant Physiol.* **1989**, *91,* 1288- 1295.

59 Schmidt, R.D., in: *Proceedings of the World Conference on Biotechnology for the Fats & Oils Industry*, Applewhite, T.H. (ed.), *A.O.C.S.* **1988**, pp 169–172.

60 Harvey, B.L., Downey, R.K., *Can. J. Plant Sci.* **1964**, *44*, 104–111.

61 Downey, R.K., Robbelen, G., Ashri, A., *Oil Crops of the World.* McGraw-Hill, New York, **1989**.

62 Sonntag, N.O.R., *Inform*, **1991**, *2*, 449–463.

63 Bafor, M., Jonsson, L., Stobart, A.K., Stymne, S. *Biochem. J.* **1990**, *272*, 31–38.

64 Dutta, P.S., *PhD thesis.* Swedish University of Agric. Sci. Sweden, **1991**.

65 Slocombe, S.P., Cummins, I., Jarvis, P., Murphy, D.J., *Plant Mol. Biol.*, **1992**, *20*, 151–156.

66 Meeusen, R.L. in: *Agricultural Biotechology: Opportunities for International Development*, Persley, G.J. (ed.), CAB International, Wallingford, U.K., **1990**, pp 108–122.

67 Bollinger, W.H. in: *Agricultural Biotechnology: Opportunities for International development.* Persley, G.J. (ed.), CAB International, Wallingford, U.K., **1990**, pp 378- 395.

68 Altman, D.W. *Current Opinion in Biotechnology*, **1993**, *4*, 177–179.

69 Scowcroft, W.R., in: *Agricultural Biotechnology; Opportunities for International Development.*, Persley, G.J. (ed.), CAB International, Wallingford, U.K., **1990**, pp 177- 179.

70 Anon. *Oil World Forecast*, **1988**.

71 Corley, R.H.V., Lee, C.H., Law, I.H., Wong, C.Y. *The Planter*, **1986**, Kuala Lumpur *62*, 233–240.

72 Cheah, S.C., Ooi, L.C.L., Rahimah, A.R., Maria, M. *Proc. Miami Biotechnology Winter Symposium on Advances in Gene Technology: Feeding the World in the 21st Century*, Miami, USA, **1992**, p5.

73 Jones, L.H., in: *Agricultural Biotechnology; Opportunities for International Development.* CAB International, Wallingford, U.K., **1990**, pp 25–30.

74 Haumann, B.F., *Inform*, **1992**, *3*, 1080–1093.

75 Dotson, K. *Inform.* **1992**, *3*, 242–265.

76 U., N. *Jap. J. Bot.* **1935**, *7*, 389–452.

7 Processing of Novel Oil Crops and Seed Oils

J.T.P. Derksen, B.G. Muuse and F.P. Cuperus

7.1 Introduction

The development of optimal processing technologies for oil crops and seed oils is crucial for the industrial acceptance of new seed oils and, therefore, for the introduction into agriculture of designer oil crops. This is particularly true when the novel oil crops are not biotechnological alterations of existing species, but comprise novel, hitherto uncultivated species.

This chapter reviews established processing technologies for seed oil winning and refining, particularly in relation to the new crops. Some novel biotechnological developments, such as enzymatic oil splitting, in the oleochemical processing of seed oils to economically attractive specialties or industrially interesting commodities are also discussed.

In 1990 the total world production of vegetable oils amounted to approximately 61 million metric tons (MT), of which almost 80% originated from only 5 crops: soybean, oil palm, rapeseed, sunflower and coconut [1]. This limited number of currently available oilseed and oilfruit crops has spurred industrial interest in the development of new crops, which are optimized for specific applications (see Chapter 2,6 & 9). The new crops of interest contain a higher percentage of a desirable fatty acid or a lower percentage of undesirable fatty acids and crops that contain unique, unusual fatty acids. Oleochemical industries in particular have expressed their interest in new fatty acids with unusual properties and functionalities for non-food applications, since current sources contain no more than approximately 10 different types of fatty acid. Such unusual fatty acids could on the one hand replace raw materials from petrochemical origins with renewable resources, and on the other hand expand the existing range of raw materials available and potentially lead to novel end products (see Chapter 2). Moreover, consumer products made from renewable resources may also carry an appealing environment-friendly or "green" label.

The development of new crops to provide renewable resources for industry is also very much welcomed by agriculture. Not only can a new or increased demand for agricultural products be beneficial to farmers incomes, it could also diversify the existing range of cash crops. In particular the introduction in agricultural practice of plant species that were not previously exploited as crops can lead to a broadening of existing narrow crop rotation schedules, thus reducing the need for pesticides to maintain production levels (see Chapters 6 & 9).

In recent years the development of new oil crops has received a new impetus when the EEC decided to fund research on the introduction of these crops in Europe. One of these research projects, VOICI (Vegetable Oils for Innovation in Chemical Industries) was initiated in 1991 as part of the EEC ECLAIR (European Collaborative Linkage of Agriculture and Industry through Research) framework program. This ECLAIR-VOICI project was expanded in the Netherlands by additional funding from the Dutch Ministry of Agriculture, Nature Management and Fisheries (NOP, National Oilseeds Program) to evaluate a broader range of oil crops than provided for in the VOICI project (Table 7.1). Characteristic of both the ECLAIR-VOICI and the NOP programs

Table 7.1. Selected potential oilseed crops

Oilseed crop	Seed yield (tons/ha)	Oil content (% dry weight)	Major fatty acid	FA content (% of total)	Industrial applications
Crambe abyssinica (Crambe)	2.5–3.5	26–39	erucic acid	55–60	plastic, additives, lubricants, cosmetics, fabric softeners
Limnathes alba (Meadowfoam)	0.5–1.0	17–29	very-long-chain FA	95+	lubricants, cosmetics
Dimorphotheca pluvialis (Cape Marigold)	1.2–1.7	18–26	dimorphecolic acid	58–65	coatings, pharmaceuticals, flavors, lubricants, polymers
Calendula officinalis (Marigold)	1.5–2.5	19–24	calendic acid	58–63	paints, coatings
Euphorbia lagascae (Spurge)	1.0–1.5	45–52	vernolic acid	59–65	polymers, plasticizers

is the multidisciplinary approach: participants specialized in such diverse fields as plant breeding, plant physiology, weed control, plant diseases, harvest mechanization, oilseed processing, animal feed technology, oleochemistry as well as in market research, work together in a joint effort towards the introduction of new oil crops.

It has been recognized that a major bottleneck in introducing a new plant species as an agricultural crop is the limited knowledge on the proper handling and processing of the agricultural products to industrial raw materials and in the characterization of these raw materials. Although through plant breeding a large number of plant varieties can be generated, the industry is not so much interested in a field of attractive plants *per se*. What needs to be added for the successful introduction of a new plant species or variety as a novel renewable resource is the know-how to process the plant products to useful and economically attractive specialties or commodities.

Within the framework of the NOP and ECLAIR-VOICI programs, it is the task of the Agrotechnological Research Institute (ATO-DLO) in Wageningen, The Netherlands, to study the processing of novel oil crops. It may be superfluous to mention that very limited information is yet available in the literature on the processing of these new crops, simply because they have scarcely been studied before. Nevertheless, certain problems that can be anticipated for these new crops are analoguous to those encountered before for other crops. Therefore, this chapter presents an overview of current practices in oil crop and seed oil processing as well as some new developments in these fields. A special emphasis is given to those aspects of oil crop processing technology that particularly pertain to the novel oil crop species under study. New, environment-friendly techniques to process the new vegetable oils to industrially desirable oleochemicals are also discussed.

7.2 New oil crops

Table 7.1 presents a selection of 5 plant species that are currently being evaluated as potential oil crops in Europe. These oil crop plants were selected, for reasons of agronomic feasibility as well as of industrial interest and market opportunities, from a wide range of plant species screened, both in Europe [2–4] and in the USA [5,6]. The seed oils from the plants selected all have features that make them 'unusual', when compared to existing oil crops. The applications of these and other novel seed oils are discussed in more detail in Chapter 2.

7.2.1 *Crambe abyssinica* Hochst. ex Fries and *Limnanthes alba* Benth

These seed oils contain large amounts of very-long-chain fatty acids. *Crambe abyssinica* contains up to 60 % erucic acid (Δ13c-docosenoic acid), exclusively positioned on the 1,3-positions of its triacylglycerols. The presence of erucic acid makes this crop an alternative to High Erucic Acid Rapeseed (HEAR), but has the advantage of a consistently higher erucic acid content. Currently, the major application of erucic acid is as its erucamide derivative in polymer production as an anti-block or slip agent. Many other applications are foreseen for erucic acid and its hydrogenated derivative behenic acid, in such fields as detergents, lubricants and cosmetics [7,8]. *Limnanthes alba* contains over 95% fatty acids of the C_{20} and C_{22} type, 63% being Δ_5- and Δ_{11}-eicosenoic acids [9]. *Limnanthes alba* seed oil already finds use as a base oil in cosmetics but also has a large potential market for use in lubricants.

7.2.2 *Dimorphotheca pluvialis* (L.) Munch

This seed oil contains more than 60% of a hydroxydiene fatty acid: dimorphecolic acid (Δ_9-hydroxy, 10t,12t-octadecadienoic acid) (Figure 7.1). In contrast to the more familiar hydroxy fatty acid ricinoleic acid (Δ_{12}-hydroxy, 9c-

OH

Lesquerolic acid (14-Hydroxy,11c-eicosenoic acid)

OH

Ricinoleic acid (12-Hydroxy,9c-octadecenoic acid)

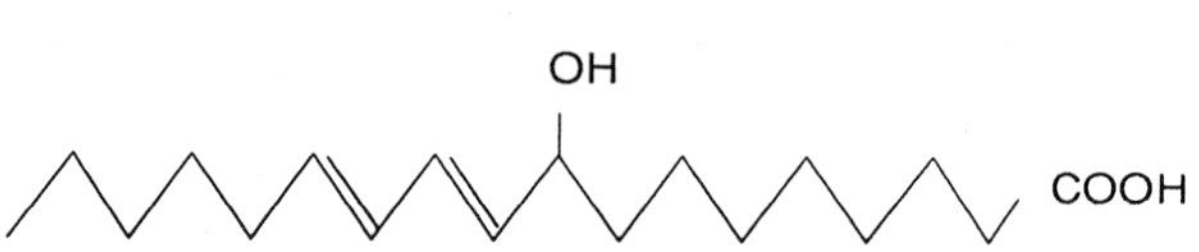

OH

Dimorphecolic acid (9-Hydroxy-10t,12t-octadecadienoic acid)

Figure 7.1. Structural formulas of natural hydroxy fatty acids.

octadecenoic acid) from castor oil, or to the newer hydroxy fatty acid lesquerolic acid (Δ14-hydroxy, 14c-eicosenoic acid) from *Lesquerella* species, dimorphecolic acid contains a conjugated diene-moiety, α-positioned with respect to the hydroxyl functionality (see Fig.7.1). This feature renders this fatty acid much more reactive than the two other hydroxy fatty acids mentioned. This reactivity may be exploited to obtain innovative oleochemicals, but also requires careful seed and oil handling and processing. Hydroxy fatty acids in general, and ricinoleic acid in particular, have many non-food applications, including polymers, inks, coatings, lubricants, plasticizers etc. [for an overview, see 10- 12]. It may be expected that the new hydroxy fatty acids from *Dimorphotheca pluvialis* and from *Lesquerella spp.* will expand this area.

7.2.3 *Calendula officinalis* L.

The oil of this plant contains fatty acids with a conjugated triene- funtionality (calendic acid or Δ_8t,10t,12c-octadecatrienoic acid) similar to tung oil from *Aleuritis fordii*. This oil presents an alternative source for these fatty acids, highly valued in the paint, coatings and cosmetics industries.

7.2.4 *Euphorbia lagascae* Sprengel

This oil contains up to 65% of vernolic acid (Δ_{12},13-epoxy-9c-octadecenoic acid), a low viscosity epoxy fatty acid [9]. This fatty acid also occurs in *Vernonia* species, new oil crops currently under investigation in the USA [13]. Applications for this fatty acid are similar to those for epoxidized soybean oil and can be found in paint and coatings industries and as stabilizers or plasticizers in polymers.

7.3 Oilseed extraction

The first step in the processing of any type of oil crop is the extraction of the oil. In an oilseed extraction plant there are three general approaches available to get a maximal recovery of oil from the seeds at the highest possible quality.

These are full mechanical pressing or expelling, full solvent extraction and a combination of these two: prepress-solvent extraction. Which of these techniques is used depends chiefly on the oil content of the seeds. The three processes mentioned are dealt with seperately below, but first we will briefly discuss the important issue of oilseed pretreatment.

7.3.1 Seed pretreatment

A careful pretreatment of the oilseeds is very important to maximize oil yield. This does not only include a careful seed cleaning to remove all foreign material, such as stones, from the raw seed material, but also the drying of the seeds. Depending on the seed type it may also be necessary to remove fibers (cottonseed) or seed hulls (e.g., soybeans, sunflower or rapeseed). Removal of the seed hulls will not only increase oil yield but also produce a press cake with a lower fiber and higher protein content. This results in a high-value press cake for animal feeding purposes. After cleaning, the seeds are usually cracked or flaked, especially when a solvent extraction step is included in the oil winning.

A recent development in the pretreatment of oilseeds is the use of an extruder (or expander) to aid oil recovery both in mechanical expelling [14] and in solvent extraction [15,16]. In the extruder the flaked seed mass is conveyed through a barrel, where it is compressed under high pressure. When steam is injected in the barrel the increasing temperature of the seed mass results in the inactivation of trypsin inhibitors and of protein and lipid degrading enzymes. When the seed mass is discharged from the barrel it expands due to the presence of compressed steam, resulting in highly porous 'collets'. The advantages of using these collets rather than flaked seeds for oil extraction are several. Mechanical pressing of extruder collets significantly increases the throughput of the expeller at a pressing efficiency of 70% [14]. In solvent extraction, the use of extruder collets increases the bulk density of the oilseed mass, thus increasing the extractor loading, and, therefore, the extraction capacity, by 15–30% [16]. Also the extraction time is reduced, due to improved mass transfer characteristics of the oil from the seeds to the solvent. The improved drainage from the porous collets decreases the solvent carry-over and thus the energy required to desolventize the seed meal. At present, extruders are available for use in oilseed pretreatment with capacities (based on soybeans) of up to 1400 metric tons per day [16].

Cell-wall degrading enzymes (cellulases, hemicellulases, pectinases) are also increasingly used as a method of oilseed pretreatment [17,18]. Due to the action of these enzymes the cell-walls become more permeable and the oil bodies release their contents more readily. Enzyme pretreatment has shown to decrease solvent extraction times by half and to increase mechanical pressing efficiencies to 90% [18]. However, the costs associated with this new methodology are still a major bottleneck with respect to the pretreatment of conventional oilseeds.

7.3.2 Mechanical pressing

The most straightforward way to recover oil from oilseeds is by mechanical pressing (expelling or crushing) in a screw press [19]. This is the method of choice when the seeds or fruits contain a high amount of oil, typically 40 % (w/w) or more, such as peanut, sunflower, sesame seeds or olives. In particular when the oil is intended for human food use, mechanical pressing has distinct advantages: no solvents are required that are potentially harmful to man or environment, and oil refining can usually be eliminated. In fact, for certain vegetable oils that are highly praised for their specific flavor, e.g., olive oil and sesame seed oil, refining is undesirable due to the loss of flavor notes.

Mechanical pressing of oilseeds (or oil bearing fruits such as olives) is performed at as low as possible temperatures, to prevent loss of flavor or degradation of thermally labile polyunsaturated fatty acids (linoleic, linolenic acid). If necessary the shaft of the expeller can also be cooled to prevent heating of the oilseed mass during pressure build-up. The oil thus produced may be labeled virgin oil. Keeping the temperature low, however, has a negative effect on total oil recovery. Therefore, one either has to compromise between oil yield and quality, or collect oil at increasingly high press temperatures in separate fractions, with differing qualities. The latter is common practice for, e.g., olives.

In general, one can say that the quality of cold-pressed oil is very good. There is little loss of oil components, the color is usually good and the oil contains a low amount of gums (lecithin or phospholipids). Refining is therefore usually not required. The oil yield, however, is typically limited to around 80%, depending on such factors as oilseed species, seed pretreatment, seed moisture content, seed preheating, press temperature, expeller pressure, etc. [20]. Heat treatment of oilseeds increases the oil yield, but also increases its content of non-triacylglycerol compounds such as phospholipids, pigments and sterols [21]. The presence of increased amounts of phospholipids and tocopherols has a favorable effect on the oxidative stability of the crude oil [22]. For a mathematical model of mechanical oil expression, see [23].

No matter how high the press temperature or the expeller pressure, it is impossible to recover all the oil from the press cake. Depending on the oilseed and the press conditions between 6 and 10% of oil, on a dry seed weight basis, will typically stay behind in the press cake. This represents a loss in oil extraction efficiency and decreases the value of the press cake when a defatted seed meal is required for use as a protein supplement in animal feed. The press cake may, therefore, be subjected to a solvent extraction step to recover the last remaining oil.

7.3.3 Solvent extraction

Oil crops that contain a relatively low amount of oil ($< 25\%$ w/w) are typically extracted with organic solvents. The most obvious example is soybeans, which contain approximately 20% oil. For many decades hexane has been the solvent of choice for commercial extraction, mainly for reasons of availability (price), oil solubility, water mixing behavior, boiling point and heat of evaporation. When the soybeans (or any other type of seeds or fruits for that matter) are properly pretreated, a virtually complete extraction of the oil can be achieved with hexane as a solvent.

On an industrial scale, oil extraction is accomplished by pumping hexane on a moving bed of flaked or otherwise pretreated seeds. The seeds are allowed to drain and the oil-rich hexane phase (or 'miscella') is collected. The hexane must now be recovered from the miscella, by heating under vacuum and by steam stripping of the oil. An interesting new development is the application of membrane technology in solvent recovery [24,25]. The use of ultrafiltration and reverse-osmosis membranes to separate triacylglycerols from hexane can potentially lead to substantial savings in energy costs [26]. The residual hexane in the seed flakes must also be recovered. This is done by treating the seed meal with steam in a so-called "desolventizer-toaster" unit. For soybeans this unit also provides the seed meal with the nutritional characteristics required for use in animal feed formulations.

Even though the hexane is recovered very efficiently from the miscella and from the seed meal, up to 0.15% may be lost [27]. This may not look like much, but at a typical throughput of an extraction plant of 2000 tons per day it could represent as much as 3 tons of solvent per plant per day. Due to continuous hexane price increases as well as safety, environmental and health concerns, a lot of effort is put in identifying suitable alternative solvents [27,28]. An ideal solvent would be one that has a high solvent power and selectivity for triacylglycerols, can be easily removed from meal and oil, has a low flammability, is stable, non-reactive, non-toxic and of high purity [28]. To date, many solvents have been evaluated according to these criteria, as reviewed by Johnson and Lusas [28]. These include: petroleum-ethers [29], trichloroethylene, methylenechloride [30], alcohols (ethanol and isopropanol [31,32], ketones (e.g., acetone), fluorocarbons, pressurized gases (butane, propane, supercritical carbon dioxide) as well as various solvent mixtures [33]. Unfortunately, all solvents tested have certain drawbacks when used in oilseed extraction and an ideal solvent has still to be found.

Lately there has been a renewed interest in using isopropanol as an extraction solvent [34]. A major advantage of using isopropanol rather than hexane is its lower toxicity, reducing both health and environmental concerns. Its lower flammability compared with hexane will also contribute to an increased industrial safety. An additional advantage for certain crops such as cottonseed is the ability of isopropanol to co-extract antinutritional factors (gossypol, aflatoxins) from the seed meal [35]. However, there are also some problems associated with the use of this solvent.

Firstly, the solvent power of isopropanol for triacylglycerols is much lower than that of hexane, and depends very strongly on the extraction temperature and the fatty acid composition of the seed oil [35].

Secondly, the heat of vaporization of isopropanol is more than twice as high as that of hexane, while also its boiling point is also higher. This results in a higher energy input required for extraction and solvent recovery. However, the limited triacylglycerol solubility in isopropanol at low temperatures can be exploited to achieve a phase separation between the bulk oil and the solvent, thus facilitating solvent recovery from the miscella. The advent of (pervaporation) membrane systems to separate the solvent from the oil phase may alleviate much of the energy input problem in isopropanol recovery.

Thirdly, since isopropanol readily mixes with water, the oil extraction is usually performed with the isopropanol-water azeotrope, consisting of 87.8% isopropanol and 12.2% water (w/w). To prevent dilution of the azeotrope with water from the seeds and thus reducing solvent power, a careful control of seed water content is required. Prior drying of the seeds to, e.g., 3% moisture may be recommended. To rectify the water content in the isopropanol/water mixture the use of pervaporation membrane technology may also be an interesting option [34].

Despite the above problems, however, isopropanol is the solvent that offers at present the best prospects to replace hexane as the solvent of choice in the oilseed extraction industry.

Carbon dioxide under supercritical conditions (temperature $>31\,°C$, pressure >73 bar) is another solvent that has been investigated extensively for the extraction of a wide range of oil crops, including rapeseed [36–38], soybeans [39–41], cottonseed [40,42], corn germs [43], olives [44], shea nut [45] and rice bran [46]. Particularly appealling in using supercritical carbon dioxide as an extraction solvent are the absence of toxicity, the ease of solvent recovery from miscella and seed meal by a simple pressure reduction step, and its nonflammability. The absence of toxicity of this solvent is very important to the oils for food uses but also for use in cosmetics (e.g., jojoba oil)[47]. Since the oil crops can be processed at a relatively low temperature, supercritical carbon dioxide extraction may also offer advantages in the extraction of heat-labile oils such as the polyunsaturated fatty acid-containing fish oils [47]. The very high selectivity of this solvent for triacylglycerols results in an oil which is low in non-triacylglycerol compounds, such as pigments, phospholipids and sterols, thus obviating the need for extensive oil refining [39,48]. However, the absence of phospholipids and antioxidants such as tocopherols may also reduce the oxidative stability of the oil [41,47,50]. Currently, the costs associated with supercritical fluid extraction technology and problems in constructing continuous, rather than batch extraction plants are major pitfalls in employing this technology on a large scale.

A new oil extraction solvent that has recently received interest is water, a solvent for which toxicity is hardly an issue [51,52]. The use of water in oil extraction is based upon a pretreatment of the milled and boiled seed mass with a solution of cell wall-degrading enzymes, followed by a centrifugation step to separate the emulsified oil from the seed residues. Good results of this new oil crop extraction method, with yields of up to 90%, were reported for rapeseed [51], coconut and olives [52].

7.3.4 Prepress-solvent extraction

For seeds that have an oil content too low to use full mechanical pressing and too high for economic full solvent extraction, i.e. between approximately 25 and 40%, a combination of the two techniques is usually employed [19,53]. Prepressing as a seed pretreatment step for solvent extraction thus reduces both solvent and energy required for a complete oil extraction. Prepress-solvent extraction is common practice for such oilseeds as cottonseed and rapeseed.

7.3.5 Oil winning in novel oil crops

With respect to the pretreatment of the new oil crops from Table 7.1, *Crambe abyssinica* seeds can easily be dehulled. This results in higher oil yield as well as a seed meal with a decreased fiber and increased protein content. The defatted seed meal can be used for feeds for ruminant animals. When glucosinolates are removed from the seed meal, e.g., through extraction with water, the meal can also be used as a feed for nonruminant animals [54]. The pretreatment of *Dimorphotheca pluvialis* seeds is complicated in so far as these seeds have two morphologies: cones, derived from the ray florets, and winged seeds, derived from the disc florets. In our hands the cones can be flaked easily for oil extraction purposes. The winged seeds, however, contain a large amount of fiber and are more difficult to process. The optimal pretreatment of winged seeds has not yet been found. The pretreatment of the other oil crops mentioned in the table has not yet been investigated.

Mechanical pressing as a means of oil extraction has been studied for a number of novel oil crops [55]. It can be noticed from Table 7.2 that, with the exception of *Euphorbia lagascae* seeds, oil winning by full pressing does not, in general, lead to satisfactory oil recoveries. This is partly due to a low oil content in the seeds and, for *Dimorphotheca pluvialis,* partly due to the high press temperatures neccessitated by the high viscosity of the oil at room temperature [9]. This latter oil in particular contained a lot of residual solid material ('fines')

Table 7.2. Mechanical pressing of novel oilseeds

Oilseed crop	Oil content (g/kg seed)	Temperature (°C)	Oil yield (g/kg seed)	Oil recovery (%)
Euphorbia lagascae	480	55	450	94
Dimorphotheca pluvialis				
- rods	230	125	120	52
- winged seeds	190	125	80	42
Limnanthes alba	220	60	150	68
Crambe abyssinica	320	60	230	72

Oilseeds were pressed in a Komet single screw oil expeller, model SS87G (IBG Monforts and Reiners GMBH, Mönchen-Gladbach, Germany). The expeller was equipped with a 12 mm choke and operated at a throughput of approximately 4 kg seed/hour, while the seed temperature was maintained at the temperatures indicated. Under the conditions used, 55 °C was the lowest temperature achievable without shaft cooling.

that could not be filtered off. This oil was also reported to be extremely dark in color, highly viscous, and to contain relatively high amounts of phospholipids and free fatty acids, the latter possibly due to lipase activity in the sediment [9]. *Crambe abyssinica* and *Limnanthes alba* both gave a good quality seed oil upon full pressing [9,55,56], albeit at a low yield. It was found that *Euphorbia lagascae* seeds could be very easily cold-pressed at high recoveries and at good oil quality [9].

Seed oil recovery by solvent extraction has been described in the literature for several new oil crops:

Crambe abyssinica seeds were successfully extracted by Carlson and co-workers in a commercial extraction facility using hexane as a solvent [56]. It was found to be important in processing the crambe seeds to preheat the seeds to 94 °C for at least 15 to 20 minutes in order to inactivate thioglucosidase activity and control sulfur content of the oil [56].

Vernonia galamensis seed oil, which contains the same epoxy fatty acid, vernolic acid, as does *Euphorbia lagascae,* was also obtained by mechanical pressing followed by hexane extraction [57]. In order to produce oil from this new oil crop with an acceptably low free fatty acid content, it was necessary to inactivate the seed lipases by a heat pretreatment of the seeds of 90 minutes at 93 °C [57].

Dimorphotheca pluvialis seeds were successfully extracted at our laboratory using hexane, pentane and isopropanol. With all three solvents a dark green oil was obtained, containing 5 - 6 % free fatty acids. This may be due to lipase action in the seeds or seed meal. An extensive heat pretreatment of the seeds may be required to inactivate seed lipase activity. However, we found that heat treatment of the oil can lead to loss of the functional hydroxydiene moiety in dimorphecolic acid and to polymerization reactions (unpublished observations).

It can be envisioned that the use of a polar extraction solvent such as isopropanol may be advantageous in extracting triacylglycerols, composed of relatively polar fatty acids ("like dissolves like"), such as hydroxy fatty acids in *Dimorphotheca pluvialis* of *Lesquerella spp.*, or epoxy fatty acids in *Euphorbia lagascae* or *Vernonia spp.* seed oils. This is currently under investigation.

Application of supercritical carbon dioxide as a solvent for the extraction of new seed oils has been succesfully demonstrated with *Oenothera biennis* (evening primrose), a gamma-linolenic acid containing oil crop [58]. Gamma-linolenic acid is a fatty acid that is especially prone to thermal and oxidative rearrangements, due to its high level of unsaturation. Using relatively low extraction temperatures from 40–60°C, good oil yields could be obtained, with no appreciable compositional changes [58].

There are some indications that when using supercritical carbon dioxide as a solvent the composition of the first extracted oil is distinctly different from oil extracted later. Since changing the pressure and temperature of the supercritical carbon dioxide leads to different solvent characteristics [59], a careful selection of the extraction conditions potentially allows one to obtain oil fractions enriched with certain desirable triacylglycerol species.

The quadrupole moment and Lewis acid-base characteristics of carbon dioxide give this molecule properties that allow specific interactions with polar compounds and functionalities [60]. This makes supercritical carbon dioxide particularly interesting to evaluate as a solvent in the extraction of thermally unstable seed oils that also possess polar functionalities, such as from *Dimorphotheca pluvialis*. At our laboratory, supercritical carbon dioxide was shown to be a good extraction solvent for *Dimorphotheca pluvialis* seed oil. In contrast to conventional solvent extraction, a light-colored yellow oil could be obtained, with reduced levels of free fatty acids and phosphorus. However, there are indications that the oil stability is also somewhat lower, possibly due to a low tocopherol content (data not shown).

Prepress-solvent extraction procedures have been succesfully employed to recover oil from several novel oil crops, such as *Crambe abyssinica* [56] and *Vernonia galamensis* [57]. Also *Lesquerella fendleri* oilseeds could be succesfully produced with a prepress-solvent extraction procedure, yielding a distinctly reddish oil, darker in color than castor oil [61].

7.4 Seed oil processing

The crude oils produced by mechanical expelling, solvent and prepress-solvent extraction procedures often contain non-triacylglycerol components that usually must be removed. Virgin edible oils, obtained by mechanical pressing, that are particularly valued for their taste, e.g., olive or sesame seed oil, are treated as little as possible. Treatment of these oils may be limited to a filtration step to remove fines. Also crude oils that are intended for the production of, for instance, fatty acids for non-food uses mostly do not require extensive further processing. For other applications the non-triacylglycerol impurities are removed in one or more refining steps, a simplified outline of which is depicted in Figure 2. In practice, the order and the desirability of the various seed oil processing steps depend highly on the oilseed crop and the level and identity of the impurities present as well as on factors such as tradition and geography. Also, processing steps are often combined. For the sake of clarity, a number of unit operations will be discussed separately.

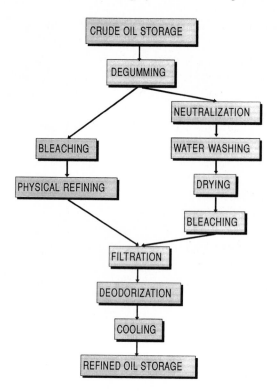

Figure 7.2. Simplified flow chart of vegetable oil processing.

7.4.1 Degumming

The first step in processing crude vegetable oils is focussed on removing the phospholipids or gums, hence "degumming". Particularly when the seed oil is produced by solvent extraction, the concentration of phospholipids in crude oil may be as high as 3% [62]. The presence of these phospholipids has harmful effects on the oil quality and stability.

Another reason to remove phospholipids from the oil can be to recover the phospholipids for further use. For example, soybean lecithin is highly valued as an emulsifier in food, feed and technical products [63,64].

The process to remove gums by the addition of water has been known for decades. In this process water is added to oil, whereupon the emulsion is kept at a temperature between 50 and 70°C for 20 - 30 min with constant stirring to allow the phospholipids to hydrate. The hydrated gums, associated with some other impurities such as some pigments, glycolipids, monoacylglycerols and metal salts, are then separated from the oil by centrifugation. The efficacy of the above protocol depends on the phospholipid species present in the oil [65,66].

To enhance the degumming process the addition of several reagents has been studied for, amongst others, Canola (rapeseed), soybean and sunflower oils [67]. These include such chemicals as citric acid, phosphoric acid, oxalic acid and acetic or maleic anhydride. Particularly good results were obtained for canola oil with the use of citric and phosphoric acid and for sunflower oil with the addition of oxalic acid or maleic anhydride. All chemicals mentioned were capable of removing 98% of the phosphorus in soybean oil.

Currently, many enhancements of the degumming process have been described in the literature, such as the "total degumming" process (TOP)[68], "super-degumming", and SNOW (Simplified Neutralization Omitting Washing), a novel alkali degumming process [69].

A novel approach in degumming practice makes use of pancreatic phospholipases to split off the fatty acid on the sn-2 position of the phospholipid molecules, thus enhancing its susceptibility to separation from the oil. It is reported that this process can decrease the phospholipid content of vegetable oils to below 5 ppm at competitive costs [70].

7.4.2 Neutralization

Crude oils may contain as much as 30% of free fatty acid, resulting from the partial hydrolysis of triacylglycerols and phospholipids. Since fatty acids render off-flavor notes, such as a soapy taste, to the oil they are usually removed during edible oil refining. Generally this is done by the addition of a measured quantity of lye and the subsequent removal of the soap stock by centrifugal separation, followed by a water washing step [71]. Contact time and temperature is dependent on the oil and on geography: in the USA a long mix process is favored, whereas in Europe a short mix process is preferred [62,71]. Degumming and neutralization is often combined in a one-step process, which is then called "caustic", "chemical" or "alkali" refining. Following the alkali neutralization a water washing and drying step may be introduced in the process, in order to remove excess alkali.

Drawbacks of the above procedure are that oil may be lost with the soap stock, that prolonged contact with alkali may result in an increased hydrolysis of the triacylglycerols, and that it consumes a large amount of energy. Therefore, a lot of research has been done on alternative methods for the deacidification of oils. The use of alcoholic solvents (methanol, ethanol and isopropanol) was tested in rice bran oil neutralization, but gave only poor results [72]. However, aqueous isopropanol could be employed to extract free fatty acids from groundnut oil [73]. The toxicologically much more acceptable solvent supercritical carbon dioxide was also successfully used to deacidify olive oil [74–75]. Another interesting oil deacidification process makes use of microbial lipases to re-esterify the free fatty acids to glycerides present in the oil [76–78].

7.4.3 Physical refining

Physical refining can typically be employed with good quality crude oils of low phosphatide content. This physical refining process, as an alternative to the chemical, alkali refining process, is used extensively in palm oil processing. During the last decade, however, it has been used more and more in the refining of other edible oils as well. Advantages lie in its simplicity, and the improved removal of particularly non-hydratable phospholipids [71,79]. Also, physical refining will remove free fatty acids as well as odors, thus a neutralization and a deodorization step can often be circumvented.

Physical refining usually comprises the addition of a small amount of phosphoric acid (0.05–0.2%) to the oil, followed by high temperature steam stripping under reduced pressure (240–270°C; 2–10 Torr)[79- 81]. In this process free fatty acids, gums and odors are removed simultaneously.

7.4.4 Bleaching

Many non-triacylglycerol components in crude vegetable oils are not removed quantitatively during the degumming and refining steps. These include pigments such as carotenoids and chlorophylls, metal salts, residual fatty acids, soaps and gums, antinutritional factors, and oxidation products. To eliminate these impurities as much as possible, the oil is usually subjected to an adsorptive treatment with bleaching earth, or "bleached" [82]. Bleaching conditions are very dependent on the type of oil processed [83] and the level and kind of impurities present. In fact, establishing the optimal conditions for bleaching may be considered more as an art than a science. (For a historical overview, see [84]).

Typically, 0.2–4.0 % adsorbent is mixed with the oil, prewarmed to 80 °C and subsequently heated to 110 °C for a maximum of 30 minutes. During this procedure the oil is preferably either blanketed with nitrogen or kept under vacuum to prevent oxidation [82]. Commercially available adsorbents are: various types of treated silica, activated carbon and various types of clay minerals [85,86]. The most popular of the latter is montmorillonite, the major mineral in bentonite ore [87]. The bentonite ore, which can be found world-wide, is mined, crushed and then acid-activated by a treatment with HCl or H_2SO_4 [87]. The efficiency of a particular bleaching earth in removing impurities in an oil is found to be very dependent on the type and level of acid-pretreatment as well as on the exact origin of the mineral [87,88].

After bleaching, the adsorbent must be thoroughly removed from the oil by filtration, to prevent fouling of the downstream processing equipment and because residual bleaching earth can act as a very strong pro-oxidant. At this point the oil should also be guarded against thermal and oxidation abuses, since now it is at its least stable [89].

An increasing problem is the disposal of spent bleaching earths, which contain 20–30% fatty materials and may sometimes be contaminated with heavy metals, such as nickel from hydrogenation steps. Governments are increasingly restricting the disposal of spent earth into landfills for environmental reasons. The higher costs associated with this waste disposal problem has spurred the industry to develop more efficient adsorbents, as well as methods to recycle bleaching earths and silicas [16].

7.4.5 Deodorization

Deodorization is a processing step which is very similar to physical refining. The oil is treated under vacuum, at high temperatures (250-270 °C) with steam to remove volatile substances from the oil that might cause off-flavors and odors [90,91]. In this process the level of free fatty acids is reduced to very low levels and also any residual volatile compounds and oxidation products are removed. The high temperature in this process causes a thermal destruction of peroxides and also results in a slight bleaching effect due to the destruction of carotenoids. New developments, such as the advent of "thin-film" deodorizer designs and heat recovery systems may increasingly lead to savings in energy costs for oil deodorizing [16].

7.4.6 Winterization

Certain seed oils, e.g., sunflower, linseed, maize (corn) and rice bran oils, contain waxes: esters of fatty acids and fatty alcohols, which originate from the seed shell [62]. These waxes present an esthetic problem in edible oils because upon storing these oils under cold conditions (e.g. in a refrigerator) a cloudy precipitate may form. Even though this precipitate has no detrimental effect on the oil quality, presents no health or organoleptic concern, and will disappear again when the oil is warmed, consumer acceptance requires that such oils be dewaxed. This is achieved by careful cooling of the oil to allow the wax-esters to crystallize, which are then removed by filtration. This process is called "winterization" or "dewaxing" [71].

7.4.7 Processing of novel seed oils

Very little information is available in literature on the processing of novel seed oils. Nevertheless, in the light of the available processing technology a few comments and expectations may be presented here regarding their treatment.

The refining of *Crambe abyssinica* seed oil does not appear to present any particular problems [93]. This oil can be processed in an analogous manner to High Erucic Acid Rapeseed, as may *Limnanthes alba* seed oil.

Processing of the hydroxy fatty acid-containing *Lesquerella fendleri* seed oil requires some extra attention. It is reported that in particular the bleaching of this oil may present some problems. Treatments with charcoal and bleaching clay reduce the dark-reddish color of this oil somewhat but the residual color still remains somewhat darker than that of refined castor oil [61].

The dark-green color, indicative of a high chlorophyll content, of *Dimorphotheca pluvialis* seed oil may also present a serious problem in bleaching. Preliminary experiments in our laboratory have not yet led to a satisfactory bleaching protocol; the most successful approach is still to prevent the extraction of the pigments by using supercritical carbon dioxide as an extraction solvent. Also the high amount of gums and free fatty acids may be a problem in refining, particularly since the increased temperatures required in refining and also the presence of lye may have adverse effects on the integrity of the triacylglycerols and fatty acids (unpublished observations). However, since the dimorphecolic acid from this oil is interesting mainly for its use as a raw material in oleochemistry, extensive refining may not be an absolute requirement. Development of new enzyme and membrane technologies to recover the fatty acid from a feed stream that does not necessarily have to be composed of pure triacylglycerols may still enable a large scale use of this oil.

The seed oil from *Euphorbia lagascae* can be easily pressed and is of good quality with little impurities [9]. Therefore, depending on the application, extensive refining of this oil appears not to be necessary.

7.4.8 Economics of new oil crops

Since only very limited information is available on the processing of novel oil crops, let alone on financial aspects of their oil winning and refining, it is extremely difficult to put price tags on these oils. Nevertheless, based on estimates of yield and production costs, Kleinhanss and Heins of the German Research Institute "Landbauforschung Völkenrode", in Braunschweig, calculated the following costs for a number of new seed oils (price level 1986): *Crambe abyssinica*, 2.69 DM/kg; *Coriandrum sativum* (petroselinic acid-rich oil), 5.86 DM/kg; *Euphorbia lathyris* (high-oleic acid oil), 2.01 DM/kg; and *Cuphea*, 5.94 DM/kg [94]. Of course, these prices reflect only the costs at one moment in time and may thus only serve to get a general idea. Still, at the moment these costs are somewhat higher than those of conventional seed oils, which range from 0.55 DM/kg for palm oil, 1.75 DM for rapeseed oil (non-subsidized), 2.25 DM/kg for high-oleic sunflower oil, to 5.20 DM/kg for olive oil [94]. Nevertheless, it must not be forgotten that conventional oil crops already have a long history of crop improvement and production optimization. It is expected that the production costs for these novel seed oils will drop after further yield increases are gained through plant breeding efforts and development of new processing technologies, specifically tailored to these new crops. In addition, for certain new oils and fatty acids, e.g., from *Dimorphotheca pluvialis* there are at present no alternatives available, except perhaps through elaborate and complicated organic chemical synthesis from petrochemical feed stocks. This makes direct price comparisons very difficult, if not impossible. Moreover, when these unique fatty acids are utilized in niche markets like specialty chemicals, the crude seed oil price is less of a concern than when they are marketed as bulk, commodity oils or chemicals.

7.5 Seed oil processing in oleochemistry

7.5.1 Oleochemical raw materials

Unlike in human or animal nutrition, unprocessed vegetable oils are currently seldom used in non-food applications. Most of the vegetable oils are first processed to a limited number of base compounds or intermediates, i.e., fatty acids, fatty acid methyl esters, fatty amines and fatty alcohols, that may then be further processed to a large array of end-products (see Figure 7.3 and Chapter 2 for a more detailed discussion of this topic).

Of the base compounds, fatty acids are quantitatively the most important oil processing intermediates, accounting for 50–60% [94- 96]. They are produced by oil splitting (triacylglycerol hydrolysis) yielding glycerol as a side product, which finds application in, e.g., cosmetics, pharmaceutics, tobacco industry, esters, resins and polymers [97]. The biggest end-use markets for fatty acids are in soaps (metal salts) and coatings [96].

Fatty acid methyl esters are produced by direct transesterification of the oil with methanol (transmethylation). Currently, most of the fatty acid methyl esters are intermediates in the production of fatty alcohols and fatty esters [98]. An increasing interest is seen in the development of the potentially large end-use market for fatty acid methyl esters as motor fuels ("bio-diesel")[99–100].

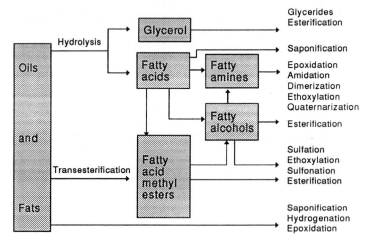

Figure 7.3. Processing routes in oleochemistry.

At present, a substantial portion of fatty alcohols is still synthesized from petrochemical feed-stocks by "Oxo" or "Ziegler" processes [101]. However, they are increasingly prepared by selective hydrogenation of "natural" fatty acids and fatty acid methyl esters [95]. Although some fatty alcohols find an end-use market in the cosmetic and toiletries industry, more than 70% is derivatized to surfactants, such as alcohol ethoxylates, alcohol sulfates and alcohol ethoxysulfates, in the detergent industry [94,95].

Fatty amines can be prepared from fatty acids or fatty alcohols and further derivatized to many end-products that find uses in laundry (e.g. fabric softeners) and household products as well as in a wide range of industrial processes [for an overview see 95]. The average annual increase in the use of fatty amines is estimated at 3.7% between 1990 and 2000.

7.5.2 Oil splitting

Considering the fact that fatty acids are key intermediates in the processing of seed oils for non-food applications, a brief discussion of oil splitting is required.

By far the most important process world-wide for the production of fatty acids from fats and oils is the Colgate-Emery process, which employs steam at high temperatures and high pressures (250–300°C, 50–70 bar) to split triacylglycerols in fatty acids and glycerol. A lower temperature alternative for chemical oil splitting, the Twitchell process, makes use of sulfuric and sulfonic acids as catalysts. However, this latter process can lead to severe corrosion in the process equipment, uses environment-unfriendly catalysts and produces a rather poor quality fatty acid [102].

The high temperatures required in the Colgate-Emery process demand a high energy input. Even though much of the energy is reclaimed through heat-exchangers, processing costs have increased dramatically, as a result of the energy price increases in the seventies. This has spurred the interest of oleochemical industries in milder, low-temperature oil-splitting technologies.

Probably the most versatile mild oil-splitting technology makes use of enzymes (lipases) as environment-friendly catalysts. Lipase (or triacylglycerol hydrolase, EC 3.1.1.3) is currently available on a large (multi-tonne) scale and at relatively low cost due to the development of industrial, microbial lipases particuarly for the detergent industry. At present, a large number of lipases are available that are either non-specific, i.e. hydrolyse fatty acids on all three acyl positions in a triacylglycerol molecule, or 1,3-specific, i.e. hydrolyse only fatty acids on the α-and α'-positions. The latter enzymes are particularly useful for the specific hydrolysis of highly asymmetric triacylglycerols, as in *Crambe abyssinica* [103] but also for the production by transesterification of asymmetric triacylglycerols for applications such as cocoabutter equivalents [104].

Since the operating conditions of enzymatic oil splitting are very mild (atmospheric pressure, room temperature), this process is ideally suited to hydrolyse heat-labile oils. One category of oils that may particularly benefit from lipase-catalysed oil splitting includes expensive polyunsaturated fatty acid-containing seed oils, such as from the novel oilseed crops *Calendula officinalis* [9](conjugated trienes), and *Borago officinalis* and *Oenothera spp.* (γ-linolenic acid, a dietary and pharmaceutically interesting α-linolenic acid isomer)[105,106]. Another category of oils for which enzymatic hydrolysis may very well be the only alternative to obtain intact fatty acids, comprises seed oils that contain functionalized fatty acids such as those from *Dimorphotheca pluvialis* (hydroxy-diene fatty acids), *Lesquerella fendleri* (hydroxy-monoene fatty acids) and *Euphorbia lagascae* (epoxy fatty acids). In particular the recovery of intact dimorphecolic acid from the very heat-labile *Dimorphotheca pluvialis* seed oil by lipase-catalysed oil splitting has already been succesfully demonstrated [55,103].

7.5.3 Lipase bioreactors

To allow the use of one batch of enzyme in multiple rounds of hydrolysis as well as to circumvent the downstream separation of seed oil hydrolysate and enzyme catalyst, lipases are preferably immobilized on solid supports. Many types of enzyme support have been employed with varying degrees of success [107], the most versatile of which makes use of hollow-fiber membranes [55,103, 108–110]. A schematic diagram of this latter type of bioreactor set-up is given in Figure 7.4. In such a hollow-fiber membrane bioreactor module the lipase is adsorptively immobilized on the lumen side of the hydrophilic membrane fiber. This membrane also functions as the interface between the water phase on the shell side, supplying the hydrolysis water, and the oil phase on the lumen phase, which supplies the substrate oil. On this interface the hydrolysis reaction takes place. In our hands, enzyme activity half-lives in this type of bioreactor of more than 4 months could be obtained (data not shown).

The concept of immobilized lipase-containing membrane bioreactors in oil splitting has already been proven for many types of lipases and plant oils. Both non-specific [109,110] and *sn*-1,3-specific lipases [103] have been employed.

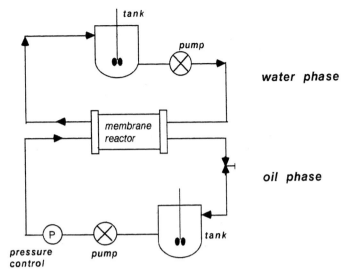

Figure 7.4. Schematic diagram of a hollow-fiber membrane bioreactor set-up for enzymatic hydrolysis of vegetable oils.

The membrane module at the heart of the bioreactor set-up contains lipase, adsorptively immobilized on the inner wall of the hollow fibers. The substrate oil phase flows through the lumen of the hollow fibre membranes, whereas the water phase is supplied to the outer (shell) side of these hollow fibers. The hydrolysis reaction takes place at the interface of oil and water, which is stabilized by the membrane.

Although common vegetable oils have been enzymatically hydrolysed in test systems [109–110], it may be expected that the real economic benefit of a mild and/or selective oil hydrolysis lies in the splitting of heat-labile oils for relatively small scale, high value-added applications.

7.5.4 Downstream processing

If the bioreactor contains immobilized non-specific lipases, e.g., from *Candida cylindraceae,* the seed oil triacylglycerols are completely hydrolysed to fatty acids and glycerol. Since the hydrophilic membrane support is highly permeable to glycerol, this product is removed from the reaction mixture. This not only allows an easy recovery of this side product, but, more importantly, draws the thermodynamic equilibrium of the hydrolysis reaction to the right, thus allowing this reaction to proceed towards completion. During experiments up to 96% complete hydrolysis of the triacyglycerols has been observed [109,110].

When 1,3-specific lipases are employed, such as from *Rhizopus javanicus, Rhizopus delemar* or *Mucor miehei,* only negligible amounts of glycerol are formed. This implies that in this case practically all the reactants remain in the organic phase. Since a thermodynamic equilibrium has been established, the hydrolysis reaction will not be complete but will result in a mixture of fatty acids, tri-, di-, and monoacylglycerols. To allow the reaction to proceed towards completion, and produce only free fatty acids and monoacylglycerols as end-products it is essential to remove continuously one of the products from the reaction. A possible in-line removal of the fatty acids desired can be achieved by using a fractional crystallization step. At our laboratory we have opted to study the in-line removal of dimorphecolic acid from the continuous 1,3-specific hydrolysis of *Dimorphotheca pluvailis* seed oil. This can be achieved by an in-line cold trap, to selectively precipitate the high- melting dimorphecolic acid by fractionated crystallization. Thus, an increased degree of hydrolysis and an equilibrium shift towards more complete hydrolysis could be observed [111,112].

The development of in-line downstream processing technologies is not only relevant to achieve an equilibrium shift in 1,3-specific enzymatic hydrolysis reactions. At present, fatty acids are predominantly purified by fractional vacuum distillation [113]. However, not only do distillation steps require a substantial energy input, but heat-labile fatty acids, such as dimorphecolic acid, cannot be purified by vacuum distillation, due to thermal breakdown (results not shown). Therefore, the use of in-line crystallization methods or membrane technology for selective fatty acid removal also facilitates the purification of reaction products.

7.6 Conclusions

There are still many questions to be answered with respect to the processing of novel oil crops. This is, of course, hardly surprising, since these crops have been studied for no more than a few years, whereas conventional oil crops have been through many decades of processing optimization. Nevertheless, the knowledge available in this field should accelerate tremendously the development of oil crops and seed oil processing technologies, specifically tailored to the needs of the new crops. This could lead to the introduction into agriculture of these crops within time-frames of 2–15 years, depending on the plant species.

The development of environment-friendly technologies for the production of oleochemicals from renewable resources is a process that is particularly encouraging, especially in the light of the growing concern for the environment and the decreasing supplies of petrochemical resources, anticipated for the beginning of the 21st century.

Acknowledgements

The work presented in this chapter was supported in part by grants from the EC (ECLAIR-VOICI program) and from the Netherlands Ministry of Agriculture, Nature Management and Fisheries (National Oil Program). The authors gratefully acknowledge the helpful comments of Dr. H. Smit of Karlshamns B.V., Koog a/d Zaan, The Netherlands.

References

1 Kaufman, A.J., Ruebusch, R.J., *INFORM* **1990**, *1* 1034- 1048.

2 Hondelmann, W., Radatz, W., *Fette, Seifen, Anstrichmittel* **1982**, *84*, 73–75.

3 Hondelmann, W., Dambroth, M., *Plant Research and Development* **1990**, *31*, 38–49.

4 Meier zu Beerentrup, H., Röbbelen, G., *Angew. Botanik* **1987**, *61*, 287–303.

5 Princen, L.H., *Economic Botany* **1983**, *37*, 478–492.

6 Princen, L.H., Rothfus, J.A., *J. Am. Oil Chem. Soc.* **1984,** *61*, 281–289.

7 Sonntag, N.O.V., *INFORM* **1991,** *2*, 449–463.

8 Nieschlag, H.J., Wolff, I.A., *J. Amer. Oil Chem. Soc.* **1971,** *48*, 723–727.

9 Muuse, B.G., Cuperus, F.P., Derksen, J.T.P., *Ind. Crops Prod.* **1992,** *1*, [in press]

10 Vignolo, R., Naughton, F., *INFORM* **1991,** *2*, 692–703.

11 Naughton, F.C., *J. Amer. Oil Chem. Soc.* **1974,** *51*, 65–71.

12 Achaya, K.T., *J. Amer. Oil Chem. Soc.* **1971,** *48*, 758–763.

13 Carlson, K.D., Schneider, W.J., Chang, S.P., Princen, L.H., in: *New sources of fats and oils*: E.H. Pryde, L.H. Princen & K.D. Mukherjee (Eds.), AOCS monograph no. 9, **1981**; pp. 297–318.

14 Nelson, A.I., Wijeratne, W.B., Yeh, S.W., Wei, T.M., Wei, L.S., *J. Amer. Oil Chem. Soc.* **1987**, *64,* 1341–1347.

15 Pavlik, R.P., Kemper, T.G., *INFORM* **1990,** *1*, 200–205.

16 Carlson, K.F., Scott, J.D., *INFORM* **1991,** *2*, 1034–1060.

17 Fullbrook, P.D., *J. Amer. Oil Chem. Soc.* **1983**, *60*, 428A-430A.

18 Anonymous, *J. Amer. Oil Chem. Soc.* **1986,** *63*, 969–970.

19 Tandy, D.C., in: *Introduction to Fats and Oils Technology*: Wan, P.J. (ed.) Champaign IL, USA, AOCS Monograph, **1991**; pp. 59–84.

20 Vadke, V.S., Sosulski, F.W., *J. Amer. Oil Chem. Soc.* **1988,** *65*, 1169–1176.
21 Prior, E.M., Vadke, V.S., Sosulski, F.W., *J. Amer. Oil Chem. Soc.* **1991**, *68*, 401–406.
22 Prior, E.M., Vadke, V.S., Sosulski, F.W., *J. Amer. Oil Chem. Soc.* **1991**, *68*, 407–411.
23 Mrema, G.C., McNulty, P.B., *J. Agric. Engng Res.* **1985**, *31*, 361–370.
24 Kuk, M.S., Hron, R.J., G. Abraham, *J. Amer. Oil Chem. Soc.* **1989**, *66*, 1374–1380.
25 Koseoglu, S.S., Lawhan, J.T., Lusas, E.W., *J. Amer. Oil Chem. Soc.* **1990**, *67*, 315–322.
26 Koseoglu, S.S., Engelgau, D.E., *J. Amer. Oil Chem. Soc.* **1990**, *67*, 239–249.
27 Hron, R.J., Koltun, S.P., Graci, A.V., *J. Amer. Oil Chem. Soc.* **1982**, *59*, 674A-684A.
28 Johnson, L.A., Lusas, E.W., *J. Amer. Oil Chem. Soc.* **1983**, *60*, 181A-194A.
29 Zajic, J., Brat, J., Svajgl, O., *Fat Sci. Technol.* **1990**, *92*, 610–614.
30 Johnson, L.J., Farnsworth, J.T., Sadek, N.Z., Chamkasem, N., Lusas, E.W., Reid, B.L., *J. Amer. Oil Chem. Soc.* **1986**, *63*, 647–652.
31 Baker, E.C., Sullivan, D.A., *J. Amer. Oil Chem. Soc.* **1983**, *60*, 1271–1277.
32 Warner, K., Baker, E.C., *J. Amer. Oil Chem. Soc.* **1984**, *61*, 1861–1864.
33 Khor, H.T., Chan, S.L., *J. Amer. Oil Chem. Soc.* **1985**, *62*, 98–99.
34 Lusas, E.W., Watkins, L.R., Koseoglu, S., *INFORM* **1991**, *2*, 970–976.
35 Lusas, E.W., Watkins, L.R., Koseoglu, S.S., Rhee, K.C., *83rd Annual Meeting of the American Oil Chemists Society, May 13,* **1992**, Toronto, Canada.
36 Lee, A.K.K., Bulley, N.R., Fattori, M., Meisen, A., *J. Amer. Oil Chem. Soc.* **1986**, *63*, 921–925.
37 Fattori, M., Bulley, N,R., Meisen, A., *J. Amer. Oil Chem. Soc.* **1988**, *65*, 968–974.
38 Temelli, F., *J. Food Sci.* **1992**, *57*, 440–442.
39 Friedrich, J.P., List, G.R., Heakin, A.J., *J. Amer. Oil Chem. Soc.* **1982**, *59*, 288–292.
40 Snyder, J.M., Friedrich, J.P., Christianson, D.D., *J. Amer. Oil Chem. Soc.* **1984**, *61*, 1851–1856.
41 List, G.R., Friedrich, J.P., *J. Amer. Oil Chem. Soc.* **1985**, *62*, 82–84.
42 List, G.R., Friedrich, J.P., Pominski, J., *J. Amer. Oil Chem. Soc.* **1984**, *61*, 1847–1849.
43 List, G.R., Friedrich, J.P., Christianson, D.D., *J. Amer. Oil Chem. Soc.* **1984**, *61*, 1849–1851.
44 Bondioli, P., Marlani C., Lanzani A., Fedeli, E., Mossa, A., Muller, A., *J. Amer. Oil Chem. Soc.* **1992**, *69*, 478–480.
45 Turpin, P.E., Coxon, D.T., Padley, F.B., *Fat Sci. Technol.* **1990**, *92*, 179–184.
46 Ramsay M.E., Hsu J.T., Novak, R.A., Reightler, W.J., *Food Technol.* **1991**, 98–104.
47 Latta, S., *INFORM* **1990**, *1*, 810–820.
48 Friedrich, J.P., E.H. Pryde, *J. Amer. Oil Chem. Soc.* **1984**, *61*, 223–228.
49 List, G.R., Friedrich, J.P., *J. Amer. Oil Chem. Soc.* **1985**, *62*, 82–84.

50 List, G.R., Friedrich, J.P., *J. Amer. Oil Chem. Soc.* **1989**, *66*, 98–101.

51 Olsen, H.S., *Asean Food Conference 1988, Thailand*, NOVO, Denmark, **1988**.

52 Christensen, F.M., *INFORM* **1991**, *2*, 984–987.

53 French, D.P., *J. Amer. Oil Chem. Soc.* **1988**, *65*, 1409–1416.

54 Carlson, K.D., Tookey, H.L., *J. Amer. Oil Chem. Soc.* **1983**, *60*, 1979–1985.

55 Derksen, J.T.P., Muuse, B.G., Cuperus, F.P., Van Gelder, W.M.J., *Ind. Crops Prod.* **1992**, *1*, [in press]

56 Carlson, K.D., Baker, E.C., Mustakas, G.C., *J. Amer. Oil Chem. Soc.* **1985**, *62*, 897–905.

57 Ayorinde, F.O., Carlson, K.D., Pavlik, R.P., McVety, J., *J. Amer. Oil Chem. Soc.* **1990**, *67*, 512–518.

58 Favati, F., King, J.W., Mazzanti, M., *J. Amer. Oil Chem. Soc.* **1991**, *68*, 422–427.

59 Randolph, T.W., *Trends Biotechnol.* **1990**, *8*, 78–82.

60 McHugh, M.A., in: *Biotechnology and Food process engineering*: H.G. Schwartzberg & M.A. Rao, (eds.) Marcel Dekker, New York, **1992**, pp. 203–212.

61 Anonymous, USDA monograph. October **1991**.

62 Haraldsson, G., *J. Amer. Oil Chem. Soc.* **1983**, *60*, 203A-208A.

63 Pardun, H., *Fat Sci. Technol.* **1989**, *91*, 45–58.

64 Schneider, M., *Fat Sci. Technol.* **1992**, *94*, 524–533.

65 Kanamoto, R., Wada, Y., Miyajima, G., Kito, M., *J. Amer. Oil Chem. Soc.* **1981**, *58*, 1050–1053.

66 Ragan, J.E., Handel, A.P., *J. Amer. Oil Chem. Soc.* **1985**, *62*, 1568–1572.

67 Smiles, A., Kakuda, Y., MacDonald, B.E., *J. Amer. Oil Chem. Soc.* **1988**, *65*, 1151–1155.

68 Dijkstra, A.J., Van Opstal, M., *J. Amer. Oil Chem. Soc.* **1989**, *66*, 1002–1009.

69 Dijkstra, A.J., *Lecture at 48th DGF Annual Meeting, September 7–10*, Essen, Germany, **1992**.

70 Buchold, H., *Lecture at 48th DGF Annual Meeting, September 7–10*, Essen, Germany, **1992**.

71 Duff, H.G., in: *Introduction to Fats and Oils Technology:* Wan, P.J. (ed.) Champaign IL, USA, AOCS Monograph, **1991**; pp. 85–94.

72 Kim, S.K., Kim, C.J., Cheigh, H.S., Yoon, S.H., *J. Amer. Oil Chem. Soc.* **1985**, *62*, 1492–1495.

73 Shah, K.J., Venkatesan, T.K., *J. Amer. Oil Chem. Soc.* **1989**, *66*, 783–787.

74 Brunetti, L., Daghetta, A., Fedeli, E., Kikic, I., Zanderighi, L., *J. Amer. Oil Chem. Soc.* **1989**, *66*, 209–217.

75 Goncalves, M., Vasconcelos, A.M.P., Gomes de Azevedo, E.J.S., Chavez das Neves, H.J., Nunes da Ponte, M., *J. Amer. Oil Chem. Soc.* **1991**, *68*, 474–480.

76 Bhattacharyya, S., Bhattacharyya, D.K., *J. Amer. Oil Chem. Soc.* **1989**, *66*, 1469–1471.

77 Bhattacharyya, S., Bhattacharyya, D.K., Chakraborty, A.R., Sengupta, R., *Fat Sci. Technol.* **1989**, *91*, 27–30.

78 Ducret, A., Pina, M., Montet, D., Graille, J., *Biotechnol. Lett.* **1992**, *14*, 185–188.

79 Stage, H., *J. Amer. Oil Chem. Soc.* **1985**, *62*, 299–308.

80 Tandy, D.C., McPherson, W.J., *J. Amer. Oil Chem. Soc.* **1984**, *61*, 1253–1258.

81 Forster, A., Harper, A.J., *J. Amer. Oil Chem. Soc.* **1983**, *60*, 217A-223A.

82 Hastert, R.C., in: *Introduction to Fats and Oils Technology:* Wan, P.J. (ed.) Champaign IL, USA, AOCS Monograph, **1991**; pp. 95–104.

83 Zschau, W., *INFORM* **1990**, *1*, 638–644.

84 Zschau, W., *Fat Sci. Technol.* **1987**, *89*, 184–189.

85 Boki, K., Kubo, M., Wada, T., Tamura, T.. *J. Amer. Oil Chem. Soc.* **1992**, *69*, 232–236.

86 Taylor, D.R., Jenkins, D.B., Ungermann, C.B., *J. Amer. Oil Chem. Soc.* **1989**, *66*, 334–341.

87 Taylor, D.R., Jenkins, D.B., *Society of Mining Engineers of AIME - Transactions* **1988**, *282*, 1901–1910.

88 Morgan, D.A., Shaw, D.B., Sidebottom, M.J., Soon, T.C., Raylor, R.S., *J. Amer. Oil Chem. Soc.* **1985**, *62*, 292–299.

89 Wiedermann, L. Erickson, D., *INFORM* **1991**, *2*, 200–213.

90 Gavin, A.M., in: *Introduction to Fats and Oils Technology:* Wan, P.J. (ed.) Champaign IL, USA, AOCS Monograph, **1991**; pp. 137–164.

91 Dudrow, F.A., *J. Amer. Oil Chem. Soc.* **1983**, *60*, 224A-226A.

92 Duff, H.G., in: *Introduction to Fats and Oils Technology:* Wan, P.J. (ed.) Champaign IL, USA, AOCS Monograph, **1991**; pp. 105–113.

93 Salunkhe, D.K., Desai, B.B., CRC press Inc., Boca Raton FL, **1986**.

94 Kleinhanss, W., Heins, G., *Ökonomishe Analyse der Wettbewerbsfähigkeit der Produktion und Verwendung pflanzlicher Öle und Fette als Industriegrundstoff.* Landbauforschung Völkenrode, Sonderheft 100, **1989**.

95 Kaufman, A.J., Ruebusch, R.J., *INFORM* **1990**, *1*, 1034–1048.

96 Warwel, S., *Fat Sci. Technol.* **1992** *94*, 512–523.

97 Steinberger, U., Preuss, W., *Fat Sci. Technol.* **1987**, *89*, 297–303.

98 Meffert, A., *J. Amer. Oil Chem. Soc.* **1984**, *61*, 255–258.

99 Von Bunk, A., Ziebell, W., Espig, G., *Der Tropenlandwirt*, **1990**, *91*, 5–17.

100 Korte, V., Hemmerlein, N., Richter, H., Report TV 8837, Dr.-Ing h.c. F. Porsche AG, Entwicklungszentrum Weissach, Abteilung EFA2, Germany, **1991**.

101 Johnson, R.W., in: *Fatty acids in Industry*: R.W. Johnson and E. Fritz, (eds.) Marcel Dekker Inc. New York. **1989**; pp. 217–231.

102 Sonntag, N.O.V., in: *Fatty acids in Industry:* R.W. Johnson and E. Fritz, (eds.), Marcel Dekker, Inc. New York, Chapter 2, **1989**; pp. 23–72.

103 Derksen, J.T.P., Boswinkel, G., Van Gelder, W.M.J., in: *Lipases, structure, mechanism and genetic engineering*: L. Alberghina, R.D. Schmid and R. Verger (eds.) GBF Monograph 16, VCH Publishers, New York **1991**. pp. 377–380

104 Bloomer, S., Adlercreutz, P., Mattiasson, B., *J. Amer. Oil Chem. Soc.* **1990**, *67*, 519–524.

105 Muuse B.G., Essers, M.L., Van Soest, L.J.M., *Neth. J. Agricult. Sci.* **1988**, *36*, 357–363.

106 Christie, W.W., *Fat Sci. Technol.* **1991**, *92*, 65–66.

107 Malcata, F.X., Reyes, H.R., Garcia, H.S., Hill, G.G., Amundson, C.H., *J. Amer. Oil Chem. Soc.* **1990**, *67*, 890- 910.

108 Kloosterman, J., Van Wassenaar, P.D., Bel, W.J., *Fat Sci. Technol.* **1987**, *89*, 592–597.

109 Pronk,W., Kerkhof, P.J.A.M.,Van Helden, C.,Van 't Riet, K., *Biotechnology and Bioengineering* **1988**, *32*, 512–518.

110 Pronk,W.,Van der Burgt, M., Boswinkel, G.,Van 't Riet, K., *J. Amer. Oil Chem. Soc.* **1991**, *68*, 852–856.

111 Derksen, J.T.P., Boswinkel, G.,Van Gelder,W.M.J.,Van 't Riet, K., Cuperus, F.P. in: Agricultural Biotechnology in focus in the Netherlands,Vol 2: D.H.Vuijk (ed.) PUDOC Wageningen **1993** (in press).

112 Derksen, J.T.P., Krosse, A.-M., Tassignon, P., Cuperus, F.P., in: *Proceedings 6th Forum for Applied Biotechnology, September 24–25, 1992*, Brugge, Belgium (in press).

113 Huibers, D.T.A., Fritz, E., in: *Fatty acids in Industry*: R.W. Johnson and E.Fritz (eds.) Marcel Dekker Inc. New York. **1989**; pp. 85–112.

8 Release of Transgenic Oil Crops

P.J. Dale and J.A. Scheffler

8.1 Introduction

Traditional methods of plant breeding have steadily improved the productivity of oil crops and have led to varieties that are better adapted to specific environments, are more resistant to pests and diseases, and have improved quality and quantity of crop product. Although plant breeding will undoubtably continue to make a significant contribution to crop improvement, the genes available are limited to those from species that are sexually compatible with the crop to be improved. Various methods have been employed during the history of plant breeding to facilitate cross pollination with distantly related plant species, including variations in pollination technique, ovary and embryo culture, and this has provided a valuable source of novel genes [1]. There are, nevertheless, frequently problems of sexual sterility in wide hybrids, and the failure of chromosomes to pair at meiosis, making it difficult or impossible to obtain the desired gene or genes in a good agronomic genetic background. Even when chromosome pairing occurs, it may take several backcross generations to remove unwanted genes contributed by the gene donor species, and tight linkage between desirable and undesirable genes can make this unattainable in practice.

With the advent of recombinant DNA methods, which allow DNA to be cut precisely at particular base pair sequences, and the development of transformation methods for the direct incorporation of DNA into plants (see chapters 5 and 6), it is possible to introduce specific genes into plants that have been isolated from a wide range of organisms. This provides the opportunity to extend the gene-pool available for crop improvement to all life forms, including unrelated plant species, microbes, viruses and even animals. Table 8.1 gives examples of genes currently in use for crop plant transformation, and their origins.

Table 8.1. Types and origins of genes being inserted into crop plants

Trait conferred by gene	Gene product	Source of gene	Target plant(s)[1]
Pest resistance			
Various insect pests	Bt insecticidal protein	*Bacillus thuringiensis*	*Gossypium hirsutum* cotton[23]
Various insect pests	Trypsin inhibitor protein	*Vigna unguiculata*	*Nicotiana tabacum* tobacco[24]
Viral disease resistance			
Tomato mosaic virus	Viral coat protein	Tobacco mosaic virus	*Lycopersicon esculentum* tomato[25]
Potato leaf roll virus, virus X, virus Y	Viral coat protein	Corresponding potato virus	*Solanum tuberosum* potato [26,27]
Fungal disease resistance			
Alternaria longipes (Brown spot fungus)	Chitinase	*Serratia marcescens*	*Nicotiana tabacum* tobacco[28]
Rhizoctonia solani (damping off, seedling blight etc)	Bean endochitinase	*Phaseolus vulgaris*	*Nicotiana tabacum* tobacco[29] *Brassica napus* rape[29]
Herbicide resistance/tolerance			
glyphosate	Analogue of EPSP synthase	Various plant and microbial genes	*Glycine max* soybean[30] *Gossypium hirsutum* cotton[31] *Linum usitatissimum* flax[32]
sulfonylurea	Acetolactate synthase	*Arabidopsis thaliana*	*Zea mays* corn[33] *Brassica napus* rape[34]
glufosinate	Phosphinothricin acetyltransferase	*Streptomyces hygroscopicus*	*Brassica napus* rape[35] *Beta vulgaris* sugarbeet[36]
bromoxynil	Bromoxynil specific nitrilase	*Klebsiella ozaenae*	*Gossypium hirsutum* cotton[37]

Table 8.1. Continued

Food processing/quality			
Improved storage	Antisense polygalacturonase	*Lycopersicon esculentum*	*Lycopersicon esculentum* tomato[37,38]
Flower color	Dihydroflavonol 4-reductase (DFR)	*Zea mays* *Petunia hybrida* petunia[39]	
Increased stearic acid	Antisense stearoyl-ACP desaturase	*Brassica rapa*	*Brassica napus* rape[40]
Increased mannitol	Mannitol dehydrogenase	*Escherichia coli*	*Nicotiana tabacum* tobacco[22,41]
Increased methionine	Seed coat protein	*Bertholletia excelsa*	*Glycine max* soybean [42] *Brassica napus* rape [43]
Increased starch content	ADP-glucose pyrophosphorylase	*Escherichia coli*	*Solanum tuberosum* potato[44]
Flavor enhancer	synthesized monellin	Artificially synthesized	*Lycopersicon esculentum* tomato[45]
Specialty chemicals			
Increased lauric acid	Lauroyl-ACP thioesterase	*Umbellularia californica*	*Brassica napus* rape [46]
Serum albumin	Human serum albumin	*Homo sapiens*	*Solanum tuberosum* potato[47]
Enkephalins	Leu-enkephalin	chimeric gene, part from *Homo sapiens* & *Arabidopsis thaliana*	*Brassica napus* rape [48]
Cyclodextrins	Cyclodextrin glycosyltransferase	*Klebsiella pneumoniae*	*Solanum tuberosum* potato[49]
Male sterility system	Ribonuclease and ribonuclease inhibitor	*Bacillus amyloliquefaciens*	*Brassica napus* rape[50,51]
Biodegradable thermoplastic	Polyhydroxybuty-rate (PHB)	*Alcaligenes eutrophus*	*Arabidopsis thaliana* [52]

[1] Only selected examples are given, not a complete listing.

8.2 Release of transgenic plants from contained conditions

Following the creation of transgenic plants by one of the transformation methods currently available, the plants are first evaluated in the laboratory and glasshouse to determine whether the transgenes are being expressed and if the phenotype is modified in the way intended. The environment under these contained conditions is essentially artificial, so eventually it becomes necessary to assess the performance of transgenic plants in a field environment similar to

Table 8.2. Field releases of transgenic oilseed crops

Crop species	Plant character	Gene product
Brassica napus (oilseed rape)	Glufosinate herbicide resistance	Phosphinothricin acetyltransferase[53]
	Insect resistance	Bt insecticidal protein[54]
	Increased methionine in meal	Seed storage protein[55]
	Increased stearic acid in oil	Antisense stearoyl ACP desaturase[54]
	Male sterility	Ribonuclease[1]
Glycine max (soybean)	Glufosinate herbicide resistance	Phosphinothricin acetyltransferase[54]
	Glyphosate herbicide resistance	Altered EPSP synthase[54]
	Increased methionine in meal	Seed storage protein[42]
Gossypium hirsutum (cotton)	Bromoxynil herbicide resistance	Bromoxynil specific nitrilase[54]
	Glyphosate herbicide resistance	Altered EPSP synthase[54]
	Sulfonylurea herbicide resistance	Acetolactate synthase[54]
	Insect resistance	Bt insecticidal protein[54]
Linum ussitatissimum (flax)	Sulfonylurea herbicide resistance	Acetolactate synthase[56]
Zea mays (maize, corn)	Glufosinate herbicide resistance	Phosphinothricin acetyltransferase[54]
	Bromoxynil herbicide resistance	Bromoxynil specific nitrilase[54]
	Sulfonylurea herbicide resistance	Acetolactate synthase[54]
	Modified protein	Wheat germ agglutin[54]
	Male sterility	Ribonuclease[1]
	Insect resistance	Bt insecticidal protein[54]

[1] Jan Leemans, Plant Genetic Systems, Gent, pers. comm.

that used in agricultural practice. For example, there have been various kinds of genes introduced into oil crops; some of these have been assessed in small field experiments and others are progressing towards being approved for commercial use in agriculture (Table 8.2).

Before transgenic plants can be grown under field conditions it is necessary, under national regulations and by international agreement, to carry out a risk assessment to determine the possible consequences of that release.

8.3 Risk assessment

What is risk assessment, and why is it practiced for transgenic plants and not for plants genetically modified by traditional breeding methods? The reasoning is not that transgenic plants are believed to be innately more hazardous than non-transgenic plants, but that modern methods of genetic engineering provide a much wider gene- pool and hence diversity of genes from which to choose. In traditional plant breeding (and similarly in natural populations) many millions of new gene combinations are generated each year, and even though it is not possible to predict the precise consequences of those gene combinations, the range of variation falls within bounds that are familar to the plant breeder. There are times however, when traditional breeding methods give recombinant lines with increased and potentially harmful levels of secondary products such as high glucosinolate levels in *B. napus* seeds [2] or high glycoalkaloid content in potato tubers [3], but procedures have been adopted to identify and eliminate these undesirable genetic combinations.

Recombinant DNA technology and plant transformation provide a means of introducing into plants, genes that we have no direct experience of in their new genetic background. Risk assessment is a process of applying the available data and expertise to make a judgement on what the consequences might be of growing a particular transgenic phenotype under field conditions, and whether they would be different from growing the corresponding unmodified plant.

The factors considered in risk assessment are summarized in Table 8.3. Questions can broadly be grouped into those about the gene inserted and those about the crop itself. It is clearly necessary to consider the interaction between the two: the extent to which the transgenes alter the biology of the crop.

Table 8.3. Risk assessment can be divided into questions about the crop species and those about the genes inserted. Factors relating to the genes are addressed for each new gene inserted; factors about the crop will frequently be constant for that crop, but may be modified by the transgenes inserted.

CROP FACTORS
Sexual compatibility with related species
Likely fate of hybrids: establishment, fertility, seed production, selection
Pollen longevity
Method of pollen dispersal
Distance pollen will travel and give successful hybridization
Method of reproduction: vegetative, seed
Method of seed dispersal
Seed longevity in soil
Method of vegetative propagation: tubers, stolons, rhizomes
Persistence and dispersal of vegetative parts
Persistence of crop species in agricultural habitats
Invasiveness of crop species in natural habitats

GENE FACTORS
Source of the gene
Source of the regulatory DNA sequence
Location of transgenes in the plant genome: nucleus, organelle
Number of copies inserted
Homozygous or heterozygous condition
Structural stability of construct
Expression stability of transgenes
Structure and expression stability over vegetative and sexual generations
Level of expression and tissue specificity
Effect of the transgenes on recipient plant
Effect on tolerance to natural or imposed stress
Effect on persistence in agricultural habitats
Effect on invasiveness in natural habitats
Effect on allergenic properties of crop
Effect on organisms associated with the crop
Safety for human/animal consumption
Consequences if the gene is transferred to related species
Consequences if the gene is transferred to other organisms

8.4 Concerns about the use of transgenic plants

Two principal concerns are expressed about the use of transgenic crops. First, that the genes inserted might make the crop more persistent in agricultural habitats, more invasive in natural habitats or might modify the crop to make it undesirable to life or to the environment. Second, that the transgenes might somehow be transferred to other organisms, particularly to sexually compatible wild relatives including weed species.

Questions about the effect of the transgene on plant phenotype can be addressed directly by a careful analysis of the transgenic plant along with its non-transgenic counterpart [4]. The comparison can be made in several environments, to estimate if there have been changes in characters which might influence persistence or invasiveness or whether there have been undesirable changes in other plant characters. Experiments of this type have been carried out in the United Kingdom PROSAMO (Plannned Release of Selected and Modified Organisms) Program, where transgenic and non-transgenic genotypes have been planted in natural habitats, and the transgenes used in those experiments have shown no significant effect on plant survival and persistence [5].

The opportunities for transgenes to be transferred from crops species to their wild relatives depend on whether they are sexually compatible, have overlapping spatial distribution and flowering times, and have a means of transferring pollen from one to the other. The hybrid plants must be viable, become established in the environment, be fertile and set viable seeds. Their fate after this will depend on subsequent environmental conditions and selection pressures [6].

Cross pollination between traditionally bred crop plants and related weeds is believed to have led to the transfer of genes to weed species and enabled them to become more like the crop they infest. *Oryza perennis*, a weed relative of rice (*Oryza sativa*), has become more like rice through cross pollination followed by intense selection for rice-like seed and vegetative characteristics. Even though genetic exchange may have contributed to the evolution of better adapted weeds, it is important not to underestimate the ability of weed species to adapt independently of genetic exchange with the crop. Mimicry of lentil seed morphology (*Lens esculentum*) by the unrelated weed, common vetch (*Vicia sativa*), is controlled in vetch by a single recessive gene which makes the seeds flatter and more like lentil. Barnyard grass (*Echinocloa crus-galli*) is sexually incompatible with rice but weed strains have developed which, to the untrained eye, are indistinguishable from it [7,8].

One of the main considerations in any assessment of the consequences of releasing transgenic plants is the nature of the gene inserted. Novel classes of genes will become available in future as more is understood about biosynthetic processes and as more genes are isolated, but only four classes of genes will be considered here as examples.

8.4.1 Herbicide resistance

Several genes conferring resistance to different herbicides are now available and a high proportion of the transgenic plants evaluated outside containment to date have contained one of these (Table 8.2). The persistence of herbicide resistant crop varieties in agricultural habitats is not likely to be affected by the insertion of a particular resistance gene, unless the control of volunteers in subsequent cropping is normally carried out by using the same herbicide. In practice there is usually a choice of several herbicides for this purpose. It is also unlikely that herbicide resistant crop plants will be more invasive in wild habitats because herbicides are not normally used under these conditions.

The transfer of herbicide resistance to wild relatives, especially weeds, is an important consideration and would be evaluated during the risk assessment procedure. Again, there is usually a choice of several herbicides, with different modes of action, for weed control. If an agricultural system depends on the use of one herbicide, then the consequences of gene transfer to weeds could be more serious. The consequences of using several different resistance genes in a crop species is also important to consider, because of the possibility of weeds becoming resistant to several herbicides.

8.4.2 Pest and disease resistance

The introduction of genes conferring resistance to important pests and diseases may give the crop plant a selective advantage in wild habitats compared with its non-transgenic counterpart, and could provide wild relatives with selective advantage through gene exchange. This possibility would be considered as part of the risk assessment procedure. It is worth noting, however, that many of the pest and disease resistance genes currently present in conventionally bred varieties originated from wild relatives. Although the mechanisms of disease and pest resistance may be different between transgenes and conventionally introduced genes, the principles and consequences of gene transfer from crop plant to wild species should be similar. Another aspect to consider is the effect of the resistance genes on the pest or disease organism. Much is already known about the capacity of these organisms to adapt and "break down" resistance introduced by breeding. Transformation and molecular biology provide opportunities to introduce several resistance mechanisms which should be more difficult to overcome, but in practice it may be desirable to develop crop tolerance strategies, rather than complete resistance, to decrease selection pressure on the pathogen [8,9]

8.4.3 Antibiotic resistance

During the transformation process only a low proportion of the treated plant cells successfully integrate functional copies of transgenes. Because of this, it is generally necessary to use an antibiotic resistance marker gene in the transgene construct which allows transformed cells to grow preferentially on a selective culture medium containing the antibiotic (usually kanamycin). The consequence of this is that the regenerated plants carry the resistance gene and produce the antibiotic-deactivating protein. This protein is broken down along with other proteins during decay of plant material in the soil or during the digestion process if eaten.

Some authorities regulating the release of transgenic plants have raised the question of whether an antibiotic resistance gene within transgenic plants could be transferred to microorganisms in the soil or in animal digestive systems, and thereby spread these genes to organisms that do not normally possess them. Several sources of evidence indicate that this does not present a hazard. First, there is no substantiated evidence that DNA can be transferred from plants to microorganisms that are associated with them. There is a preliminary report that the club root organism, *Plasmodiophora brassicae,* is able to take up DNA from the *Brassica* plant it infects [10], but this needs to be verified. If there is DNA transfer, it would be necessary to determine its fate and whether it ever becomes integrated into the fungal genome and hence could be passed on to future generations. Second, there is no evidence of gene transfer between plant material being digested and gut microflora. Third, genes for antibiotic resistance are common in nature; they are widespread in bacteria present in the digestive systems of animals and humans [11], and also in the soil [12]. It is also important to note that the antibiotics themselves will only be used in the laboratory to generate transgenic plants and will not be applied to crops in the field [13,14]. There is no prospect, therefore, of antibiotic residues occurring on crops, or of antibiotic selection being imposed on microbial populations, or of giving a selective advantage to antibiotic resistant plants.

Even though there is no good evidence that antibiotic resistance genes will cause any hazard in transgenic crop plants, some regulatory authorities have advocated the minimal construct principle: that a construct should be the minimum size to modify the plant phenotype in the way desired. Various strategies have been proposed for producing transgenic plants without antibiotic resistance genes or of eliminating them before a transgenic variety is released. The methods include, a) cotransformation with *A. tumefaciens* and separation of the antibiotic resistance gene from the gene of interest by genetic recombination in the sexual progeny [15], b) introduction of a mechanism which excises the antibiotic resistance gene [16,17] and c) the use of microinjection which can be capable of high transformation efficiencies, making antibiotic resistance genes unnecessary [18].

8.4.4 Oil crops for industrial use

Erucic acid has virtually been eliminated from rapeseed varieties by traditional breeding, because animals fed on large amounts of it were reported to show retarded growth and changes in various internal organs [19]. Although high erucic acid rapeseed oil has never been shown to be hazardous to human health, all rapeseed used for animal and human consumption is now from low erucic acid varieties. High erucic acid varieties are still grown for industrial use so it is necessary to use good genetic isolation and agricultural practice to ensure that varieties grown to produce edible oils do not become contaminated by those containing high levels of erucic acid.

Genetic isolation and good practice will also be required to ensure that there is no contamination from transgenic varieties modified to produce other oils for industrial use. It may be feasible to include a visual marker such as flower color or a selectable marker gene on the construct used to modify oil biosynthesis, so that varieties for industrial use can be identified easily.

8.5 Widespread use of transgenic crops

The first transgenic varieties are now in large scale evaluation trials and it is likely that transgenic crops will become widespread in the coming decade [20]. Because transgenic varieties that are registered in one country will be available for use by farmers in another country, it is important that there is international harmonization on risk assessment and release procedures; and that they are thorough, careful, and equitable for commerce and research. Harmonization of procedures within the European Economic Community is now being advanced in response to the EEC directive No. 90/220, although not quickly enough according to some parties (see Chapter 6). Standardization in a worldwide context is being coordinated through the Organisation for Economic Co-operation and Development (OECD) [21].

Public perception about the use of transgenic plants in agriculture, and consumption of their products, is crucially important. There is currently an international debate on various issues relating to the widespread use of transgenic crops in agriculture such as the necessity (or not): (1), to label the products of transgenic varieties [14]; (2), to use the smallest possible DNA construct to modify plant phenotype; and (3), to allow antibiotic resistant genes to be present in transgenic commercial crops [13,14]. It is important that members of the public are informed about the principles and opportunities from modern methods of genetic modification.

There are significant advantages to be gained from the improved control of pests and diseases, and from an ability to modify plants in ways helpful to maintaining world food supply [22]. The modification of oils for a range of agricultural and industrial uses, is also likely to make a significant contibution to providing a sustainable supply of resources from agriculture. However, the translation of the achievements of modern biotechnology from the laboratory to the farm, and eventually into shops, supermarkets and industry, relies on the acceptance of the products by the public. The commercialization of such products will depend upon an effective, but fair regulatory framework to evaluate their release and utilization.

References

1 Simmonds, N.W. *Principles of Crop Improvement*, London: Longman, **1979**.
2 Thompson, K.F., Hughes,W.G. *Oilseed rape*: Scarisbrick,D.H.,Daniels,R.W.,(eds.) London: William Collins Sons & Co Ltd, **1986**; pp. 32–82.
3 National Research Council USA, *Field Testing Genetically Modified Organisms: Framework for Decisions*, Washington DC: National Academy Press, **1989**.
4 Dale, P.J., McPartlan,H.C., *Theor. Appl. Genet.* **1992**, *84*, 585–591.
5 Crawley, M.J. in: *Proceedings of the 2nd International Symposium on The Biosafety Results of Field Tests of Genetically Modified Plants and Microorganisms*: Casper, R., Landsmann, J. (eds.) Braunschweig, Germany: Biologische Bundesanstalt für Land- und Forstwirtschaft, **1992**; pp. 43–52.

6 Dale, P.J. *Plant. Physiol.* **1992**, *100*, 13–15.

7 Barrett, S.C.H. *Econ. Bot.* **1983**, *37*, 255–282.

8 Gould,F., *Amer. Scientist* **1991**, *79*, 496–507.

9 Gould, F., Martinez-Ramirez, A., Anderson, A., Ferre, J., Silva, F.J., Moar,W.J. *Proc. Natl. Acad. Sci. USA* **1992**, *89*, 7986–7990.

10 Bryngelsson, T., Gustafsson, M., Green, B., Lind, C., *Physiol. Mol. Plant Path.* **1988**, *33*, 163–171.

11 Smith, H.W. *New Zealand Veterinary Journal* **1967**, *15*, 153–166.

12 Drahos, D.J., Hemming, B.C., McPherson,S., *Biotechnol.* **1986**, *4*, 439–444.

13 Flavell, R.B., Dart, E., Fuchs, R.L., Fraley, R.T., *Biotechnol.* **1992**, *10*, 141–144.

14 Bryant, J., Leather, S., *Trends in Biotechnol.* **1992**, *10*, 274–275.

15 De Block, M., Debrouwer, D., *Theor. Appl. Genet.* **1991**, *82*, 257–263.

16 Dale, E.C., Ow, D.W., *Proc. Natl. Acad. Sci. USA* **1991**, *88*, 10558–10562.

17 Russell, S.H., Hoopes, J.L., Odell, J.T., *Mol. Gen. Genet.* **1992**, *234*, 49–59.

18 Neuhaus, G., Spangenberg, G., Mittelsten Scheid, O., Schweiger, H-G. *Theor. Appl. Genet.* **1987**, *75*, 30–36.

19 Vles, R.O., Gottenbos, J.J., in: *Oil Crops of the World*: Röbbelen, G., Downey, R.K., Ashri, A.(eds.) New York: McGraw-Hill, **1989**, pp. 63–86.

20 Chasseray, E., Duesing, J., *Agro Food Industry hi-tech.* **1992**, *3*, 5–10.

21 O.E.C.D., *Recombinant DNA Safety Considerations*, Paris: Organisation for Economic Co-operation and Development, **1986**.

22 Fraley, R. *Biotechnol.* **1992**, *10*, 40–43.

23 Perlak, F.J., Deaton, R.W., Armstrong,T.A., Fuchs, R.L., Sims, S.R., Greenplate, J.T., Fischhoff, D.A., *Biotechnol.* **1990**, *8*, 939–943.

24 Boulter, D., *Outlook on Agric.* **1989**, *18*, 2–6.

25 Nelson, R.S., McCormick, S.M., Delannay, X., Dube, P., Layton, J., Anderson, E.J., Kaniewska, M., Proksch, R.K., Horsch, R.B., Rogers, S.G., Fraley, R.T., Beachy, R.N., *Biotechnol.* **1988**, *6*, 403–409.

26 Kaniewski,W., Lawson, C., Sammons, B., Haley, L., Hart, J., Delannay, X.,Tumer, N.E., *Biotechnol.* **1990**, *8*, 750–754.

27 Lawson,C., Kaniewski,W., Haley, L., Rozman, R., Newell, C., Sanders, P.,Tumer, N.E., *Biotechnol.* **1990**, *8*, 127–134.

28 Jones, J.D.G., Dean,C., Gidoni, D., Gilbert, D., Bond-Nutter, D., Lee, R., Bedbrook, J., Dunsmuir, P., *Mol. Gen. Genet.* **1988**, *212*, 536–542.

29 Broglie, K., Chet, I., Holliday, M., Cressman, R., Biddle,P., Knowlton, S., Mauvais, C.J., Broglie, R., *Science* **1991**, *254*, 1194–1197.

30 Gasser, C.S., Fraley, R.T. *Scientific American* **1992**, *266*, 34–39.

31 John, M.E., Stewart,J.McD. *Trends in Biotechnology* **1992**, *10*, 165–170.

32 Jordan, M.C., McHughen, A. *Plant Cell Rep.* **1988**, *7*, 281–284.

33 Fromm, M.E., Morrish, F., Armstrong, C.,Williams, R., Thomas, J., Klein,T.M., *Biotechnol.* **1990**, *8*, 833–839.

34 Miki, B.L., Labbe, H., Hattori, J., Ouellet,T., Gabard, J., Sunohara, G., Charest, P.J., Iyer,V.N., *Theor. Appl. Genet.* **1990**, *80*, 449–458.

35 De Block, M., De Brouwer, D.,Tenning, P., *Plant Physiol.* **1989**, *91*, 694–701.

36 D'Halluin, K., Bossut, M., Bonne, E., Mazur, B., Leemans, J. Botterman, J., *Biotechnol.* **1992**, *10*, 309–314.

37 Mitten,D.H., Redenbaugh, M.K., Sovero, M., Kramer, M.G., in: *Proceedings of the 2nd International Symposium on the Biosafety Results of Field Tests of Genetically Modified Plants and Microorganisms*: Casper, R.,Landsman,J.,(eds.) Braunschweig, Germany: Biologische Bundesanstalt für Land- und Forstwirtschaft, **1992**; pp. 179–184.

38 Sheehy, R.E., Kramer, M., Hiatt, W.R., *Proc. Natl. Acad. Sci. USA* **1988**, *85*, 8805–8809.

39 Meyer, P., Linn, F., Heidmann, I., Meyer, H., Niedenhof, I., Saedler, H. *Mol. Gen. Genet.* **1992**, *231*, 345–352.

40 Knutzon, D.S.,Thompson, G.A., Radke, S.E., Johnson,W.B., Knauf,V.C., Kridl, J.C., *Proc. Natl. Acad. Sci. USA* **1992**, *89*, 2624–2628.

41 Tarczynski, M.C., Jensen, R.G., Bohnert, H.J., in: *Third International Congress of Plant Molecular Biology Abstracts*: Hallick,R.B.(ed.) Tucson,Arizona:Department of Biochemistry,University of Arizona, **1991**; abs.1517.

42 Townsend, J.A., Thomas, L.A., Kulisek, E.S., Daywalt, M.J., Winter, K.R.K., Grace, D.J., Crook,W.J., Schmidt, H.J., Corbin,T.C., Altenbach, S.B. in: *American Society of Agronomy Abstracts*: Madison,Wisconsin: American Society of Agronomy, **1992**; p. 198.

43 Guerche, P., De Almeida, E.R.P., Schwarztein, M.A., Gander, E., Krebbers, E., Pelletier, G., *Mol. Gen. Genet.* **1990**, *221*, 306–314.

44 Barry, G.F., Stark, D.M., Muskopf, Y.M., McKinnie, R.E., Timmerman, K.P., Kishore, G.M., in: *American Society of Agronomy Abstracts*: Madison,Wisconsin: American Society of Agronomy, **1992**; p. 187.

45 Penarrubia, L., Kim, R., Giovannoni, J., Kim, S.-H., Fischer, R.L., *Biotechnol.* **1992**, *10*, 561–564.

46 Voelker,T.A.,Worrell, A.C., Anderson, L., Bleibaum, J., Fan, C., Hawkins, D.J., Radke, S.E., Davies, H.M., *Science* **1992**, *257*, 72–73.

47 Sijmons, P.C., Dekker, B.M.M., Schrammeijer, B.,Verwoerd,T.C., van den Elzen, P.J.M., Hoekema, A., *Biotechnol.* **1990**, *8*, 217–221.

48 Vandekerckhove, J., Van Damme, J., Van Lijsebettens, M., Botterman, J., De Block, M., Vandewiele, M., De Clercq, A., Leemans, J., Van Montagu, M., Krebbers, E., *Biotechnol.* **1989**, *7*, 929–932.

49 Oakes, J.V., Shewmaker, C.K., Stalker, D.M., *Biotechnol.* **1991**, *9*, 982–986.

50 Mariani, C., De Beuckeleer, M.,Truettner, J., Leemans, J., Goldberg, R.B. *Nature* **1990**, *347*, 737–741.

51 Mariani, C., Gossele,V., De Beuckeleer, M., De Block, M., Goldberg, R.B., De Greef,W., Leemans, J. *Nature* **1992**, *357*, 384–387.

52 Poirier, Y., Dennis, D.E., Klomparens, K., Somerville, C., *Science* **1992**, *256*, 520–523.

53 Oelck, M.M., Phan, C.V., Eckes, P., Donn, G., Rakow, G., Keller,W.A., in: *GCIRC Eighth International Rapeseed Congress*: McGregor, D.I.,(ed.) Saskatchewan, Canada: Organizing Committee, **1991**; pp. 292–297.

54 USDA Environmental Release Permits, U.S. Department of Agriculture Animal and Plant Health Inspection Service Biotechnology, Biologics and Environmental Protection, Biotechnology Permit Unit, 6505 Belcrest Road, Hyattsville, Maryland, USA March, **1992**.

55 Krebbers,E., Rüdelsheim, P., De Greef,W.,Vandekerckhove, J. in: *GCIRC Eighth International Rapeseed Congress*: McGregor, D.I.,(ed.) Saskatchewan, Canada: Organizing Committee, **1991**; pp. 716–721.

56 McHughen, A., Holm, F. *Euphytica* **1991**, *55*, 49–56.

9 Future Perspectives for Oil Crops

D.J. Murphy

9.1 Introduction

Oil crops have always been an important component of any balanced agricultural system. They provide 20–40% of the total calorific intake of most human societies as well as supplementing the diet of domestic animals. With the seemingly relentless rise in human populations, particularly in developing countries, continued increases in the yield and production of oil crops will be required simply to maintain current levels of dietary intake. The provision of food for human societies should always be the primary goal of global agriculture, with the feeding of domestic animals and non-food uses taking second place. In the last decade of the 20th century, we have the luxury of a global agricultural surplus, albeit unevenly distributed and with many pockets of largely man-made famine. This surplus is, however, unlikely to persist as more and more pressures are brought to bear on our agricultural systems. Not only will agricultural products be required to feed a lot more people, they will eventually be the only source of hydrocarbons and a potential source of renewable "biofuels". In the preceding Chapters of this monograph, the major oil crops have also been described with respect to their history, current status and potential for improvement through breeding. The mechanisms of oil synthesis and utilisation and its manipulation by modern biotechnology have been examined, together with a brief account of how such advances are being monitored and regulated in field releases of new crop varieties. The purpose of this Chapter is to look forward to the future of oil crops during the coming two centuries in the light of present agricultural, scientific, geological and demographic trends.

9.2 Designer oil crops

In 1992, the first two announcements were made of the successful use of genetic engineering to modify the fatty acid profile of an oil crop [1,2]. Field trials of these plants are now underway in the USA and are planned for Europe in the near future. In both cases, rapeseed was the crop in question, largely because it is the oil crop that is most amenable to the transformation/regeneration procedures involved in gene transfer into plants (see Chapter 6). Many of the other major oil crops have, however, now been transformed and it is likely that by the end of the decade, genetically-engineered soybean, sunflower and linseed varieties will also be available. This means that biotechnologists will soon have the capacity to make available a huge spectrum of designer oil crops, with oil compositions tailored specially for individual or general industrial, edible or pharmaceutical applications. The ways in which society will utilise this resource will depend largely on the economic and political factors obtaining at that time.

Current pressures from huge agricultural surpluses in the developed world obviously favor the non-edible uses of crops. To these pressures must be added the sometimes contradictory influences of international free trade and the desire to protect (often uneconomic) local or national agricultural practices. With populations largely static, or even in decline, it is unlikely that there will be significant increases in demand for edible vegetable oils in developed countries. The slight increases due to the shift away from animal fat to plant-derived oils can be balanced by a general reduction in overall dietary fat intake within the social groups in which these trends are typically made manifest. It is likely, therefore, that a significant proportion of the agricultural capacity of developed countries will be available for the production of non-edible crops for a considerable time into the future. Providing that costs and particularly inputs can be kept relatively low, designer oil crops could provide an alternative, stable and readily manipulable production system for basic or partially-refined industrial materials.

A great attraction of annual oil crops over perennial tree crops is their versatility with respect to changing production requirements. For example, a chemical company using lauric acid in one of its processes may contract farmers to produce 100,000T of high-lauric rapeseed oil on an annual basis. It may be that the company then finds that it can substitute a high-capric oil in its industrial process at a considerable cost saving. In this case, the company is changing its requirements at short notice. Annual oil crops have the flexibility to respond each year to such demand-led fluctuations in production requirements, whereas perennial tree crops would take decades for a similar response. From the viewpoint of the industrial user, this ready adaptability of annual oil crops is likely to offset, at least in part, their lower yields, compared to the perennial oil crops. In the above example, the chemical company would simply alter its contracts and could require the same farmers to grow a high-capric rapeseed variety instead of a high-lauric variety. This would have virtually no effect on the farmers who would purchase the new variety of seed and grow and harvest the rapeseed crop in exactly the same way as before. Even the crushing and refining processes would be virtually identical. The only difference would be a greatly enhanced value or utility to the downstream industrial user. Such flexibility of oil production is likely to have a considerable attraction for both large and small industrial users alike.

Although a relatively optimistic tone has been taken here with respect to the future of designer oil crops, it is important to recognise the factors which may obstruct or delay their introduction into general agriculture. A major concern, as outlined in Chapter 6, is the complex and lengthy regulatory procedures that exist in some countries for the field testing of transgenic crops (see also Chapter 8). The existence of quite different rules in different countries or trade blocs will tend to lead to the most rapid progression of designer oil crop development in those countries with the most favourable regulatory environments. There is a danger here of Europe falling behind the USA unless its regulations are clarified and simplified in the near future.

Public opinion is another important factor that can mitigate against the introduction of any innovatory process. Well-organised and articulate groups in the USA are currently campaigning against the marketing of the first genetically engineered food product, the Flavr Savr©[1] tomato. Their actions have included consumer boycotts, picketing of supermarkets and an active publicity campaign. It is less likely that industrial or pharmaceutical products derived from genetically engineered oil crops would evoke the same emotional response. Nevertheless, these actions and the publicity derived therefrom may result in a general adverse perception for any product of modern biotechnology. It will be necessary for scientists, farmers and industrialists to participate in a constructive and educational dialogue with the general public in order to explain the benefits of designer oil crops and to address some of the real concerns felt by people from many different walks of life.

In addition to these more general concerns, there are numerous minor, but still important, practical problems with the widespread use of designer oil crops. A very real difficulty may be in distinguishing the different seed varieties on the farm or in the crushing plant.

To take the example of rapeseed, only a relatively few genes will need to be manipulated in order to produce the kind of diversity of products shown in Figure 9.1. The introduction of these genes will only affect oil quality and the

---

[1] Flavr Savr is copyrighted by Calgene Inc., USA.

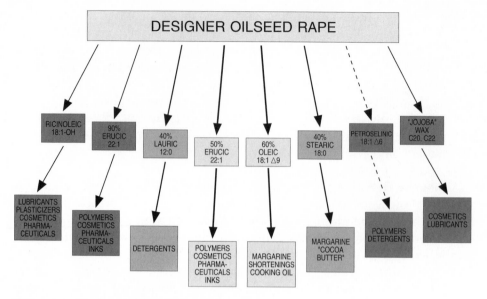

Figure 9.1. Designer Oilseed Rape Varieties

Clear boxes. At present there are two types of rapeseed variety, with respect to oil content. By far the most common variety is the edible, "Canola" type, containing about 60% oleic acid. A relatively minor, but industrially useful variety is high-erucic acid rapeseed (HEAR), which contains about 50% erucic acid.

Lightly shaded boxes. In 1992, the preliminary development of two new genetically engineered varieties was announced [1,2]. These are, respectively, a 40% stearic and a 40% lauric acid variety. These two "designer" varieties are currently undergoing field trials in the USA and elsewhere.

Darkly shaded boxes. At least four other genetically engineered varieties are under development in Institutes and companies in the UK, USA and Germany. These include a high-petroselinic, a very-high erucic (90% +), a ricinoleic and a jojoba wax variety. All four varieties are targeted at medium- to high-value industrial markets.

rape plants and seeds will otherwise be indistinguishable from one another in much the same way that the present high-erucic and zero-erucic rapeseed cultivars look identical.

The availability of ten identical-looking rapeseed cultivars could give a lot of scope for confusion and mix-ups on the farm. This could occur at several levels. Seed batches may be mixed up before or during sowing so that two genetically different, but virtually indistinguishable varieties are sown together. Batches of harvested seed may also be mixed either on the farm or at the crushers. Less likely, but still possible, is the mixing of different batches of crude or refined oils or the inadequate cleaning of machinery, resulting in trace or higher levels of contamination of one oil by another. This could be serious if an edible-grade oil were to be contaminated by even very small amounts of industrial or pharmacologically-active oils. Such contamination, although potentially problematic is controllable by the stringent segregation of all different seed and oil batches at every stage of their existence, from the seed supplier, to the farmyard and onwards to the crushing mill and beyond.

Clearly, attention would have to be given to the adequate labelling at all stages of the process from seed purchase, to sowing, growth, harvesting, storing, transporting, crushing and refining. There is very limited potential for the use of genetic engineering to "color code" seeds or flowers, and anyway, this is likely to be prohibitively expensive. Washable sprays could be used instead to color code the seeds. This would need to be supplemented by a relatively cheap and easy assay system to check the oil profiles of batches of seeds of unknown or uncertain provenance. The present analytical methods based on gas chromatography are relatively sophisticated and cannot readily be adapted to such uses. In the past, a large number of "dip- stick" tests have been devised to measure parameters from soil pH to sugar level in the urine. More recently, advances in immunological methods have allowed a similar antibody-based dip-stick test to be developed to measure such complex factors as the concentration of alfatoxins in samples of peanut butter. It will be a considerable challenge, but not impossible, for lipid chemists and engineers to come up with a similar, cheap, rapid and reliable method for testing the fatty acid compositions of seeds and oils.

A different threat is that posed by cross-hybridisation of one cultivar by pollen from a different cultivar. Here, rapeseed is at an advantage compared to its vegetable *Brassica* relatives and many other oil crop species, because its pollen is self-compatible. This means that pollen from one rapeseed plant can fertilize the same plant. The result is a much lower rate of cross-fertilization. Since rapeseed pollen is relatively heavy and is not transported efficiently by wind, the plant relies on insects such as bees for pollination. Pollinating insects do not, on average, fly very far and it is probable that a simple geographic isolation distance of 100–200 metres will prevent almost all cross-pollination of adjacent rapeseed fields. Rather than relying on geographical isolation alone, researchers are now considering the genetic manipulation of rapeseed so that each plant will *only* accept its *own* pollen, i.e. the plants are genetically isolated from one another. The development of such varieties will be expensive and therefore depends upon the future extent of designer rapeseed cultivation on farms and on the success of alternative methods of reducing cross-pollination.

Designer oil crops are now a reality and offer many attractive commercial possibilities for the future. It is important to remember, however, that they are only one aspect of the future direction of oil crops. From the above discussion, it can be seen that they have several drawbacks and, as mentioned in Chapter 6, account should also be taken of the desirability of maintaining agricultural diversity. This is true, not only for long-term and sometimes intangible ecological and aesthetic reasons, but also for sound agronomic or commercial reasons. Monocultures require much maintenance to keep them pest and disease-free. The build up of rapeseed disease is so serious that the crop can often only be grown in the same field about once every four years. For these reasons it is prudent to consider the ways in which modern biotechnology can assist and accelerate the development of alternative oil crops.

9.3 Alternative oil crops

Alternative oil crops are plants which produce potentially useful seed oils but which are not yet in widespread commercial cultivation. Several specific examples of such potential oil crops are given in Chapter 2. In some cases, their lack of popularity is due to a resistance to change by growers and/or users, but most frequently, it is because these new crops have associated problems which retard their further development. Often these problems are straightforward to describe but much more complex to solve. Examples range from premature pod-shattering in *Cuphea* to indeterminate seed set in coriander.

When a potential new oil crop is chosen for domestication, it is almost inevitable that the overall seed yield and oil yield will be suboptimal and that there will be a number of disease and pest problems to be overcome. Historically, these tasks lie within the realm of the plant breeder and previous experience with crops such as rapeseed and sunflower has shown how dramatic progress can often be made in resolving such problems. (see Chapters 1 & 3). Unfortunately, such work takes a great deal of time and effort and is therefore expensive. Modern biotechnology can assist by making a range of new techniques available to the breeder. These range from tissue culture and embryo rescue, to induced mutation and genome mapping (see Chapters 3 & 6). Although these methods can accelerate and widen the scope of breeding programs, the process is still relatively expensive and unlikely to yield elite cultivars of the alternative crops for at least a decade or more. It is therefore important to select a relatively small number of alternative oil crops for intensive improvement programs.

Four alternative temperate oil crops at various stages of domestication plus two established crops are shown in Figure 9.2. High-oleic rapeseed is the variety now grown as an edible oilseed throughout Europe and North America. High-linolenic linseed is one of the few established industrial oilseeds grown in

DIVERSIFICATION OF OILSEED INDUSTRIAL CROPS
IN TEMPERATE REGIONS

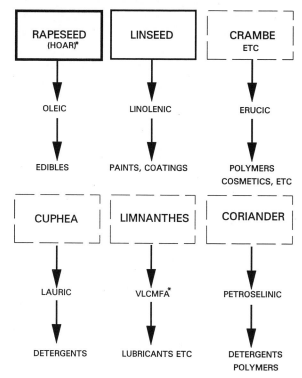

Figure 9.2. Alternative Oil Crops
The cool temperate maritime climate of Northern Europe supports two major oilseed crops, i.e. edible-grade rapeseed and industrial- grade linseed (□). There are numerous examples of other plant species with useful oil profiles that are available for domestication. Four such examples, all of which have been considered as potential industrial crops, are shown above (∷).
*HOAR, high-oleic-acid rapeseed; VLCMFA, very-long-chain- monounsaturated fatty acid.

temperate regions. *Crambe* is a distant relative of rapeseed which produces a high-erucic oil, but unlike high-erucic rapeseed, it will not interbreed with edible rapeseed varieties. It has been grown on a small scale in the USA, but problems such as low yield and high glucosinolate levels in the seed meal must be solved before it can become a major oil crop. *Cuphea* produces a high-lauric oil but has serious problems of pod- shattering and pollen self-incompatibility that have defeated the best efforts of plant breeders for over 15 years. *Limnanthes* (meadowfoam) produces long-chain industrial oils but, even after more than a decade of cultivation in the Western USA, it continues to suffer from low yield and pollen self-incompatibility. These factors result in less than half the financial return to growers compared to that required to make the *Limnanthes* crop a commercial success. Finally, coriander produces a high-petroselinic oil of potential industrial use. Unfortunately, it too suffers from low yield and indeterminate flowering and seed set which have hampered its domestication, despite many years of work in Germany.

An example of the progress that can be made by the combination of classical breeding and modern biotechnology can be found in *Cuphea*. Over three decades ago, it was found that several species of the genus *Cuphea* were rich sources of medium chain-length oils. Such oils can be obtained commercially only from tropical oil crops, such as coconut and oil palm, and the suppliers therefore have a monopoly. The development of a temperate oilseed source of medium chain-length oils, including lauric acid, was undertaken in Germany and the USA in order to break this monopoly [3]. Although some progress in the domestication of various *Cuphea* species was made, the outstanding problem of premature pod shattering remained unresolved until recently. It was only in 1992 that the first reports of non-shattering varieties emerged. This has now given the first promise of a commercially viable crop and has resulted in the extension of modern biotechnological methods to *Cuphea*. It should also allow the rapid development of several *Cuphea* species, each one being the source of a different medium chain-length oil, e.g. capric, lauric or myristic. It is estimated that, given an oil yield of 0.6T/ha, the European and North American demand of medium chain-length oils could be met by producing 2.8M ha *Cuphea* spp [3]. To put this figure into context, the current area under conventional rapeseed cultivation in the EEC alone is about 2.4Mha [4]. The anticipated revenue from the seed oil alone is US $1.12billion. Present uses of such oils, particularly as dietary supplements in infant feeds and for medicinal purposes, are severely restricted by the cost of synthesising them from fractionated coconut and palm kernel oils. With *Cuphea* as a less expensive source, an increase in the demand for medium chain-length oils may therefore be expected.

It is factors such as these which are driving the domestication of alternative oil crops. Over the coming decades it is likely that between four and ten new temperate oil crops will be domesticated and available for widespread cultivation. Some will have relatively small niche markets, e.g. evening primrose as a source of γ-linolenic oils, but others may eventually rival the currently dominant oil crops like soybean, palm, rapeseed and sunflower. In this regard, it is instructive to note that only 50 years ago all four of these major oil crops were very minor crops indeed (See Chapter 1).

Besides the temperate oil crops, there are several major (coconut, oil palm) and minor (cocoa, castorbean) oil crops grown in non- temperate regions. Most of these are tree crops and cannot be cultivated outside the tropics. One exception is the castorbean plant which is an annual oil crop producing a valuable high-ricinoleate industrial oil. Attempts are underway in the UK and USA to transfer the gene for the ricinoleate character from castorbean to rapeseed [5]. An alternative strategy is to adapt the castorbean itself so that it will grow in cooler climates. Given a warm, sunny summer, castorbean will produce a reasonable yield of seeds before the onset of the first frosts in latitudes as far north as southern France. With some further development it may be possible to grow castorbean in a similar geographical range to sunflower or soybean. Other non-temperate alternative oil crops include jojoba, which produces a long-chain oil with an expanding market as a lubricant and cosmetic, and *Vernonia*, which produces an epoxy oil used in resins and coatings. Jojoba has been under development for about 20 years and suffered from high initial expectations that were not realised. Nevertheless, it now has a relatively secure future as a niche oil, although as a perennial plant, breeding developments will take a long time before they result in real improvements in the crop. *Vernonia* and other novel oilseeds like *Lesquerella* and *Euphorbia* are under development, mainly in the USA, but are many years from commercialisation.

9.4 Global hydrocarbon reserves

At present, oil crop-derived industrial feedstocks must complete with fossil-derived raw materials. In considering the future of oil crops, it is therefore germane to assess the extent and rate of depletion of non-renewable hydrocarbon reserves. Terrestrial hydrocarbon deposits were mostly laid down from fossilised vegetable matter during the Carboniferous period some 300 million years ago. These deposits are now recoverable as petroleum, natural gas, and coal. The process of hydrocarbon deposition continues to this day. Recent deposits are often recovered in a non-compressed form as peat, but the latter cannot even begin to satisfy the present global demand for hydrocarbons, either as fuels or industrial raw materials. The major hydrocarbon deposits are being used up at an ever increasing rate, despite their non-renewability and the prospect that they will be exhausted within the foreseeable future.

Recoverable crude petroleum reserves are estimated at 120 billion T and those of natural gas at 100 billion TOE (tonnes oil equivalent) [6]. As these readily-recoverable reserves are depleted, there will be economic incentives to improve extraction technologies and this should raise the combined total of petroleum and gas reserves to at least 450 billion TOE [7]. Recoverable coal reserves are estimated at 500 billion TOE but further reserves (e.g. in Antarctica) and improved mining methods to win currently inaccessible coal may increase this figure somewhat [8]. This gives a total of the available global hydrocarbon reserves of at least 950 billion TOE.

What are these hydrocarbons used for? The answer is, mostly as fuel in the production of electricity. Coal, petroleum and gas are also burned directly to provide thermal energy, e.g. to heat houses and premises, but this is a relatively minor use compared to electricity generation. In 1988, fossil fuels were responsible for two thirds of the world electricity production of 10.4×10^{15} Wh. Hydro-power provided about 20% and nuclear power 15% of electricity production [6,9]. The current global primary energy consumption is over 8 billion TOE per year and over the next century, this may rise to 16 billion TOE as the world population increases from 5 to 10 billion over the same period [10]. Assuming that annual energy consumption only rises to 10–12 billion TOE and that fossil fuels are required for two thirds of this, the annual demand for fossil fuels will be 7–10 billion TOE. A simple calculation shows that the estimated global reserves will be exhausted within about 100 years. Even at present rates of consumption, the reserves will disappear by the year 2180. These calculations have not included the non-fuel uses of the hydrocarbon reserves. These are quite significant, particularly in the case of petroleum, which is the raw material for an entire industry producing lubricants, paints, plastics, textiles, etc. While there are alternative, renewable sources of fuel, albeit not as convenient or cheap as fossil sources, there are no other sources of renewable petroleum-like hydrocarbons - except for oil crops.

9.5 Renewable fuels

9.5.1 Non-biological fuels

With the total depletion of fossil hydrocarbon reserves only a matter of time, it is important to consider alternative, renewable sources of energy. The main contenders here are hydro and nuclear power which already supply about one third of the world's electricity. Although there is some potential for increasing hydropower generation, this is largely constrained by the loss of often valuable land to reservoirs. It is estimated that the full development of the global hydropower potential would cost at least 1% of the world land area [10]. Nuclear fission is not strictly a renewable source of energy, but fast-breeder reactors are 60 times more sparing of uranium fuel than thermal reactors. Nuclear fission based upon uranium fuels has the capacity to power the equivalent of five times the current demand for electricity for well over 1500 years [10]. The development of nuclear fission as a source of energy has received a series of setbacks following accidents in the USA, Europe and the former USSR. To a great extent, these are due to poor design, inadequate funding and infrastructure, or human error. A risk- benefit analysis of nuclear power shows that, given adequate levels of reactor design, containment and safety stand-ards, it compares well with coal or oil-derived power. It is less "clean" than hydropower or the other minor renewable sources such as solar or wind power. Nevertheless, it is likely that there is little alternative to the continued use of nuclear power if future generations wish to maintain the high levels of energy consumption that are a feature of affluent countries in the late 20th century.

Minor sources of renewable fuel, such as solar power, wind power, wave energy, geothermal energy and nuclear fusion have been reviewed recently by Gilland [10]. The conclusion is that, although it is theoretically possible for renewable energy sources to satisfy our current demands, it is unlikely that this potential will ever be realised. It would require a huge investment in hardware such as solar collectors, dams, wind generators etc, and large areas of the world land mass would need to be sacrificed. To quote the author;

> "Unless a technological breakthrough in nuclear fusion is achieved in the near future, the bulk of the power will have to be supplied by fossil fuels and fast reactors, and the fast reactor will be the only means of maintaining power production when the inevitable decline in fossil fuel production sets in." [10].

9.5.2 Biological fuels

Biological materials are already used as energy sources, particularly in less developed societies. Examples include crops, grasses and trees. Timber is the largest source of bioenergy and has the greatest potential for future exploita-tion. Current roundwood production is about 3.2 billion m^3, of which half is used as fuel, to give an energy of 400M TOE. It is estimated that the maximum sustainable annual roundwood cut is 5.2 billion m^3 to give 650M TOE [11]. This would satisfy about 8% of current global primary energy demands. Of course, this does not take into account the transportation costs, since the calorific value per tonne of wood is considerably lower than that of coal or oil. Heavy, bulky wood cargoes would be very expensive to ship to power stations from logging areas. Therefore, wood can never be considered as a major fuel source, e.g. for electricity generation. The potential of maize or sugar conversion to alcohol can also be dismissed very quickly. It would require the *entire* world's cropland to supply less than 40% of our primary energy demand by this method [10].

Can oil crops provide sufficient renewable energy to meet our future needs? Temperate oilseed crops can provide 1–2T oil per hectare each year while tropical tree crops can yield 5T/ha. In contrast, roundwood forests can yield 125m³/ha or 160 TOE/ha but can only be harvested about once every 10–20 years; i.e. their average equivalent annual yield is about 1T/ha. Therefore, the cultivation for oil crops of an area equivalent to that occupied by closed forest, i.e. 2.4 billion ha [12], could yield 1–5 times the energy potential of the roundwood stands. This might supply 8–40% of our current energy consumption but at the cost of removing most or all of the world's prime agricultural land from food production. Clearly, therefore, oil crops cannot come anywhere near to addressing the future renewable energy needs of the planet. Even as part of a portfolio of biofuels, they can never provide more than a small fraction of future energy requirements. There is also the important point that, unlike the raw materials for nuclear, hydro, solar or tidal power, biological raw materials like plant oils and wood can also serve as feedstocks for a plethora of useful products.

9.6 The future of oil crops

The current global food surpluses have given us an important breathing space in which to assess the future needs that we will have for our agricultural resources. At the same time, scientific advances, notably the development of modern biotechnology, have revolutionised the possibilities for the development of new edible and non-edible agricultural products. These advances will also stimulate the speedy development of novel crops and accelerate the improvement of perennial crops.

In the short term, designer oil crops, particularly those aimed at industrial use, will help to alleviate problems associated with overproduction of edible crops in developed countries. Biotechnological advances also hold the promise of doubling or trebling the yield of oil crops in developed countries (Chapter 6), providing there is the political will to promote such technology transfer. In the medium term, designer oil crops will be joined by new alternative oil crops to create a more diversified agricultural system able to respond to demand-led changes imposed by industrial users. The next fifty years will probably also see the successful use of genetic engineering to modify the high-yielding tropical oil crops to produce a whole spectrum of new edible and industrial oils. During this period, global petroleum reserves will be mostly depleted and the remaining deposits will require more difficult and costly extraction methods. The result will be huge increases in the price of petroleum and the tendency to husband it for industrial rather than fuel use. At this point, in the mid-late 21st century or maybe beforehand, it is likely that plant oils will become a more economically attractive source of hydrocarbons than petroleum.

As the 22nd century approaches, petroleum reserves will run out altogether and the remaining coal deposits would largely disappear by the end of this century. Primary energy demands would be met by a combination of nuclear power and renewable physical sources such as hydro, wind, tide and solar. Due to population pressures, it is likely that only relatively small tracts of land could be made available for the cultivation of high-yielding industrial crops. These could supply renewable hydrocarbons for industry for millennia to come, and it is most unlikely that such a valuable resource would be squandered by burning it for fuel. Instead, as our only remaining source of hydrocarbons and maybe even more importantly by then, as a major source of edible calories, oil crops will play a major role in sustaining the quality of human life well into the future.

References

1 Knutzon, D.S., Thompson, G.A. Radke, S.E., Johnson, W.B., Knauf, V.C., Kridl, J.C., *Proc. Natl. Acad. Sci.*, **1992**, *89*, 2624–2628.
2 Voelker, T.A., Worrel, A.C., Anderson, L., Fan, C., Hawkins, D.J., Radke, S.E., Davies, H.M., *Science* **1992**, *257*, 72–73.
3 Knapp, S.J. In: (Janick, J. & Simon, J.E. eds.) *Advances in New Crops* **1992**, Timber Press, Portland, Oregon.
4 Askew, M.F. *Inform* **1993**, *3*, 935–938.
5 Murphy, D.J. *Trends in Biotech.*, **1992**, *10*, 84–87.
6 BP Statistical Review of World Energy, British Petroleum, London **1989**.
7 White, D.A. In: (McLaren, D.J. & Skinner, B.J. eds.) *Resources and World Development*. Wiley, Chichester, **1987**, pp 113–128,
8 Schilling, H.-D., Wiegand, D. In: *Resources and World Development*. McLaren, D.J. & Skinner, B.J. (eds.) **1987**, pp 129–156.

9 Anonymous. *UN. Monthly Bulletin of Statistics* **March 1989.**
10 Gilland, B. *Endeavour* **1990**, *14*, 80–86.
11 Weck. J. *Die Walder der Erde* Springer, Berlin, **1957**.
12 Persson, R. *World forest resources*, Royal College of Forestry, Department of Forest Survey, **1975**, Research Note No. 17.

Index